School Specialty.
Science

cpo science

Foundations of Physics Investigations
Second Edition

ISBN: 978-1-62571-837-2
Part Number: 1576081
Printing 1—October 2016
Quad/Graphics, Versailles, KY

CPO Science
80 Northwest Boulevard
Nashua, NH 03063
(800) 932-5227
http://www.cposcience.com

Printed and Bound in the United States of America

Contributors

Principal Writers

Thomas C. Hsu, PH.D – Author

Nationally-recognized innovator in science and math education and the founder of CPO Science. Holds a Ph.D. in Applied Plasma Physics from the Massachusetts Institute of Technology (MIT), and has taught students from elementary, secondary, and college levels. Tom has worked with numerous K–12 teachers and administrators and is well known as a consultant, workshop leader, and developer of curriculum and equipment for inquiry-based learning in science and math.

Patsy Eldridge – Writer

B.S., Biology, Grove City College; M.Ed., Tufts University.

Experienced science teacher and national hands-on science trainer and presenter. As an adjunct professor for Endicott College in Beverly, Massachusetts, and the College of Charleston, developed content-intensive Physical Science courses for educators. Patsy has developed curriculum and training materials with CPO Science since 2000.

Erik Benton – Principal Investigation Editor and Writer

B.F.A., University of Massachusetts with minor in Physics

Taught for eight years in public and private schools, focusing on inquiry and experiential learning. Erik brings extensive teaching and technical expertise, ranging from elementary and adult education to wildlife research. Currently he is involved in bird population studies in Massachusetts. Erik is the CPO Science investigation writer and conducts national content presentations.

Scott Eddleman – Co-Author, Curriculum Manager

B.S., Biology, Southern Illinois University; M.Ed., Harvard University

Taught for 13 years in urban and rural settings. Developed two successful science-based school-to-career programs. Participated in a fellowship at Brown University where he conducted research on the coral reefs of Belize. Worked on National Science Foundation–funded projects at TERC. Scott has been a principal writer and curriculum developer for CPO Science since 1999.

Laine Ives – Writer

B.A., Gordon College and graduate coursework at Cornell University's Shoals Marine Laboratory and Wheelock College

Taught elementary and middle school science at an independent school and an environmental education center in New England. Taught middle and high school English overseas. Laine has worked in curriculum development at CPO Science since 2000.

Stacy Kissel – Writer

B.S., Civil and Environmental Engineering, Carnegie Mellon University; M.Ed., Physics Education, Boston College

Teaches physics, math, and integrated science at Brookline High in Massachusetts. Stacy was selected as Brookline's teacher of the year for 2007.

Senior Editor

Lynda Pennell – Executive Vice President

B.A., English; M.Ed., Administration, Reading Disabilities, Northeastern University; CAGS Media, University of Massachusetts, Boston.

Nationally known in high school restructuring and for integrating academic and career education. Served as the director of an urban school for 5 years and has 17 years teaching/administrative experience in the Boston Public Schools. Lynda has led the development for CPO Science for the past eight years. She has also been recognized for her media production work.

Editorial Consultant

Alan Hull

B.S., Emery University, Atlanta, Georgia; M.S., Georgia State University

Former high school physics and math teacher in Atlanta, Georgia. After teaching, Alan entered the school textbook publishing business where he held a variety of positions, including as a national secondary math-science consultant for major publishing companies.

Art and Illustration

Polly Crisman – Graphics Manager/Illustration

B.F.A., University of New Hampshire

Worked as a designer and illustrator in marketing and advertising departments for a variety of industries. Polly has worked in the CPO Science design department since 2001, and is responsible for organizing workflow of graphics and file management. She created the CPO Science logo and supervises the graphic design image for CPO publications and media products.

Jesse Van Valkenburgh – Illustration/Photography

B.F.A. Illustration, Rochester Institute of Technology

Worked in prepress and design. Was responsible for creative design and prepress film production for computer catalogs, brochures, and various marketing materials. Jesse completes photography and illustrations as a graphic designer for CPO Science book and media products.

Bruce Holloway – Senior Designer/Illustrator

Pratt Institute, New York, Boston Museum of Fine Arts

Expertise in illustration, advertising graphics, exhibits and product design. Commissioned throughout his career by The National Wildlife Federation's Conservation Stamp Campaign. Other commissions include the New Hampshire State Duck Stamp campaigns for 1999 and 2003.

Equipment Design and Material Support

Thomas Narro – Senior Vice President

B.S., Mechanical Engineering, Rensselaer Polytechnic Institute

Accomplished design and manufacturing engineer; experienced consultant in corporate reengineering and industrial-environmental acoustics.

Danielle Dzurik

B.S., Industrial Design, Auburn University

Focuses her efforts in product development on creating new products and improving upon older designs.

Kathryn Gavin – Purchasing and Quality Control Manager

Responsible for all functions related to purchasing raw materials and quality control of finished goods. Works closely with product development and design.

Project and Technical Support

Sara Desharnais

B.A., Creative Writing, Chester College of New England

Has worked in the publishing industry since 2004 and joined CPO Science in 2010.

Susan Gioia – Educational CPO Science Administrator

Expertise in office management. Oversees all functions necessary for the smooth product development of CPO products, including print and media.

Lynn L'Heureux

Owner of M&M Composition, LLC. Has worked in textbook composition for 10 years and specializes in math, computer, and science texts.

Science Content Reviewers

Dr. Jeff Schechter – Physicist, Boston, MA

Dr. Tim Daponte – Physics Teacher, Houston ISD, Houston, TX

Wanda Pagonis – Physics Teacher, Lylte High School, Lylte, TX

Dr. Willa Ramsay – Science Education Consultant

DaMarcus Wright – Physics Teacher, Grand Prairie ISD, Grand Prairie, TX

Beverly T. Cannon – Physics Teacher, Highland Park High School, Dallas, TX

Betsy Nahas – Physics Teacher, Chelmsford High School, Westford, MA

Neri Giovanni – Physics Teacher, Kennedy High School, San Antonio, TX

Scott Hanes – Physics Teacher, Liberty-Eylau High School, Hooks, TX

Bruce Ward – Nuclear Medical Technician, Boston, MA

Lee DeWitt – Physics Teacher, NEISD, San Antonio, TX

Steve Heady – Physics Teacher, Houston High School, Houston, TX

Dr. Michael Saulnier – Physicist, Boston, MA

Gigi Nevils – Physics Teacher, Bellaire High School, Houston, TX

Lebee Meehan – Physicist, National Aeronautics and Space Administration, Houston, TX

Dr. Manos Chaniotakis – Physicist, Massachusetts Institute of Technology, Cambridge, MA

Jay Kurima – Physics Teacher, Fort Worth ISD, Fort Worth, TX

James DeHart – Physics Teacher, West Brook High School,

David Binette – Engineering Student, Cornell University, Ithaca, NY

Rebecca DeLozier – Physics Teacher, Lewisville ISD, Shady Shores, TX

Valerie Felger – Physics Teacher, North East ISD, New Braunfels, TX

Alan Hull – Science and Math Consultant, Atlanta, GA

David Bliss – Science Teacher, Mohawk, NY

How to Read an Investigation

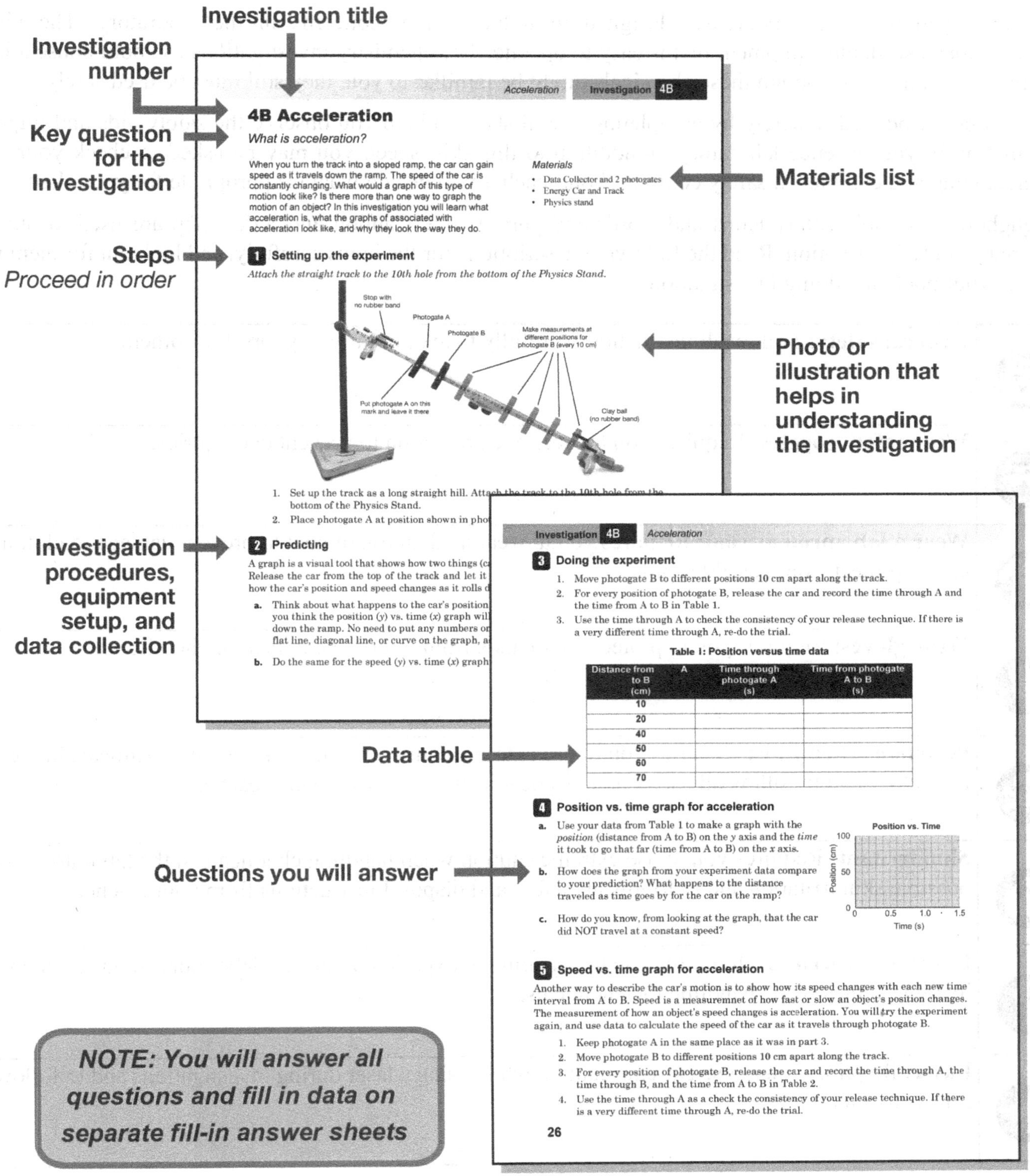

NOTE: You will answer all questions and fill in data on separate fill-in answer sheets

Lab Safety

Observing safety precautions is an extremely important practice while conducting science investigations. Using science equipment and carrying out laboratory procedures always require attention to safety. The purpose of learning and discussing safety in the lab is to help you learn how to protect yourself and others at all times.

The investigations in this book are designed to reduce safety concerns in the laboratory. The physics investigations use stable equipment that is easy to operate. The chemistry investigations use both household and laboratory chemicals. Although these chemicals might be familiar to you, they still must be used safely.

You will be introduced to safety by completing a skill sheet to help you observe the safety aids and important information in your science laboratory. In addition to this skill sheet, you may be asked to check your safety understanding and complete a safety contract. Your teacher will decide what is appropriate for your class.

Throughout this book, safety icons and words and phrases like Caution and Safety Tip are used to highlight important safety information. Read the following descriptions for the icons carefully, and look out for them when reading your book and doing investigations.

	General safety: Follow all instructions carefully to avoid injury to yourself or others.
	Wear safety goggles: Requires you to wear eye protection to prevent eye injuries.
	Wear a lab apron or coat: Requires you to wear a lab apron or coat to prevent damage to clothing and to protect from possible spills.
	Wear gloves: Requires you to protect your hands from injury due to heat or chemicals.
	Poisonous chemicals: Requires you to use extreme caution when working with chemicals in the laboratory and to follow all safety and disposal instructions from your teacher.
	Skin irritant: Requires you to use extreme caution when handling chemicals in the laboratory due to possible skin irritation and to follow all safety and disposal instructions from your teacher.
	Respiratory irritant: Requires you to perform the experiment under a laboratory hood and to avoid inhaling fumes while handling the chemicals.
	Laser: Requires you to use extreme caution while using a laser during investigations and to follow all safety instructions.

Lab safety is the responsibility of everyone! Help create a safe environment in your science lab by following the safety guidelines from your teacher as well as the guidelines discussed in this document.

Investigations

Engineering Design Challenges

1.1 Physics and Energy

What is the difference between a "live" battery and a "dead" battery?

If you weigh a fresh battery and a dead battery, you find they have the same mass. Whatever makes a battery "fresh" is not mass, but energy.

In this investigation, you will:

- charge and discharge a capacitor, which is like a battery.
- investigate how quickly a battery gains or loses energy.
- come up with an explanation for how a battery functions.

Materials List

- Electric Circuits
- DataCollector
- Super Capacitor
- Voltage probe
- Battery

1 Introduction

Many of the things we will investigate are invisible, move too quickly, or have other properties that are hard to measure directly with our ordinary senses. The purpose of an instrument is to measure things accurately—even things that are invisible, like the electricity in a wire.

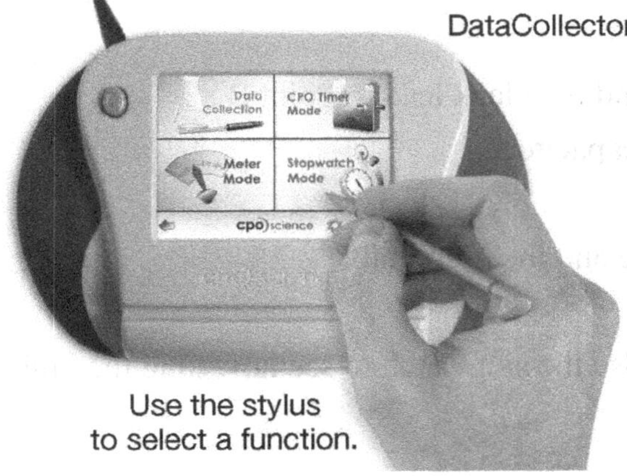

Home screen

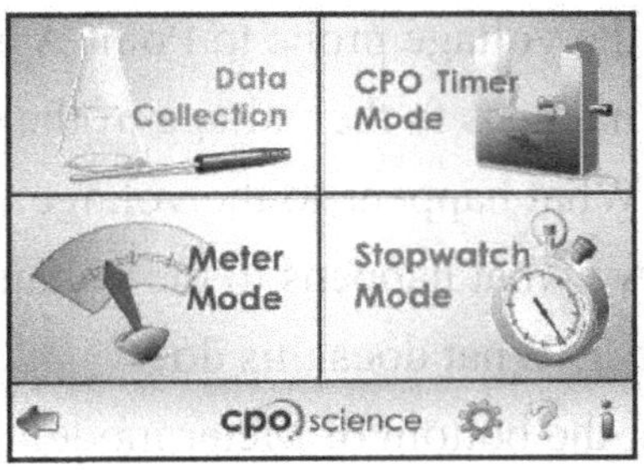

Stopwatch screen

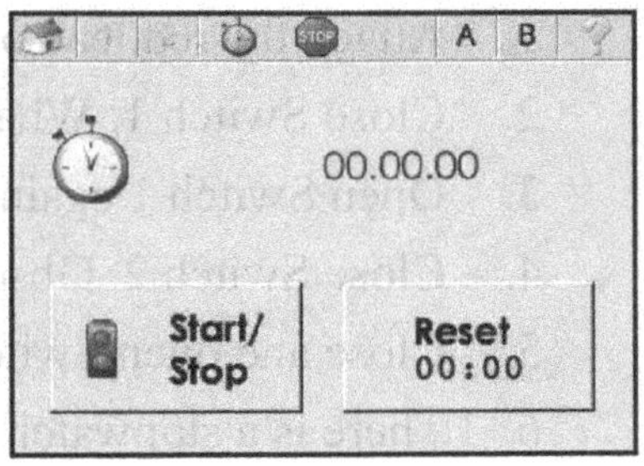

1. Connect the DataCollector to a power source and turn it on. You should see the home screen.
2. Select stopwatch mode.
3. Gently tap the Start/Stop and Reset buttons to make the stopwatch work.
4. How quickly can you start and stop the clock? Can you reach 0.10 seconds? Pass the DataCollector around and have everyone in your group try.
5. Tap the home button to get back to the home screen. Press the Meter mode button to put the DataCollector in meter mode. Connect the voltage probe.
6. Touch the ends of the voltage probe to a battery. Try reversing where you touch the red and black leads. What happens to the reading?

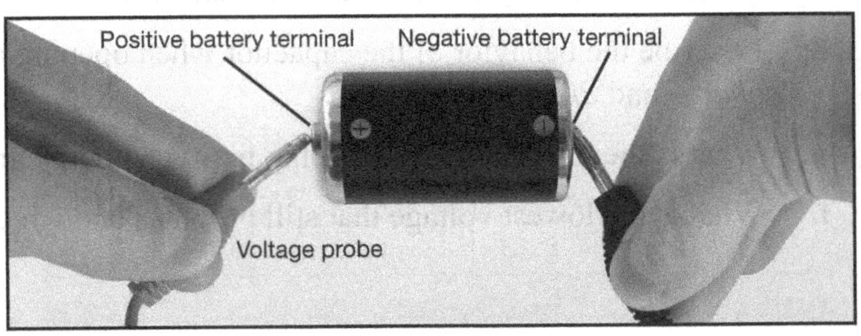

2 Making an electric circuit

Set up the circuit shown in the diagram below. Before attaching the switches to the circuit board, make sure they are both in the open position.

Connect this circuit.

Connect the voltage probe.

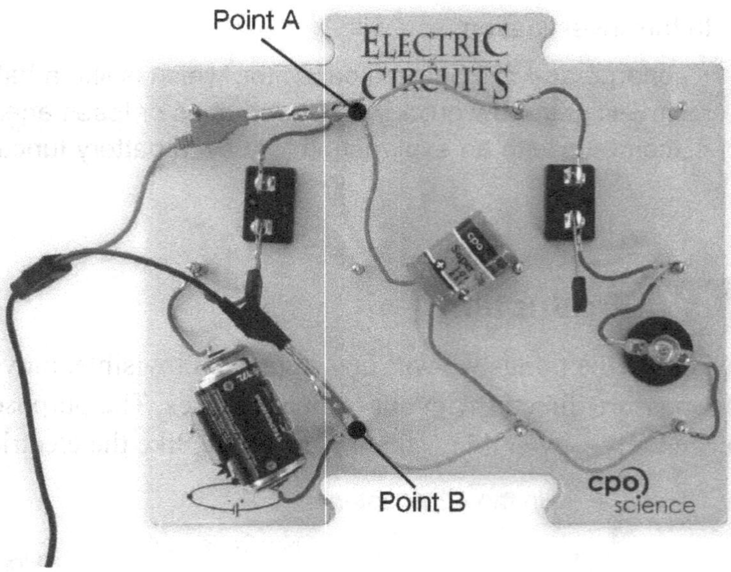

1. Attach the red lead of the voltage probe to Point A and the black lead to Point B.
2. Close Switch 1. What happens to the voltage of the capacitor?
3. Open Switch 1 again. What happens to the voltage?
4. Close Switch 2. Observe what happens to the voltage and the light bulb.
5. Close and open Switch 1. What does this do?
6. There is a stopwatch at the bottom of Meter mode. Use the stopwatch to see how long the bulb stays lit after you open Switch 1.

3 Constructing explanations

Try repeating the experiment and making observations until you can answer the following questions.

a. Describe what happens to the capacitor when you close Switch 1. Use the word *energy* in your answer. Assume the capacitor is at zero volts when you first close the switch.

b. What happens to the energy in the capacitor when you open Switch 1?

c. What happens to the energy in the capacitor when you close Switch 2? The same thing happens if Switch 2 is already closed and you open Switch 1.

d. Describe the behavior of the capacitor when opening and closing Switches 1 and 2 using the words *charge* and *discharge*.

e. What is the observable relationship between energy and the voltage measured on the DataCollector?

f. What is the lowest voltage that still provides enough electrical energy to light the bulb?

1.2 Inquiry and Optical Illusions

How do our senses perceive the universe?

The sense of sight is an extremely valuable tool for gathering information about the world around you. Sight is fundamentally about detecting patterns of light, including brightness and color. Because light can be manipulated, it is possible to see things that are not real, or at least not what they seem to be.

In this investigation, you will:

- create and explain a simple optical illusion with a prism.
- create the illusion of a color that is not really there.

Materials List

- Optics with Light & Color

1 Introduction

Light has many properties: it travels in straight lines and has the characteristic of color. Both refraction and reflection often occur when light hits a boundary between materials, such as the boundary between glass and air. The amount of light reflected or refracted depends on the angle at which you are looking relative to the surface.

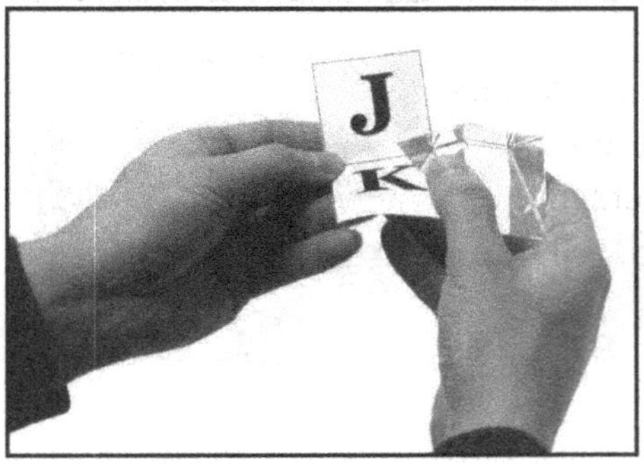

Fold the paper around the acute angle of the prism.

Look through the face as you rotate the prism.

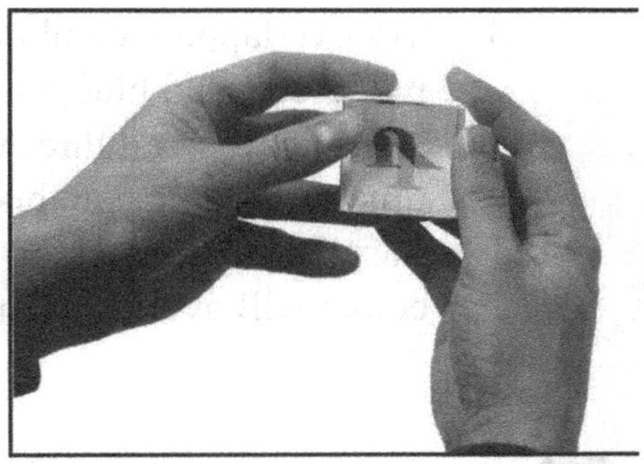

What do you observe? Are images reversed? Are images upside down?

1. Take a piece of graph paper and cut out a rectangle about 5 × 12 centimeters. Draw a line that divides the rectangle into a square 5 × 5 centimeters and a rectangle 5 × 7 centimeters. Next, write a large letter J above the line and a large letter K below the line.
2. Fold the paper along the line, and wrap it around a 45-degree corner of the prism so that the letter K slides under the flat side of the prism and the J folds up onto the angled top side.
3. Look into the prism through the uncovered face as you slowly rotate the prism up and down so you change the angle at which you look into it.
4. What do you see? Write down all the observations you can make, such as which letter you see and whether it is right side up or inverted.

2 Constructing explanations

a. Draw a diagram showing the path of the light when you see one letter. Is this reflection, refraction, or both?

b. Draw a diagram showing the path of the light when you see the other letter. Is this reflection, refraction, or both?

3 Is it a real color?

An artist mixes different colored paints to create almost any color. However, once the paint is mixed, it cannot change itself to a different color. Yet that is precisely what a video screen does—change itself to any color, 30 times per second, using light instead of paint. It can only use three colors; red, green, and blue.

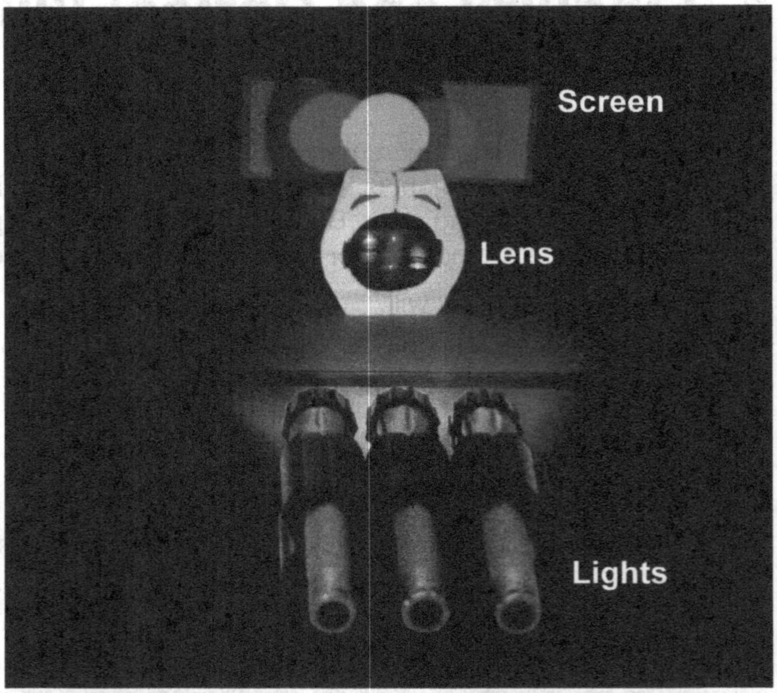

1. Arrange the flashlights, screen, and lens as shown in the photo.
2. Move the lens to get the colored circles to overlap. You may have to dim or turn off the classroom lights to see the colors accurately.
3. Try overlapping combinations of red, green, and blue, as well as all three colors together. The best results are seen when all three colors partially overlap each other at the same time. You can make this happen by detaching one of the holders and flashlights and stacking it on top of the other two in a pyramid. All three colors will overlap evenly.

Colored lights

4 Using your model

Try repeating the experiment and making observations until you can answer the following questions.

a. What is the purpose of the lens? How does it change the circles of color on the screen?

b. What color do you get when you overlap red and blue light?

c. What color do you get when you overlap blue and green light?

d. What color do you get when you overlap red and green light? Does this surprise you?

e. What "color" do you get when you overlap red, green, and blue light?

f. Is there such a thing as yellow light? If so, how is it different from the yellow color you saw in this experiment?

g. Discuss how the human eye perceives color. How does this explain the appearance of yellow?

h. The word *digital* means numbers. A digital picture is actually a set of numbers. How does a television or computer create different colors using only numbers? Discuss the answer to this question in the context of the experiment with the three flashlights.

1.3 Scientific Evidence and Sound

How does our sense of hearing shape our experience of sound?

The human ear is a remarkable sound sensor. You can easily hear sound waves that change the air pressure by one part in a million. Collecting reliable scientific evidence about sound is tricky because people are different and may hear the same physical sound in different ways.

In this investigation, you will:

- learn about the range of frequencies the ear can detect.
- interpret sound as single notes and in chords.

Materials List

- DataCollector
- Sound and Waves with speakers

1 Introduction

The average range of frequencies the human ear can hear ranges from 20 Hz to 20,000 Hz. However, there is tremendous variation between individuals. Hearing changes greatly with age and exposure to loud noises.

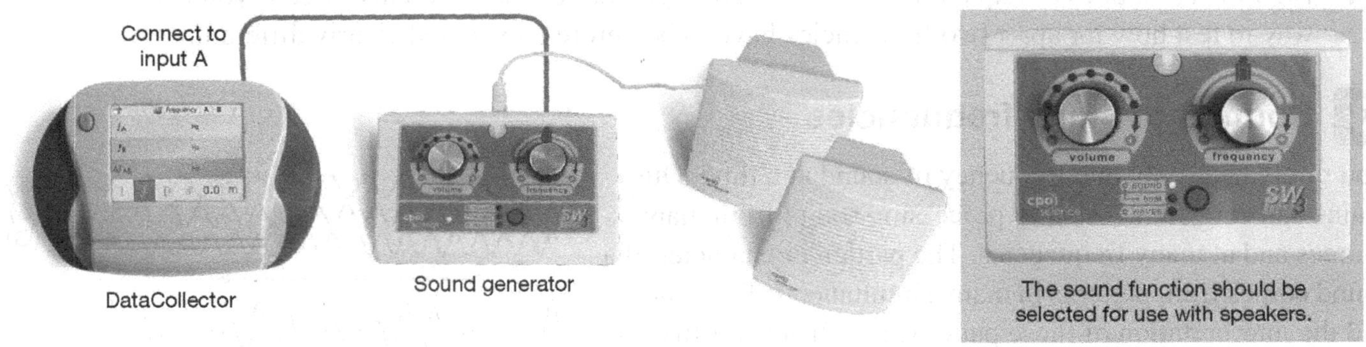

1. Connect your sound generator to your DataCollector and select Timer mode. Select the frequency (f) option on the menu bar at the bottom of the Timer mode screen. Connect a speaker to the sound generator. When you turn the DataCollector on, you should hear a sound and the screen should display a frequency near 440 Hz.
2. There are knobs for frequency and volume control. Try adjusting the frequency and see how high and low it goes.
3. See if you and your group can agree on frequencies where you hear the sound as low, medium, high, or very high frequency. Use Table 1 to record the frequencies of sound that you perceive as low, medium, high, or very high. Don't try to be too exact because the words *low*, *medium*, and *high* are not well defined. It is difficult to agree exactly on anything that is based completely on individual human perception.

Table 1: How we hear frequencies of sound

Description	Frequency (Hz)
Low	
Medium	
High	
Very high	

2 Scientific evidence

Scientific evidence should be free of opinion, and should accurately represent what actually occurs. This quality is called *objectivity.* Evidence is *objective* when it is a clear statement of facts without any human interpretation. A second quality of scientific evidence is *repeatability.* Repeatability means that any observation of the same system under the same conditions will produce the same result.

a. How objective are your classifications of low, medium, and high frequencies? What might cause these observations to contain *bias* or human interpretation?

b. How objective are the numerical measurements of frequency from the DataCollector? What bias or human interpretation appears in this evidence?

c. How repeatable are your observations based on the sounds produced by the equipment you are using?

d. Now that you have some experience with frequency, try and estimate a frequency just by listening to it. Have one person in the group set a frequency and have others in the group estimate its frequency by ear. Compare the actual measured frequencies with the estimations. How objective and repeatable is the human ability to detect and identify frequency?

e. How different does a frequency have to be before you notice the difference? See if you can devise a way to test how far apart two frequencies have to be before they sound clearly different.

3 Sounds of many frequencies

You rarely hear only one frequency of sound at a time. Our brains and ears are constantly processing sound from many sources and at many frequencies. The particular character of a sound comes from patterns in many simultaneous frequencies, and the interpretation of those patterns by a listener. Musical sounds have special relationships based on the frequencies they contain. Each musical note is a different frequency of sound. Over thousands of years, people have found combinations of frequencies that sound good together.

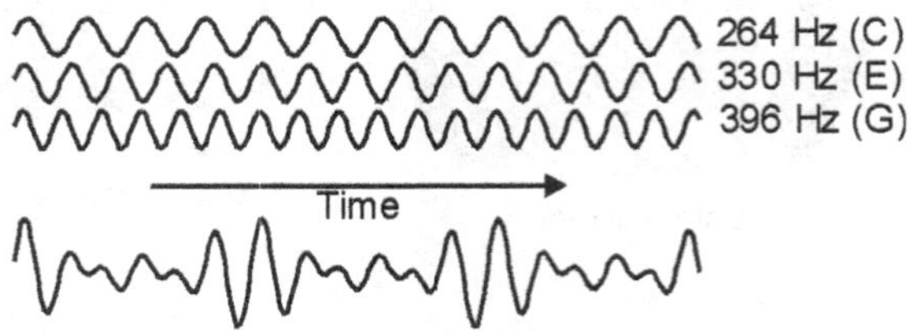

Combination of 264 Hz, 330 Hz,and 396 Hz (C major chord)

1. Turn down the volume so that you cannot hear the sound but can still read the frequency.
2. Each group in the class will be given a different frequency to tune to. Tune your frequency until you are within +/– 1 Hz.
3. Your instructor will tell each group to turn different frequenciesup and down so they can be heard together. Don't change your frequency, just adjust the volume up and down when you are asked.

4 Constructing explanations

a. Describe the sound of the first three frequencies—264 Hz, 330 Hz, 396 Hz—when heard together.

b. Describe the sound of the next three frequencies—264 Hz, 311 Hz, 396 Hz—when heard together.

c. Contrast the two sounds. Does one sound more happy or sad compared with the other? Does one sound spookier than the other? Which combination reminds you more of spring? Which of fall?

d. Which parts of this investigation asked for objective observations, and which parts asked for opinion?

2.1 Distance and Length

How do we accurately communicate length and distance?

Measurements of length and distance are important in physics. In this investigation, you will estimate the lengths of various objects and learn to precisely measure those lengths using the proper number of significant digits.

In this investigation, you will:

- practice length estimation.
- measure and describe lengths using the correct significant figures.

Materials List

- Meter stick
- Metric ruler
- Sheet of heavy paper
- Scissors and packing tape
- Empty 1-liter bottle
- Sand
- Calculator
- Paper clip
- Compass (for drawing circles)
- Map of your local area

1 Estimating

The word *estimate* means to make an educated, approximate guess. For example, you might estimate that it takes 20 minutes to drive to the store. Your estimate is likely to be based on experience, so it is not just a wild guess. The actual time might be off by a few minutes, but it is not likely to be off by 5 or 10 times that 20-minute estimate. Estimating is a very handy tool in physics because it often provides a check for whether an answer could be right or is possibly a mistake.

Estimating length requires you to be familiar with the different units of length used in the world. For a given estimate, first decide on an appropriate unit, then make the estimate. For example, to estimate your height, first choose an appropriate unit. Kilometers are not a good unit to use for a person's height because the value would be a very small fraction, in the range of 0.001–0.002 km. Values this small are hard to estimate. The millimeter is not the best unit either, because your height in millimeters would be a very large value, more than 1,000 mm. Large numbers are also hard to estimate. The meter is a good unit for estimating height because the average height of a person is between 1 and 2 meters.

a. Look at a meter stick but do not use it to measure anything yet. Imagine holding a meter stick up to your body. Estimate your height in meters.

b. Use the meter stick to measure your actual height in meters. How far off was your estimate compared with the actual measurement? Express your answer as a percentage of your actual height.

2 Estimate lengths and distances in meters

Use your sense of how long a meter is to estimate the following lengths in meters. You may want to do some of the estimations in two steps. For example, estimate the width of the pencil eraser in millimeters and then convert the value to meters.

a. the length of your classroom

b. the diameter of a pencil's eraser

c. the height of your desk or laboratory bench

d. the length of a paper clip

Reading the metric ruler (meter stick)

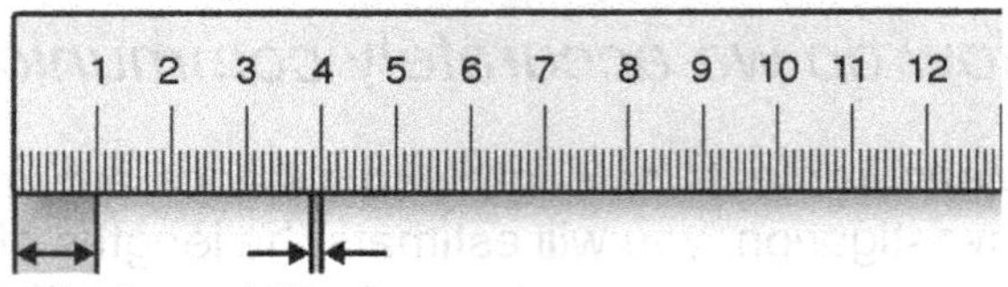

3 Significant digits and measuring length

Precise measurements are often required when estimates do not provide enough information. Consider using a metric ruler with millimeter divisions to measure the length of a pencil. At first glance, the measurement looks to be a little bit more than 9 centimeters. Should the measurement be recorded as 9 centimeters? 9.3 cm? 9.35 cm? 9.347 cm?

9.3 or 9.35 or 9.347?

When recording a measurement, you are only allowed to record the meaningful, or significant digits. **Significant digits are those that you are certain of plus one additional digit that can be meaningfully estimated.**

When the tip of the pencil is lined up with the ruler, you can definitely tell that the pencil is a little bit longer than 9.3 cm, but not quite 9.4 cm. Thus, the "9" before the decimal and the "3" after the decimal are both significant. You are certain of those digits.

Now try to make the measurement a little bit better by estimating one final digit. Look closely at where the tip of the pencil lines up with the markings on the ruler. We already noted that the tip lies between the third and fourth millimeter marks. If the pencil tip lies exactly in between the two marks, we would estimate the last digit to be a 5 and report our measurement as 9.3 and five-tenths of a millimeter, or 9.35 cm. However, if the pencil tip lies very close to but not quite touching the fourth millimeter mark, we might estimate the last digit to be an 8 or a 9 and report our measurement as 9.3 and eight-tenths of a millimeter, or 9.38 cm. Remember, the last digit is an estimate, so you do not need to worry about being absolutely right or wrong. Do your best to make a good estimate.

Both measurements, 9.35 and 9.38, have three significant digits—two are certain (the 9 and the 3) and one is estimated (the 5 or the 8). We cannot record any more digits. The number of significant digits in a measurement depends on the tool used to make the measurement.

a. Now, measure and record each length from Part 2 in meters, using the proper number of significant figures.

4 Analyzing data

a. Compare your estimates in Part 2 with your measurements in Part 3. Which estimate was closer to the actual measurement, the length of a paper clip or the length of your lab or classroom? Why?

b. Use a map to estimate the distance from your school to your home. The next time you travel from your home to school, use a car or bus odometer to measure the distance. Compare the measurement with your estimate.

5 Area

A standard sheet of paper measures 8.5 × 11 inches.

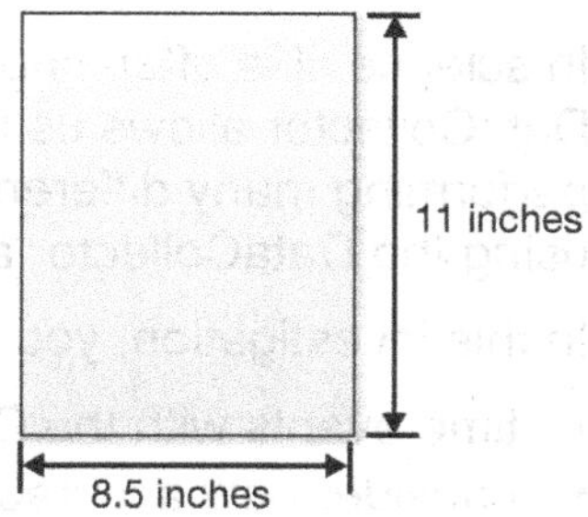

a. Calculate the area of one side of a sheet of paper in square inches.

b. Convert the length and width measurements from inches to meters.

c. Calculate the area in square meters. (*Hint*: The answer is much less than 1 m^2.)

d. Estimate the area of the floor of your lab or classroom.

e. How many sheets of paper would it take to cover your estimated area of the floor of your lab or classroom? Explain how you arrived at your estimate, and show any calculations you made.

6 Volume

A soda bottle is advertised to contain one liter of liquid. One liter is the volume of a cube that measures 10 centimeters on a side. That means there are 1,000 cubic centimeters (cm^3) in one liter (1,000 = 10 × 10 × 10).

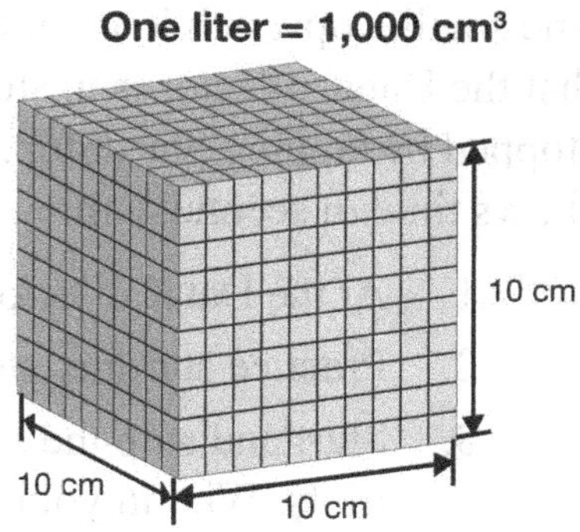

a. How many liters will fit into a cube that is 30 centimeters on a side?

b. Calculate the volume of a cylinder that has a radius of 4 centimeters and a height of 20 centimeters. The volume of a cylinder is approximately 3.14 × radius × radius × height. This relationship is often written $V = \pi r^2 h$.

c. How many liters would fit into this cylinder?

7 Volume by investigation

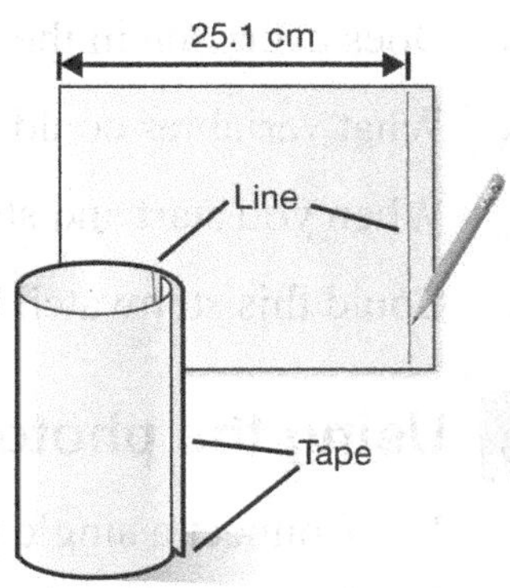

1. Take a sheet of paper and measure 25.1 centimeters along the long side. Draw a line parallel to the short side as shown in the diagram. Roll the paper into a cylinder and tape it together so the edge lines up with the line at 25.1 centimeters. This cylinder has a radius of 4 centimeters. Cut a paper circle with a radius of 4 centimeters and tape it securely to the bottom of the cylinder.
2. Take a 1 liter soda bottle and fill it with sand to the level the liquid would normally reach.
3. Pour the sand into the cylinder and measure the height it fills.

a. Calculate the height of a cylinder that has a radius (r) of 4 centimeters and a volume of 1,000 cm^3. (*Hint*: The volume of a cylinder is $\pi r^2 h$.)

b. How close did 1 liter of sand come to matching the height required to make a volume of 1,000 cm^3?

2.2 Time

How do we measure and describe time?

In science, it is often important to know how things change with time. The DataCollector allows us to make accurate, precise time measurements by performing many different functions. You will explore several of these functions using the DataCollector and a photogate.

MaterialsList
- DataCollector
- Photogate

In this investigation, you will:

- time events with the DataCollector in stopwatch mode and with a photogate.
- consider how accuracy, precision, and resolution are important to time measurements.

1 Stopwatch mode

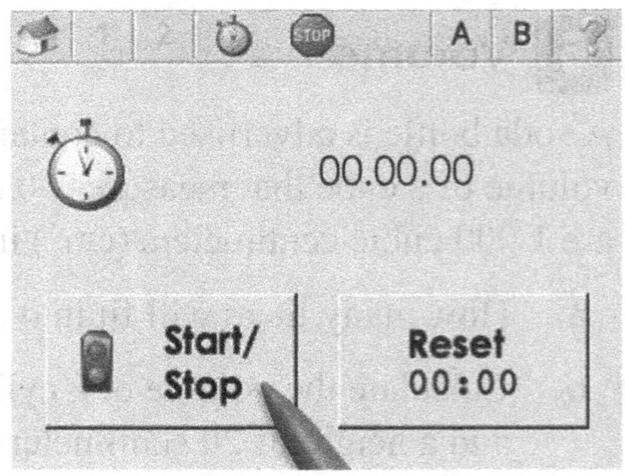

In stopwatch mode, the DataCollector is used to measure time intervals. The small stopwatch icon on the top menu bar of the screen indicates that the DataCollector is in stopwatch mode. The clock is started and stopped by tapping Start/Stop. To reset the clock, tap Reset. The display shows time in seconds, minutes, and hours.

1. At the DataCollector home screen, choose stopwatch mode.
2. Practice starting, stopping, and resetting the stopwatch.
3. Start and stop the stopwatch as quickly as you can. Each group member should perform several trials. Who in your group can get the shortest time? Find the average time for each group member.

a. Does this stopwatch measure time to the nearest tenth, hundredth, or thousandth of a second?

b. What was your average start/stop time? What is the average start/stop time for the entire class?

c. In general, what does your start/stop time represent? (*Hint*: What would you call the time it takes your body to send a signal from your brain to a muscle?)

d. Does everyone in the class have the same average start/stop time? Why or why not?

e. What variables could affect start/stop time?

f. When you start and stop the stopwatch repeatedly, can you always get the same time? Why or why not?

g. Could this stopwatch be used to time a marble falling from a table down to the floor? Why or why not?

2 Using the photogate

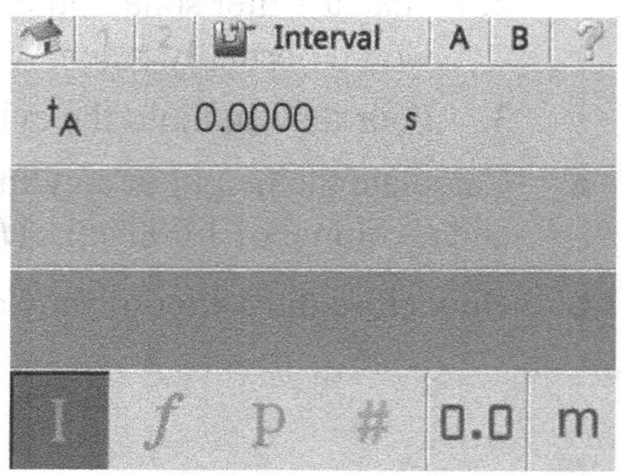

1. Connect a single photogate to the "A" input with a cord.
2. From the DataCollector home screen, choose CPO timer mode. A photogate icon at the top of the screen indicates that you are timer mode. There are four different functions for this mode displayed at the bottom of the screen. Check to see if you are already in Interval mode; the word "Interval" will be next to the photogate icon. If not, choose Interval by tapping the Interval function button "I."

3. The photogate uses an infrared beam to start and stop the DataCollector. Try blocking the beam with your finger and observe what happens to the t_A value on the screen.
4. Which group member can get the shortest time for t_A? Use your finger, a pencil, or some other object, as long as you don't do anything that would damage the photogate.

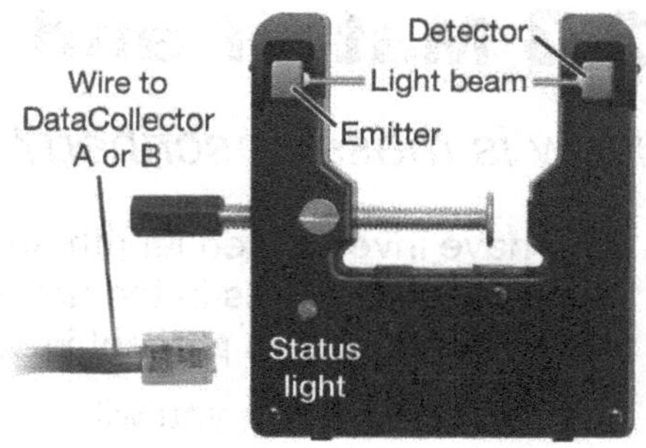

a. Describe the action you take when using a photogate to start and stop time interval measurements by the DataCollector. Be very specific. Someone who has never seen the photogate before should be able to read your answer and know what to do with the infrared beam to make the DataCollector start and stop.

b. If you block the beam several times in a row, does the DataCollector add each new measurement to the last one, or does it start at zero every time you break the beam? Your answer should provide observations for evidence.

c. Does t_A display time to the nearest tenth of a second, hundredth of a second, thousandth of a second, or some other fraction of a second?

d. What was the shortest time for t_A in your group? How does this time compare to the shortest stopwatch time your group achieved?

e. Suppose you are timing how long it takes a marble to fall from a table to the floor. Could you use the DataCollector with the photogate to find the time? Why or why not?

3 Accuracy, resolution, and precision

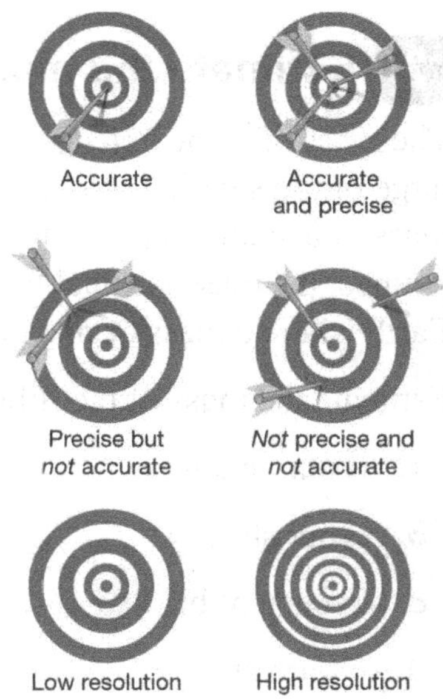

a. *Resolution* refers to the smallest interval that can be measured. Use a photogate to determine the resolution of the DataCollector. Give your answer in seconds and tell how your observations support your answer.

b. The word *accuracy* refers to how close a measurement is to the true value. Which of the following statements best describes what you know about the accuracy of time measurements made with a photogate? Give a reason for your answer.

1. The DataCollector is accurate to 0.001 seconds.
2. The DataCollector is accurate to 0.0001 seconds.
3. It is impossible to know the accuracy without more information about how the DataCollector determines one second.
4. A time of 0.0231 seconds is more accurate than a time of 26 seconds.

c. The word *precision* describes how close repeated measurements of the same quantity are. For example, an ordinary clock with hour, minute, and second hands can give you the time to the nearest second (so the resolution of the clock is 1 second). If several people look at the clock at the same time, and everyone records the same time to the nearest second, then we can say that the times are precise. If the recorded times differ by several seconds, the times are not precise. Suppose, instead of simply reading a time from a clock, you decide to measure how long it takes a toy car to move down a track. Which is likely to give you better precision—a stopwatch or photogates? Explain.

2.3 Matter and Mass

How is mass described?

You have investigated length and time. A third fundamental quantity that describes things in the universe is mass. Mass is a measure of how much matter is present in an object.

In this investigation, you will:

- estimate and measure the mass of various objects.
- review how to represent large and small numbers with scientific notation.

Materials List

- 1-liter bottle of soda (full)
- Electronic scale or triple beam balance
- Bathroom scale
- Scientific calculator
- Other objects to serve as examples of mass

1 Estimating mass

The two most common units for measuring mass are kilograms and grams. For easy reference, a 1-liter bottle of soda has a mass of about one kilogram; a large paperclip has a mass of about one gram. One kilogram equals 1,000 grams. (In the English system, a *pound* is a unit of *force*, not a unit of mass.)

a. Pick up a full (1-liter) bottle of soda to get a sense of the amount of mass in a kilogram. Next, pick up a book such as your textbook. Use the comparison to estimate the mass of the book in kilograms.

2 Estimate the mass of each object

The mass of some objects can be estimated (or conveniently measured) directly in kilograms. Others are too large or too small. For a very small object, think about how many would make up one kilogram of mass. The mass of a single object is then one kilogram divided by the number of objects. If you know the weight in pounds, the mass is equal to the number of pounds divided by 2.2. This conversion works on the surface of Earth where a mass of one kilogram has a weight of 2.2 pounds.

Estimate the mass of the following objects in kilograms.

a. your pencil

b. your shoe

c. a cement block

d. yourself

e. an automobile

f. a paper clip

3 Measuring mass

You need a very sensitive scale (an electronic scale or triple beam balance) to measure the mass of an object like a paper clip or a pencil, but you need a very sturdy scale (like a bathroom scale) to measure your mass or the mass of a cement block. Learning to choose the right measuring device is an important skill in physics.

a. Use the appropriate measuring device to measure the mass in kilograms of each object listed in Part 2.

b. Briefly discuss how your estimates compare to your actual measurements of each item.

4 Scientific notation

Scientific notation is a shorthand method used for expressing extremely large and extremely small numbers. Numbers written in scientific notation are expressed as a product of two numbers: a number between 1 and 10 (called the base) and a power of 10. For example, the number 100 in scientific notation is written as 1×10^2 because $10^2 = 100$. If the mass of an airplane is 75,000 kilograms, how is this written in scientific notation?

- To express 75,000 kilograms in scientific notation, first identify the base (number between 1 and 10). For the airplane mass, the base is 7.5.
- Next, figure out what power of 10 the base needs to be multiplied by to get 75,000. You know that $75{,}000 = 7.5 \times 10{,}000$. The number 10,000 is a power of ten because $10{,}000 = 10 \times 10 \times 10 \times 10$, or 10^4. In scientific notation, the mass of the airplane is 7.5×10^4 kilograms. To quickly choose the power of 10 you need, you can count the number of places you moved the decimal point to get to the base. In this case, you moved the decimal point four places to get from 75,000 to 7.5.

Scientific notation also works for numbers less than one. Powers of 10 that are *negative* are numbers smaller than one. For example, a quarter has a mass of 0.009 kilograms, which is smaller than one. The number 0.009 can be broken down into a product of two numbers: 9×0.001. The second number (0.001) is a power of 10. One-thousandth (0.001) can be expressed as $1 \div 1{,}000$. Since $1{,}000 = 10^3$, $1 \div 1{,}000$ is the same as $1 \div 10^3$. Mathematically, $1 \div 10^3$ is written 10^{-3}. The negative exponent means the actual number is one divided by the power of 10. Using scientific notation, the mass of a quarter is 9×10^{-3} kg. To quickly choose the power of 10 you need, you can count the number of places you moved the decimal point to get to the base. In this case, you moved the decimal three places to get from 0.009 to 9. You must not forget to include the negative sign in the power of 10, since you moved the decimal point to the right.

Table 1: Some useful powers of 10

Numbers larger than 1	Numbers smaller than 1
$10^1 = 10$	$10^{-9} = 0.000\ 000\ 001$
$10^2 = 100$	$10^{-8} = 0.000\ 000\ 01$
$10^3 = 1{,}000$	$10^{-7} = 0.000\ 000\ 1$
$10^4 = 10{,}000$	$10^{-6} = 0.000\ 001$
$10^5 = 100{,}000$	$10^{-5} = 0.000\ 01$
$10^6 = 1{,}000{,}000$	$10^{-4} = 0.000\ 1$
$10^7 = 10{,}000{,}000$	$10^{-3} = 0.001$
$10^8 = 100{,}000{,}000$	$10^{-2} = 0.01$
$10^9 = 1{,}000{,}000{,}000$	$10^{-1} = 0.1$

Words for large and small numbers

1×10^9	1 billion
1×10^6	1 million
1×10^3	1 thousand
1×10^{-3}	1 thousandth
1×10^{-6}	1 millionth

a. Rewrite your estimates from Part 2 a–f using scientific notation.

b. Look up or estimate the mass of each of these objects and write it both using scientific notation *and* as a decimal (ordinary) number. *It is difficult to write some of the masses as decimal (ordinary) numbers.*

a. an atom of hydrogen

b. an atom of uranium

c. a grain of sand

d. the space shuttle

e. an official professional soccer ball

f. the planet Earth

3.1 Position, Speed, and Velocity

How are position, speed, and velocity related?

Knowing an object's position, and measuring how fast or slow it is moving, can be helpful, but not always precise enough for science. Sometimes it is important also to know in what *direction* an object is moving.

In this investigation, you will:

- measure increases and decreases in positions values.
- measure positive and negative velocity.
- compare speed and velocity.

Materials List

- DataCollector
- SmartTrack
- Energy Car
- Graph paper
- Bubble level
- Velocity sensor

1 Position

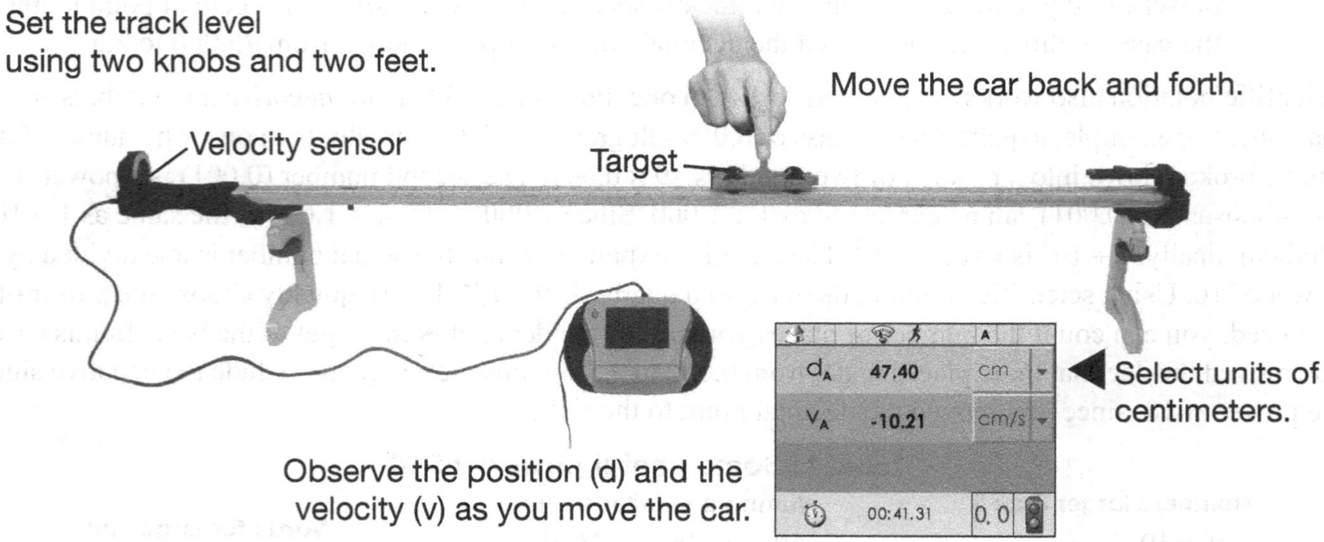

1. Attach the Distance-Velocity sensor to the track, then attach the communication cable so that one end is plugged into the sensor and the other is plugged into the A slot of the DataCollector.
2. Set up the SmartTrack so it is approximately level using the bubble level. You may need to slide a few sheets of paper under one end of the track to make it level.
3. Set the DataCollector to Meter mode. The units will default to centimeters for position, or distance (d), and centimeters/second for velocity (v).
4. Attach the target to the car and place the car at the center of the track so the target is positioned near the 50-centimeter mark, initially leaving the car at rest.
5. Move the car along the track, backward toward the sensor and forward away from the sensor. Observe the values of position and velocity. Answer the questions below using this setup.

2 Constructing explanations

a. Did you observe both increasing and decreasing positions? If so, describe what happened when the position increased and decreased.

b. Did your position value change sign? What do you think it would take to see a negative position?

c. The default zero position is at the 0-centimeter mark on the SmartTrack. If the zero were at the center of the track, would you expect see a sign change when you run this same experiment?

d. What is velocity? What determines when the velocity value changes from positive to negative?

3 Speed and velocity

1. Use the same setup of the EnergyCar and SmartTrack that you used in Part 1 of this investigation.
2. Start a new experiment by going to the home screen and tapping Data Collection.

3. Tap Setup on the menu bar at the bottom of the screen. Set the samples to 100 and the rate to 20 Hz. This will give you five seconds to move the car back and forth on the track.
4. Select Speed and cm/s as your measurement input. The other two inputs should be set to None.
5. Once again, you will begin the experiment with the car at the center of the track, but instead of moving the car back and forth randomly, create a plan with your group for how you are going to move the car for the entire five seconds. Decide in which direction and how fast you will move the car. Make at least three changes in direction, come to a complete stop at least once, and move the car fast at least once and slow at least once. Write out your plan using words and arrows indicating direction and speed. Use a step-by-step approach and describe what should be done during each step. Each step should describe some kind of change.

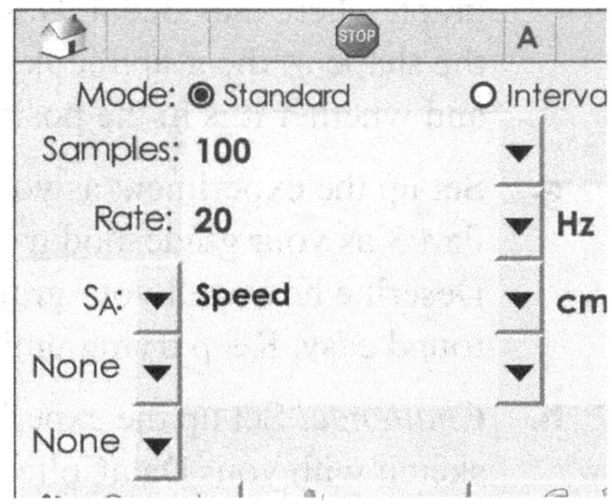

6. Hit the Go button to start the experiment.
7. In order to watch the velocity change in real time on a graph when you move your car on the track, switch to the Graph tab.
8. Move your car according to your plan and watch the graph it makes.

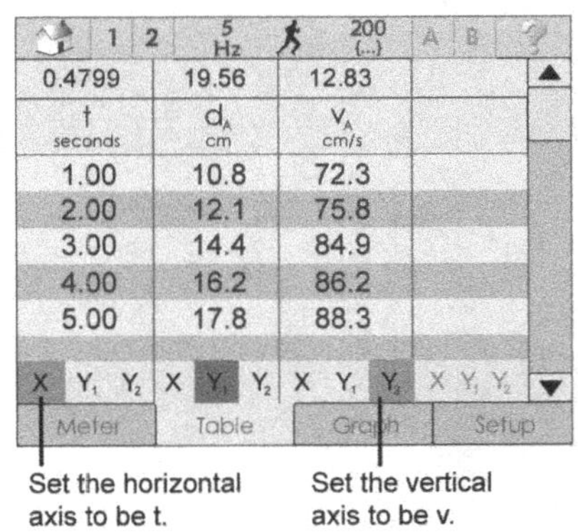

9. Take careful note of what the graph does as you move your car. Write these notes next to each instruction on your list. You will need to be able to make a connection between what is being displayed on the graph and how the car is moving.
10. To repeat the experiment with the current setup, tap the Setup tab and hit Repeat.

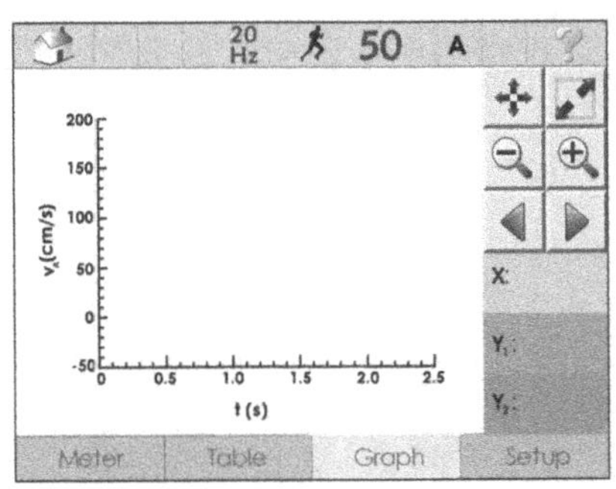

4 Analyzing the data

a. Describe the graph you see. Make a careful sketch of the graph on a piece of graph paper and label the axes.

b. When is the velocity positive? When is it negative?

c. Velocity is a change in position divided by a change in time. Speed is a change in distance divided by a change in time. What is the difference between a change in position and a change in distance?

d. Look at your sketch of the velocity vs. time graph. If you were to use this graph to make a speed vs. time graph, how would the two graphs be different? How would the two graphs be similar?

e. Use your sketch of the velocity vs. time graph to make a sketch of a speed vs. time graph. Do you think you could make an accurate velocity vs. time graph from a speed vs. time graph? Why or why not?

f. Look at your velocity vs. time graph sketch. Match up the steps from your plan with the places on the graph where they occur. Describe how each step changed on your graph. Be descriptive, and state what the shape of the graph looks like at each step: a straight line, a curve, whether it is sloping up or down, and whether it is in the positive or negative region of your graph.

g. Set up the experiment as you did for Part 3. Instead of the written steps, use the sketch you made from Part 3 as your guide, and try to match the DataCollector's graph to the sketch as you move the car. Describe how well your group was able to match the graph, what you found difficult, and what you found easy. Keep trying until you get a good match.

h. *Challenge!* Set up the experiment again. Trade sketches with another group. See if you can match their sketch with your DataCollector's graph as you move the car using their sketch as your guide. Trade sketches with as many groups as time permits. Describe your successes and challenges as you match their graphs. Make a speed vs. time sketch based on each velocity vs. time sketch you try to match.

3.2 Position, Velocity, and Time Graphs

How are graphs used to describe motion?

A graph presents information in a format that can be understood easily, once you know how to interpret it.

In this investigation, you will:

- create graphs of velocity versus position and time.
- create a predictive model for the velocity of a cart rolling down a hill.

Materials List

- DataCollector
- SmartTrack
- Energy Car
- Photogate
- Physics Stand
- Velocity sensor

1 Velocity measurements

Make photogate measurements every 10 cm down the track.

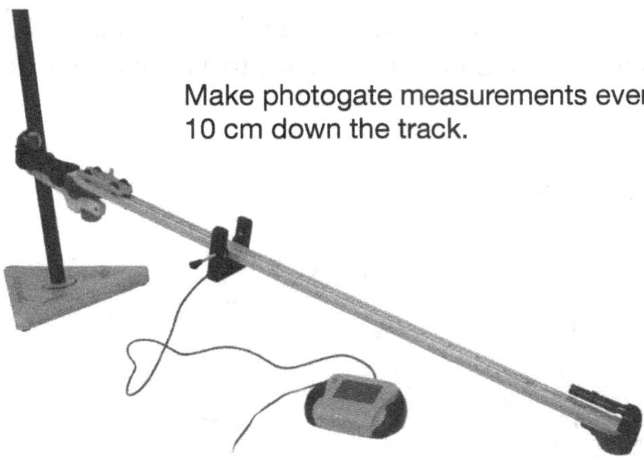

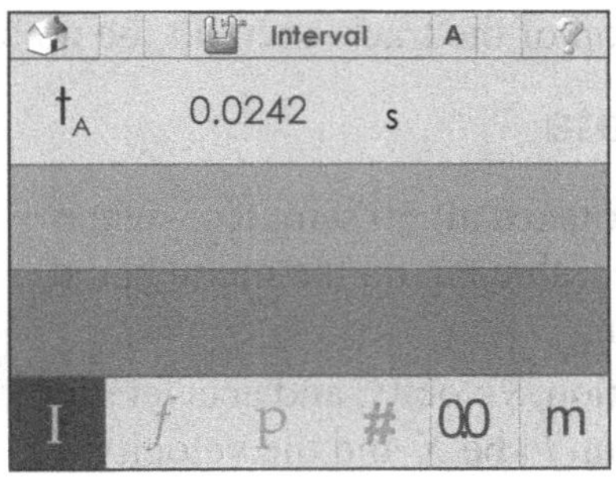

The car moves 1 cm while the light beam is blocked.

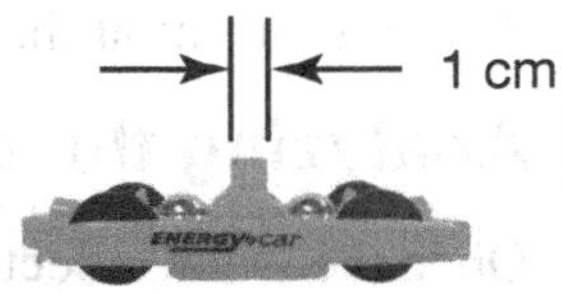

$$v = \frac{1\text{ cm}}{t_A}$$

1. Set up the track with the distance-velocity sensor attached. Use the fifth position from the bottom of the stand. Set the DataCollector to Timer mode and plug one photogate into input A.
2. Place the photogate 10 centimeters from the top of the track. This will be the location of your first trial. You will perform several trials by moving the photogate further down the track in 10-centimeter increments for each new trial. Record the photogate position and the time it takes the one-centimeter flag to break the light beam for each trial you perform in Table 1.

Table 1: Time and velocity data

Photogate position (cm)	Time through A (s)	Velocity (cm/s)

2 Analyzing the data

a. Calculate the velocity of the car in cm/s, and record the results in Table 1.

b. Do the velocities increase or stay about the same along the ramp?

c. Repeat the time measurement five times at one of the locations you measured. Calculate the velocity for each of the five trials. What is the precision of the average? Are the differences between your velocities significant? Justify your conclusion based on the estimated percent error in your measurements.

3 The velocity vs. time graph

In the next part of the investigation, you are going to generate the graph of velocity versus time for the track, and compare your data with the data you collected with the photogate.

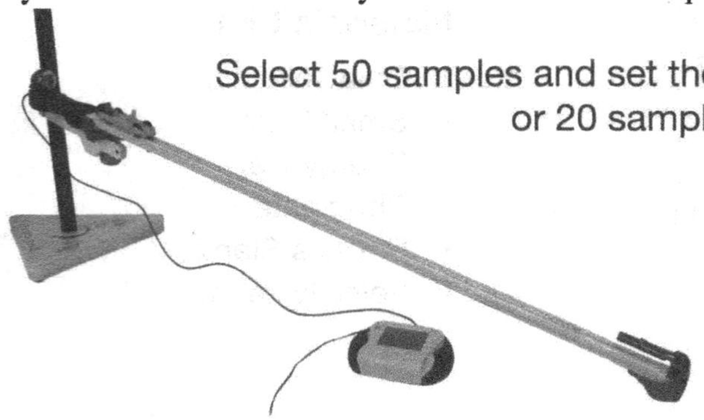

Select 50 samples and set the rate to 20 Hz, or 20 samples per second.

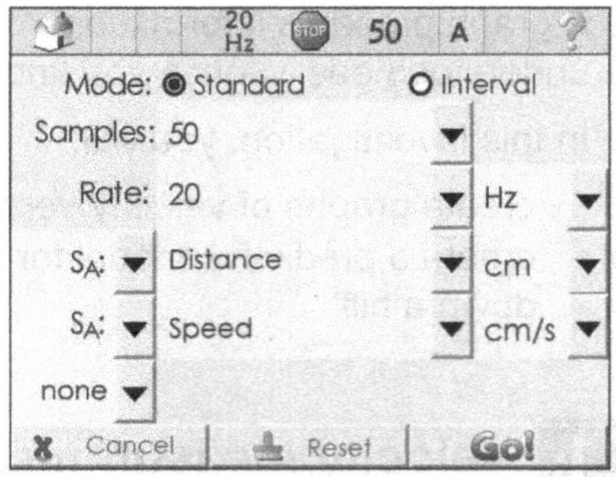

1. Switch the cord from the photogate to the velocity sensor and remove the photogate. Without changing the angle of your ramp, start a new experiment in Data Collection mode. Select the Setup tab at the bottom of the screen and set the samples to 50 and the rate to 20 Hz. Click back to Meter view to see the time, distance, and velocity.
2. Attach the target to the back of the car.
3. Put the car at the top of the track, then hit the go button, then release the car.

4 Analyzing the data

a. Once the DataCollector takes all 50 samples, switch to Table view using the Table tab on the menu bar at the bottom of the screen.

b. Scroll through the position, velocity, and acceleration data. Set the time column to be X and the velocity column to be Y1. Switch to Graph view.

c. Describe the graph you see. Is it a straight line or a curve? Does it get steeper with time or shallower?

d. Is there acceleration in the motion of the car? How do you know from the velocity vs. time graph?

e. Go back to Table view and set the position to be X and the velocity to be Y1. Return to Graph view and observe the graph you see. Is it a straight line or a curve? Does it get steeper with time or shallower?

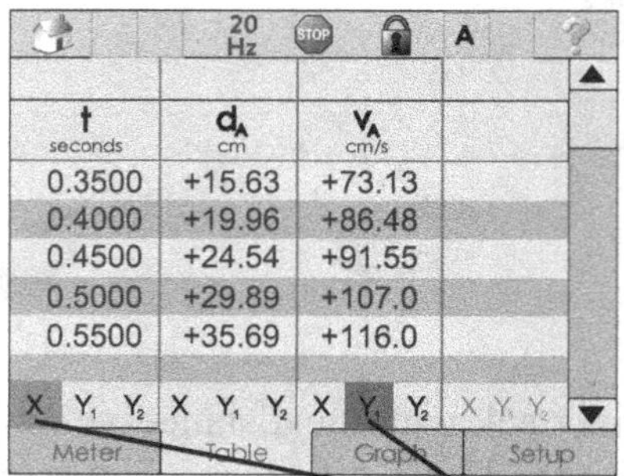

t (seconds)	d_A (cm)	v_A (cm/s)
0.3500	+15.63	+73.13
0.4000	+19.96	+86.48
0.4500	+24.54	+91.55
0.5000	+29.89	+107.0
0.5500	+35.69	+116.0

Set the horizontal axis to be t and the vertical axis to be v.

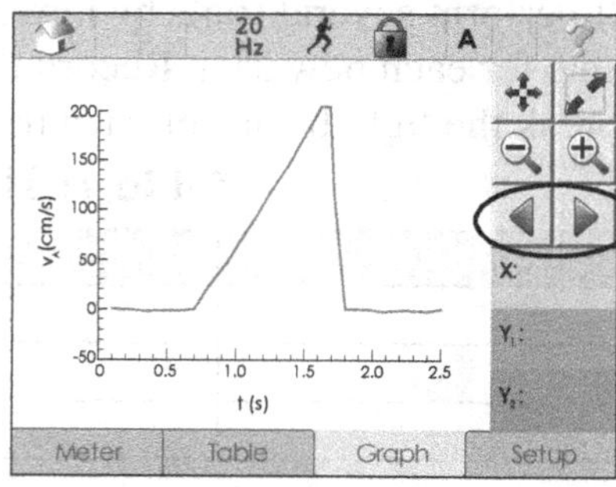

Use the cursor to read different values from the graph.

f. Move the cursor to a position on the graph that is as close as possible to a position where you made a photogate measurement of velocity. Write down the value of both position and velocity. (*Note*: To compare values, you must subtract 6.7 centimeters from the position to find the velocity at the same location for the car.)

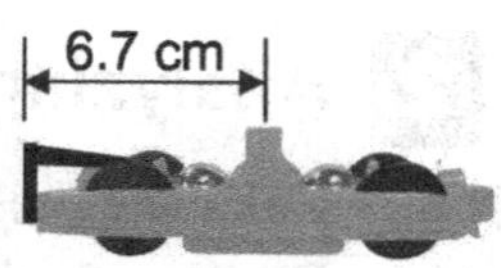

g. Compare the graph values with the photogate calculation you did at the same position in Part 2.

h. Derive a formula that will give you the photogate time if you know the velocity and the distance traveled. Use the cursor to choose a point on the graph and test your formula. (*Note*: Don't forget that you need to consider when in time your car started moving.)

3.3 Equations of Motion

How are equations used in physics?

Equations provide quantitative relationships between variables such as position, time, velocity, and acceleration.

In this investigation, you will:

- create an equation for motion that includes four variables.
- determine parameters in the equation from data.
- test the equation against real data.

Materials List

- DataCollector
- SmartTrack
- Energy Car
- Some small books
- Velocity sensor

1 Constant velocity motion

The objective of this investigation is to create constant velocity. Since there is always friction, the SmartTrack must be set at a very slight angle to compensate. If you get the angle right, the car will not move by itself, but will roll downhill with constant velocity if given a small push. You don't need the car to move very fast—any velocity between about 5 and 20 cm/s works just fine.

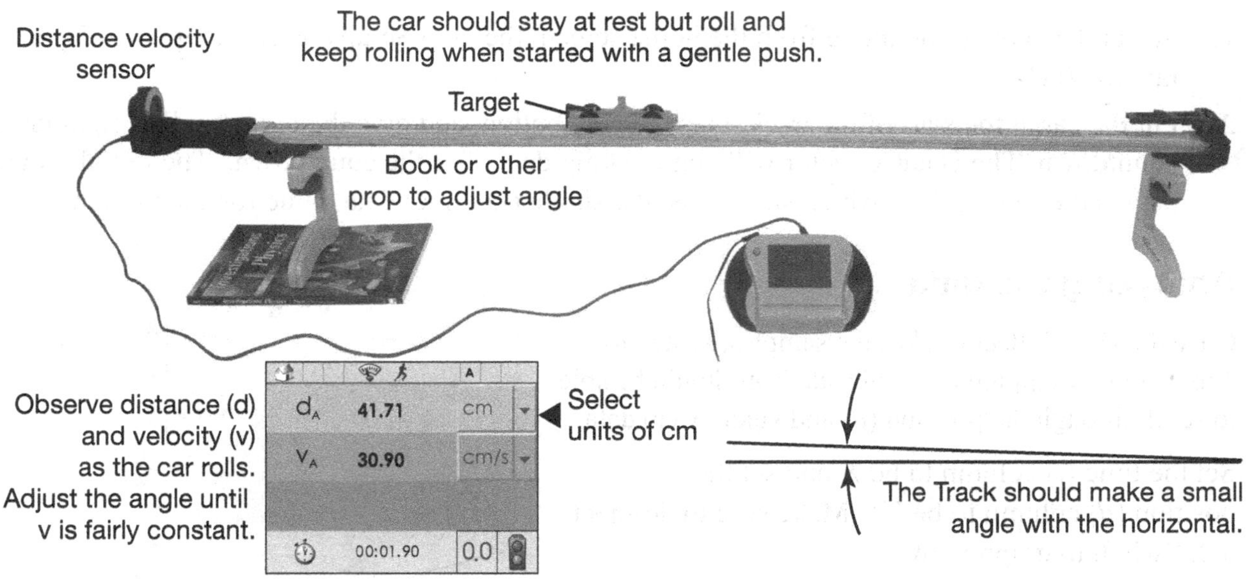

1. Set up the SmartTrack with two feet as shown in the diagram and connect the velocity sensor to the DataCollector.
2. Select Meter mode. You should see a measurement screen that shows you position (d) and velocity (v) with a stopwatch for time at the bottom. Set the units to cm.
3. Place thin objects under the free end of the track to provide a small slope. The slope should not cause the car to roll by itself, but the car should *keep* rolling if started with a small push.
4. Push the car down and let it roll freely while watching the velocity measurements. Your goal is to make the velocity as constant as you can. Adjust the angle in small increments by adding or subtracting shims under the end of the track. Record three trials of your best setup.

Table 1: Velocity data

Trial	1	2	3
Velocity (cm/s)			

2 Thinking about what you observed

a. Why did the track require a slight downhill angle to reach constant velocity?

b. What is the acceleration if the velocity is exactly constant?

c. When do you expect the acceleration to be positive? negative? zero?

3 The graph of constant velocity motion

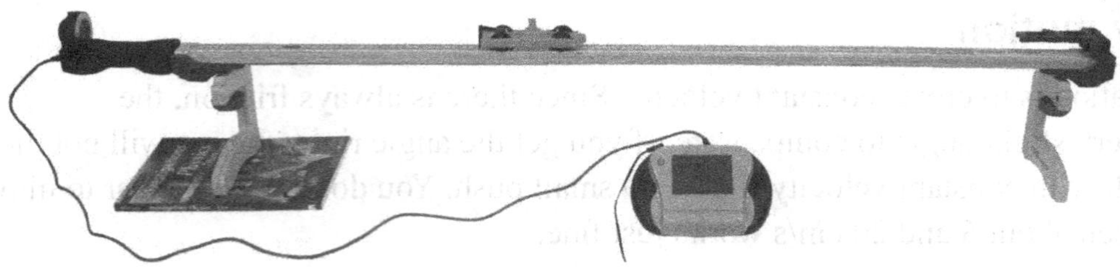

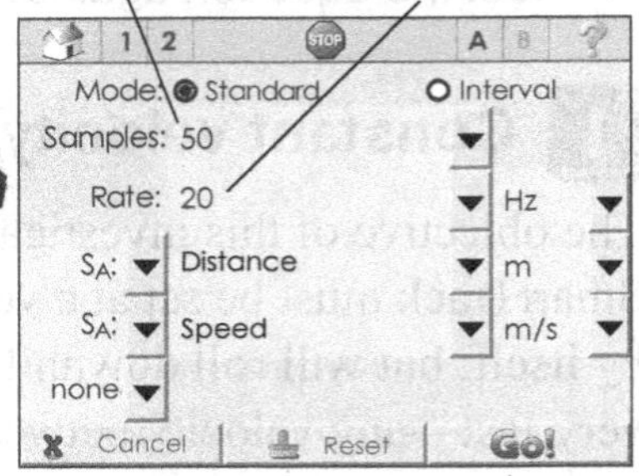

1. Select data collection mode from the home screen, then tap Setup and set samples to 50 and the rate to 20 Hz.
2. Put the car at the start of the track, press the Go button, and give the car a small push after the countdown. The DataCollector will begin taking data after the countdown. The initial view of the data is the Meter view, which shows just the single (i.e. live) data value for each input.

4 Analyzing the data

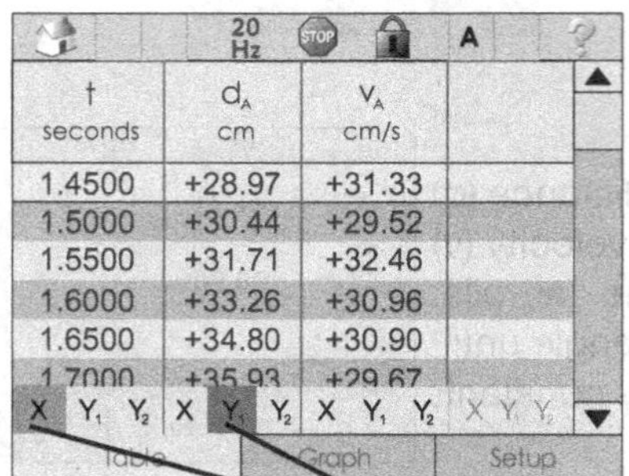

t seconds	d_A cm	V_A cm/s
1.4500	+28.97	+31.33
1.5000	+30.44	+29.52
1.5500	+31.71	+32.46
1.6000	+33.26	+30.96
1.6500	+34.80	+30.90
1.7000	+35.93	+29.67

a. Once the DataCollector takes the samples, switch to Table view by tapping the Table tab. You should be able to scroll through the position (d) and velocity (v) data.

b. Set the time (t) column to be X and set the position (d) column to be Y1. Make sure to deselect Y2. Switch to graph view.

c. Describe the graph you see. Is it a straight line or a curve?

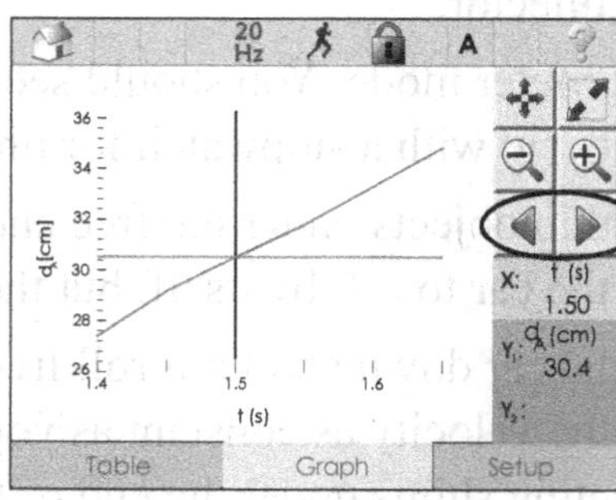

d. The equation for a straight line is $y = mx + b$. Determine the slope, m, and the y-intercept, b, from the graph using the cursor. You will have to calculate the slope from the position and time at two separate points.

e. Rewrite the equation for a straight line using the following variables: position, d; initial position, d_0; velocity, v; and time, t.

f. Compare the velocity you derive from the slope of the line in d above with the velocity you measured on the SmartTrack. Are the two values similar? By what percentage are they different? Give a possible explanation for any differences.

4.1 Acceleration

How does acceleration relate to velocity?

Acceleration is the rate at which velocity changes. The car accelerates as it rolls down the track. You can tell because the velocity changes.

In this investigation, you will:

- measure position and velocity on a ramp.
- observe positive and negative velocity.
- use the velocity data to calculate acceleration.

Materials List

- DataCollector
- SmartTrack
- Energy Car
- Physics Stand
- Velocity sensor

1 Observing acceleration

Start the car at the bottom with enough of a push that the car gets more than 3/4 of the way up the track before it rolls back down again.

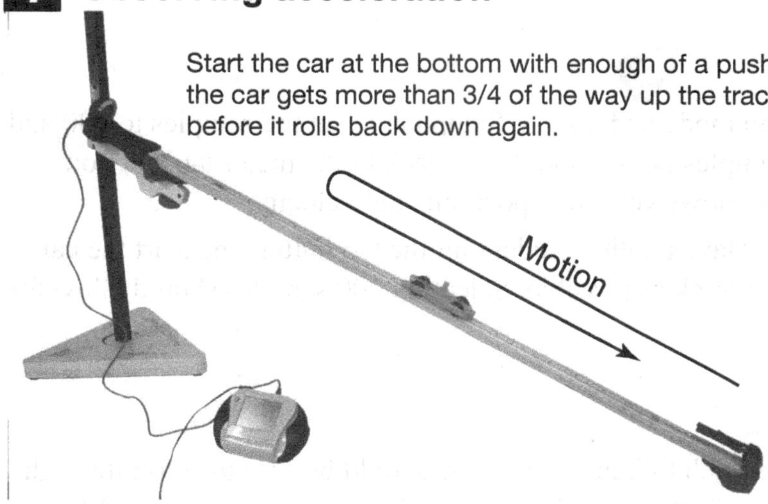

Select units of centimeters (cm).

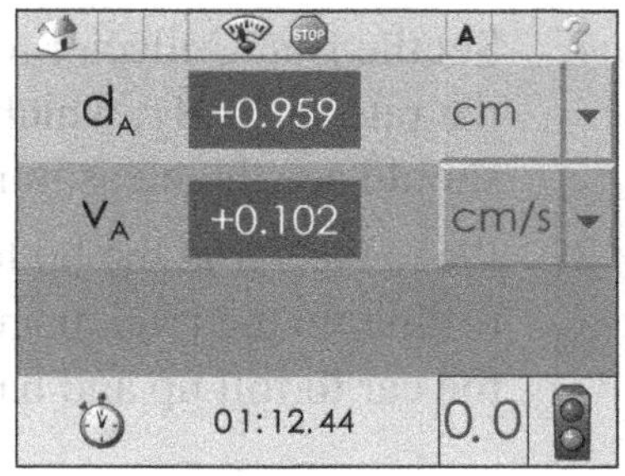

Observe the position (x) velocity (v) and acceleration (a) as the car goes up and down.

1. Set up the SmartTrack at a small angle, using the second hole from bottom of the stand.
2. Connect the velocity sensor to input A of the DataCollector. Set the DataCollector to Meter mode. Set the units to meters.
3. Put the car near the middle of the track. Move it back and forth with your hand while you watch the distance and velocity values.
4. Start the cart at the bottom and give it a short push so it rolls up the track, slows, reverses, and returns to the starting point.
5. Answer the questions in Part 2.

2 Using your model

Suppose you move the car one way, then reverse it and move it the other way.

a. When is the velocity positive? When is it negative?

b. When do you think the largest acceleration occurs? Do you think it occurs when the velocity is highest? lowest? Do you think the largest acceleration occurs when the velocity is changing the most?

c. When do you think the velocity and acceleration had the same sign?

d. When do you think the velocity and acceleration had the opposite sign?

3 Acceleration and velocity

Start the car at the bottom with enough of a push that the car gets more than 3/4 of the way up the track before it rolls back down again.

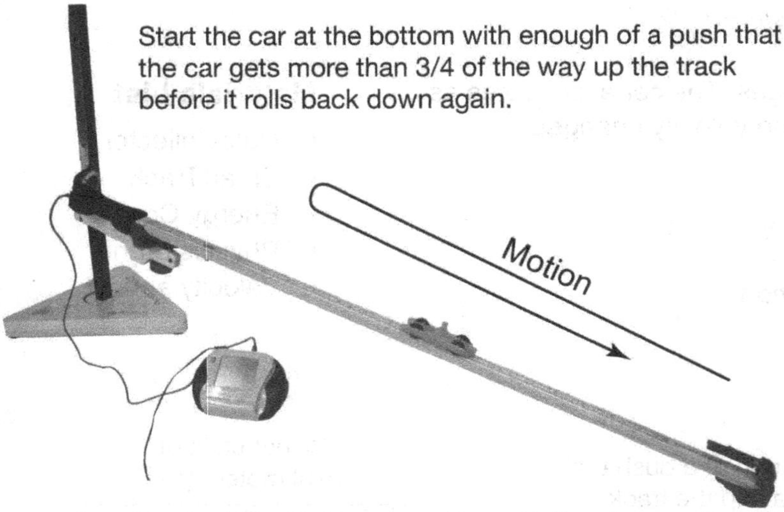

Select 100 samples and set the rate to 20 Hz, or 20 samples per second.

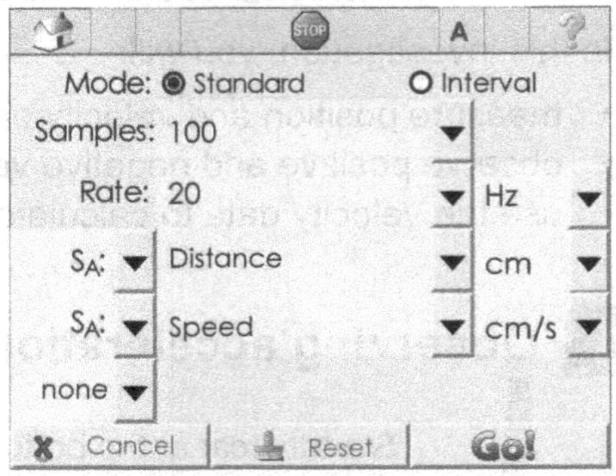

1. Set the DataCollector to Data Collection mode and use the Setup tab to set the samples to 100 and the rate to 20 Hz, which provides 20 samples per second. Click back to the meter tab and you should see the measurement screen that shows you time, position, and velocity.
2. Place the car at the bottom of the track. Have another student hit the Go button and start the car rolling up the track using as much of the track as possible. Selecting 100 samples should allow for the car to roll up and back down again.

4 Analyzing the data

a. Once the DataCollector takes the samples, switch to Table view. You should be able to scroll through the position and velocity. There may be a lot of "stopped" data both before and after the up-and-back data you really want to examine.

b. Set the time (*t*) column to be X and set the velocity (*v*) column to be Y1. Switch to graph view.

c. Describe the graph you see. Is it a straight line or a curve?

d. Explain why the velocity vs. time graph has positive and negative portions.

e. The equation for a straight line is $y = mx + b$. Determine the slope, *m*, and the y-intercept, *b*, from the graph using the cursor. You will have to calculate the slope from the velocity and time at two separate points.

f. Rewrite the equation for a straight line using the following variables: velocity, *v*; initial velocity, v0; acceleration, *a*; and time, *t*.

g. Is the acceleration positive or negative?

h. What is the meaning of the sign of the acceleration in terms of how the car moves on the track?

i. Go back to Graph view. Can you find a place where the velocity is zero but the acceleration is not zero? What was the car doing at that moment of its motion? Where on the track was the car when its velocity was zero?

j. Draw a graph representing the acceleration versus time.

k. Explain the relationship between the velocity vs. time graph and the acceleration vs. time graph. Why is one graph sloped while the other is relatively flat?

4.2 Accelerated Motion

How do we describe and predict accelerated motion?

A model for uniformly-accelerated motion includes velocity, acceleration, position, and time.

In this investigation, you will:

- determine an equation for the velocity in accelerated motion.
- use the equation to make predictions.

Materials List

- DataCollector
- SmartTrack
- Energy Car
- Photogate
- Physics Stand
- Velocity sensor

1 Uniformly accelerated motion

1. Set up the SmartTrack at an angle by attaching it to the stand at the fifth hole from the bottom.
2. Set the DataCollector to Data Collection mode from the home screen, and use the setup tab to set the samples to 50 and the rate to 20 Hz.
3. Put the car at the top of the track, and tap the Go button. The DataCollector will begin a short countdown and then will begin taking data. The initial display is the Meter view of the data where you should see time, position, and velocity.
4. Release the hold button on the SmartTrack to let the car roll down the track.

Set the SmartTrack at an angle.

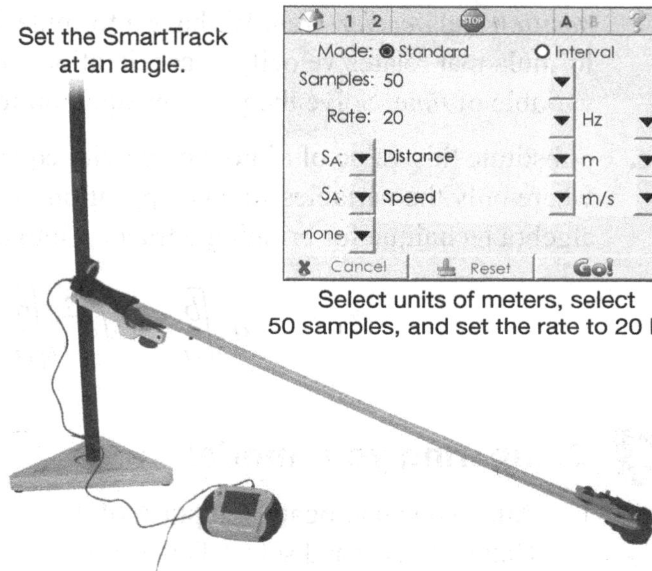

Select units of meters, select 50 samples, and set the rate to 20 Hz

2 Analyzing the data

a. Once the DataCollector takes the samples, switch to Table view. You should be able to scroll through the position and velocity data.

b. Set the time (t) column to be X and set the velocity (v) column to be Y1. Switch to Graph view.

c. Describe the graph you see. Is it a straight line or a curve?

d. The equation for a straight line is $y = mx + b$. Determine the slope, m, and the y-intercept, b, from the graph using the cursor. You will have to calculate the slope from the position and time at two separate points. Try using data points that fall within five to 30 centimeters down the track.

e. Rewrite the equation for a straight line using the following variables: velocity, v; initial velocity, $v0$; acceleration, a; and time, t.

f. Calculate the slope of two points between 30 and 55 centimeters down the track. Compare the acceleration you derive from the slope of the line at this part of the track with the acceleration you derived in d above. Are the two values similar? By what percentage are they different? Give a possible explanation for any differences.

3 Developing the equation

The equations for uniformly accelerated motion are

$$v = v_0 + at$$

$$x = x_0 + v_0 t + \frac{1}{2}at^2$$

a. Assume that the initial position, x_0, is zero. This is true if the position is measured from the flag of the car just before it starts rolling down the track, which is something you will see in the next part of the investigation. Assume also that the initial velocity, v_0, is zero. Rewrite these two equations, eliminating terms that are zero.

b. To predict the velocity of the car at any point on the track, we need a formula that relates *velocity* to *position* and *acceleration*. We have a formula that relates position to acceleration and time, and another formula that relates velocity to acceleration and time. We need to combine these to eliminate the variable of *time*. Solve the position equation for time so you have an equation in the form $t = ?$.

c. Substitute this value of t into the velocity equation and solve it to get an equation of the form: $v = ?$ where only the variables x and a appear on the right-hand side. You will need to use the following algebra technique for bringing a fraction into a square root.

$$a\sqrt{\frac{b}{a}} = \sqrt{a^2}\sqrt{\frac{b}{a}} = \sqrt{\frac{a^2 b}{a}} = \sqrt{ab}$$

4 Comparing your model

1. Attach one photogate to input A of the DataCollector and select Timer mode.
2. Using the same track setup as in Part 1, place the photogate somewhere on the track and measure its position as shown in the diagram.
3. Use the position, x, you just measured, and the value of acceleration, a, to calculate the velocity the car should have when it reaches the photogate.
4. Use the relationship $t = d/v$ to calculate how long it should take the flag on the car to break the light beam. The d in this formula is the width of the flag on the car, which is 0.01 meters (1 centimeter).

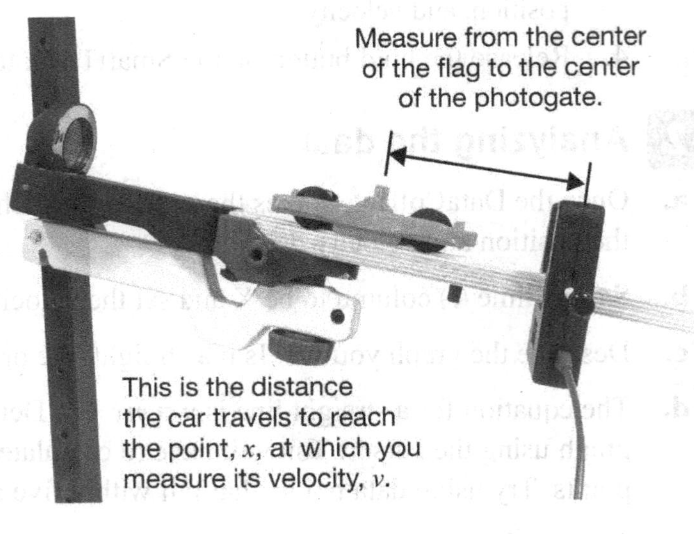

5. Let the car roll down the track, and see how close your prediction comes to the actual measured time interval.
6. Calculate the percent error between your prediction and the measurement. Take a few trials to get a sense of the accuracy of the experiment.

4.3 Free Fall

What kind of motion is falling?

We know falling objects accelerate, and that is a good place to start when examining their motion.

In this investigation, you will:

- determine an equation for the velocity in free fall.
- use the equation to make predictions.

Materials List

- DataCollector
- Physics Stand
- SmartTrack
- Energy Car
- 3 Steel marbles
- Velocity sensor
- Black plastic marble

1 Free fall

1. Set up the physics stand and the velocity sensor from the SmartTrack as shown in the diagram, putting the stand on the floor. Putting a rag in the catcher helps to eliminate noise in the data.
2. With the DataCollector on a table, select Data Collection mode from the home screen and use the Setup tab to set the samples to 20 and the rate to 20 Hz.
3. Attach the target to the Energy Car. Press the Hold button and use the target to hang the car from the pin that comes through the track. Make sure the wheels are properly set on the track.
4. Press the Go button and release the Hold button after the countdown. The DataCollector will measure the time, position, and velocity of the car as it falls.
5. The fall takes less than one second. Be careful not to push or shake the track. To get the technique right, you should perform several trials.

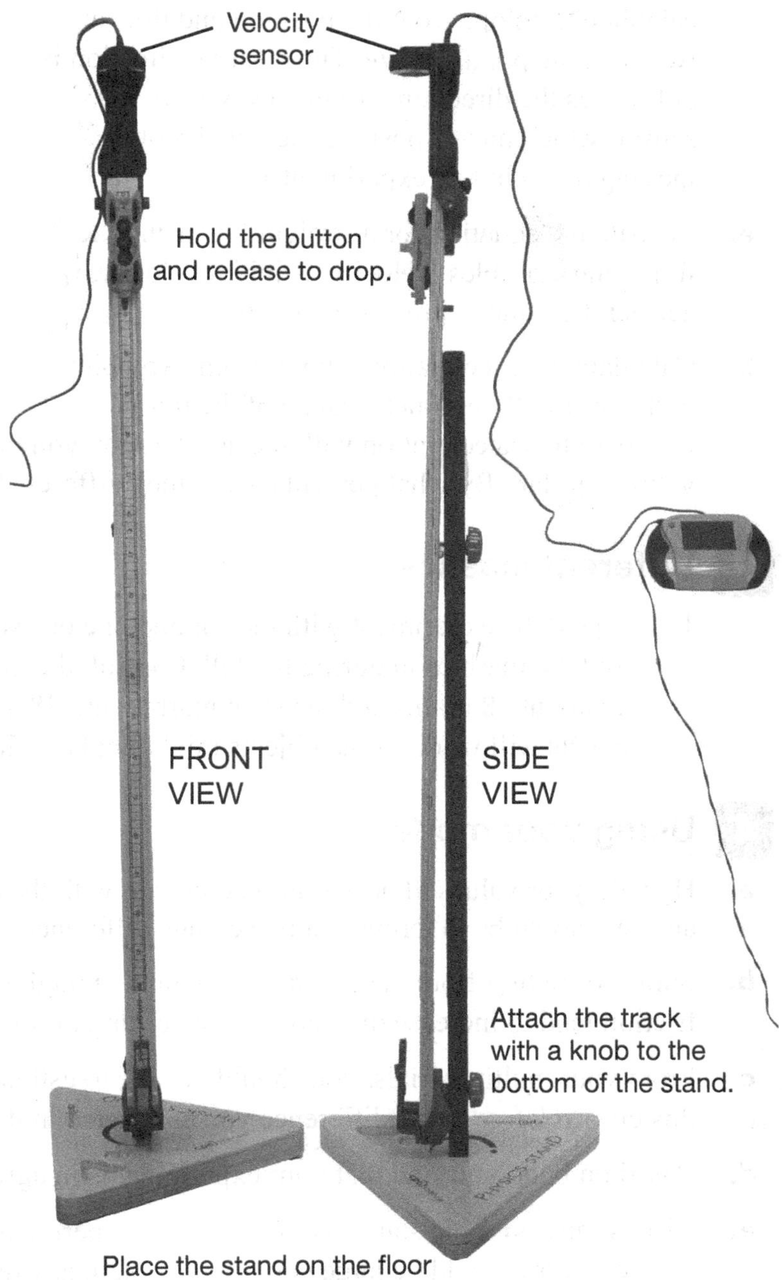

Place the stand on the floor to get enough height to make the drop.

2 Analyzing the data

a. Once the DataCollector takes the samples, switch to Table view. You should be able to scroll through the position and velocity data.

b. Set the time (t) column to be X and set the velocity (v) column to be Y1. Switch to graph view.

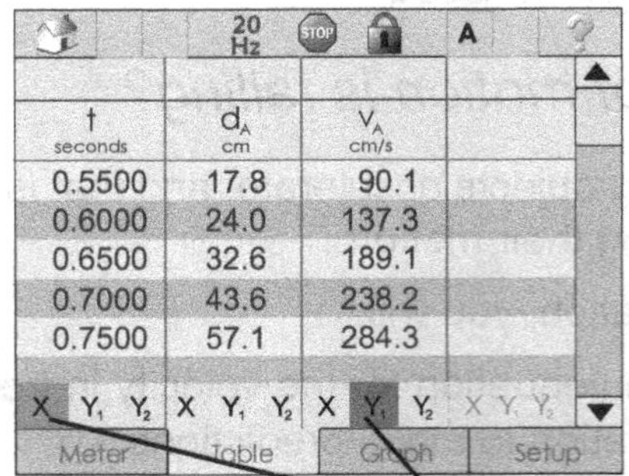

Set the horizontal axis to be t and the vertical axis to be v.

c. Describe the graph you see. Is it a straight line or a curve?

d. The equation for a straight line is $y = mx + b$. Determine the slope, m, and the y-intercept, b, from the graph using the cursor. You will have to calculate the slope from the position and time at two separate points. (*Note*: The positive direction is defined as the direction moving away from the sensor, which means positive indicated motion moving *down* in this experiment.)

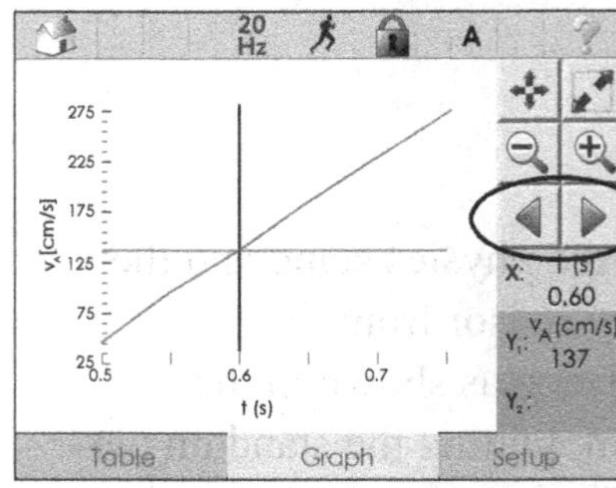

Use the cursor to read different values from the graph. In the example, the velocity is 137 cm/s at t = 0.60 s.

e. Rewrite the equation for a straight line using the following variables: velocity, v; initial velocity, v_0; acceleration due to gravity, g; and time, t.

AVERAGE ACCELERATION

$$g_{avg} = \frac{v_2 - v_1}{t_2 - t_1}$$

f. Calculate the acceleration (slope) from two points in the 30-to-60-centimeter range of the track. Compare this acceleration with the acceleration you derive from the slope of the line in d. Are the two values similar? By what percentage are they different? Give a possible explanation for any differences.

3 Different masses

1. Repeat the experiment with the car and one or two steel marbles. You need the magnet plate to hold the marbles in during the fall. Careful, the marbles can fall out on impact. The car and target are about 58 grams and the steel marbles are 28 grams. You can also try other objects. Not all objects will work. Some objects might not be reflective enough to work with the sensor.

4 Using your model

a. How do your values of acceleration compare with the accepted value of g, which is 9.82 m/s^2? Your answer should be in terms of a percentage difference between your result and the accepted value.

b. Suppose friction from the air was substantial enough to observe. What would the effect be? Would air friction tend to increase or decrease the acceleration you measured?

c. From your multiple trials, you should be able to estimate an error for your measurement. How big is this error relative to the difference you calculated in item a above?

d. Based on b, do you find that your experiment is in agreement with the accepted value of g?

e. Discuss any possible sources of error in the experiment. Make sure you assess whether each potential source of error could be large enough to explain any differences you observed.

5.1 The First Law: Force and Inertia

How does changing an object's inertia affect its motion?

Newton's first law states that objects tend to keep doing what they are doing unless acted on by an unbalanced force. This law applies both to objects at rest and objects in motion.

In this investigation, you will:

- change the Energy Car's inertia to see what effect this has on the Energy Car's motion.
- apply Newton's first law to describe the Energy Car's motion.

Materials List

- SmartTrack
- Energy Car
- Steel marbles
- Rubber band
- Track feet (2)
- Bubble level
- Velocity sensor
- DataCollector
- Electronic scale

1 Making a prediction

Newton's first law of motion applies to objects at rest and objects in motion. The first law can also be called the law of inertia. To understand why, consider what inertia is. Inertia is a property of matter that resists a change in motion. Inertia comes from an object's mass. Because of inertia, objects at rest remain at rest, and objects in motion remain in motion, unless acted upon by a net force. You will investigate the first law by launching the Energy Car on a flat track and seeing what happens when you change the car's inertia.

a. Reflect on previous observations using the Energy Car system or review Investigation 3.1. Write a hypothesis to address this question: "How will changing the inertia of the car affect the car's motion on the flat track?" Your hypothesis should follow this format: "If inertia affects the car's motion on the flat track, then when I add more marbles to the car, the car____."

2 Setting up the experiment

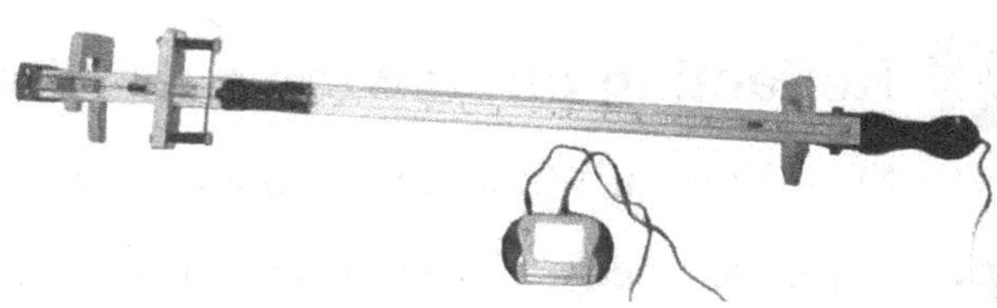

1. Attach a foot to each end of the SmartTrack so it sits level on the table. Check it with the bubble level and pieces of paper under the feet to adjust the height as necessary to make the track level.
2. Attach the velocity sensor to the start end of the SmartTrack.
3. Fasten a rubber band on the launcher and attach it to the SmartTrack. It is best to have the rubber band twist into an "x."

4. Set the adjustable screw so that the flat end is two centimeters behind the rubber band.
5. Turn the DataCollector on and then plug the velocity sensor into the DataCollector.
6. At the home window, select Data Collection mode.
7. At the Go window, choose Setup at the bottom of the screen.
8. At the Setup window, choose Standard mode, 50 samples, and 20 Hz. This will allow the DataCollector to collect 20 samples of data from the sensor each second.

9. Put the target on the EnergyCar.
10. Practice launching the Energy Car with no marbles. Once you have a consistent launch technique, get the car ready to launch, press the Go button on the DataCollector, and launch the car.
11. Make a note of the target distance value at the launch location. This value will be a reference when you look at the data table and graph.
12. Switch to Table and Graph views to study your data.

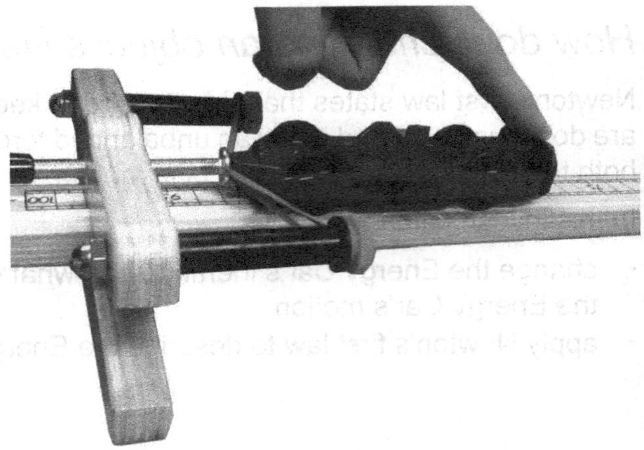

a. Describe what happens to the Energy Car's velocity as it moves along the track. Explain why this happens.

b. An object at rest remains at rest unless acted upon by an outside force. What is the outside force that acts on the car to disturb its state of rest at the start of the track?

c. An object in motion remains in motion unless acted upon by an outside force. What outside force acts on the car to change its motion as it moves along the track?

3 Planning your investigation and communicating results

a. Design an experiment to test the hypothesis you formulated in Part 1a. What is your procedure?

b. Create a data table and a graph to communicate your results.

c. Summarize your findings. Be sure to refer back to your hypothesis.

4 Reflecting on Newton's first law

a. State Newton's first law in your own words.

b. Place the Energy Car in the center of the track so it stays at rest. What do you know about the forces on the Energy Car? Identify the forces acting on it.

c. If the Energy Car is moving, and there are no unbalanced forces acting on it, does its speed increase, decrease, or remain the same? Explain.

d. Were any forces acting on the Energy Car as it rolled along the level track? Identify the forces. Explain how Newton's first law is applied to describe the motion you observed.

e. What changes occur in the forces acting on the Energy Car when the track is tilted slightly up or down? Explain how the first law is applied to describe the observed motion in the case of an uphill or downhill slope.

5.2 The Second Law: Force, Mass, and Acceleration

What is the relationship between force, mass, and acceleration?

British scientist George Atwood (1746–1807) used two masses on a light string running over a pulley to investigate the effect of gravity. You will build a similar device, aptly called an Atwood's machine, to explore the relationship between force, mass, and acceleration.

In this investigation, you will:

- measure the acceleration for an Atwood's machine of fixed total mass.
- create a graph of force versus acceleration for the Atwood's machine.
- determine the slope and *y*-intercept of your graph, and relate them to Newton's second law.

Materials List

- Physics Stand
- Double pulley
- Red safety string
- Mass hangers (2)
- Steel washers (18)
- Plastic washers (6)
- Measuring tape
- Photogate
- DataCollector
- Electronic scale or triple beam balance

1 Analyzing the Atwood's machine

To accelerate a mass, you need a net force. Newton's second law shows the relationship between force, mass, and acceleration:

NEWTON'S SECOND LAW

$$F = ma$$

Force (N), **Mass** (kg), **Acceleration** (m/s²)

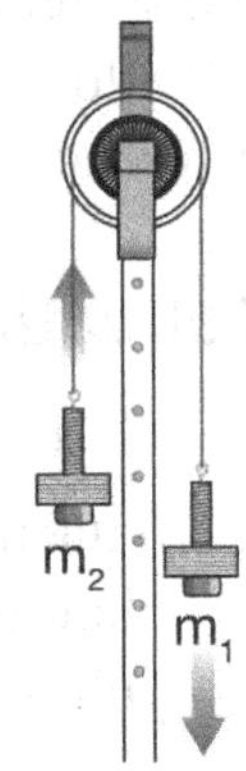

The Atwood's machine is driven by a net force equal in magnitude to the weight difference between the two mass hangers. You will vary the two masses, m_1 and m_2, but you will keep the total mass constant. As you move plastic washers from m_2 to m_1, you will use a photogate to measure the acceleration of the system. If you know the acceleration and the total mass of the system, you will be able to calculate the net force that is responsible for accelerating the system. The equation for the system's motion is a variation of the basic second law formula:

$$F_{net} = (m_1 + m_2)a$$

An ideal pulley would be frictionless and massless, and would just redirect the one-dimensional motion of the string and attached masses without interfering with the motion. However, the pulley you will use has mass and there will be some friction involved. For the purpose of this investigation, we will neglect the mass of the pulley, but we will be able to analyze the friction involved with our Atwood's machine. To represent the friction present in the system, you must subtract it from the net force, since the friction opposes the force of weight provided by m_1 and m_2;.

$$F_{net} - \text{friction} = (m_1 + m_2)a$$

It is easiest to move the friction force (*f*) to the other side of the equation, so you get:

$$F_{net} = (m_1 + m_2)a + f$$

2 Setting up the Atwood's machine

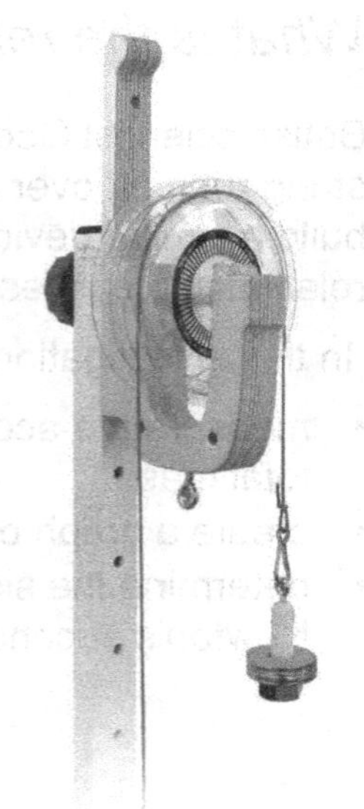

1. Set up the Atwood's machine as shown in the photograph at right. Attach the double pulley to the top of the Physics Stand. You will only use the striped pulley.
2. Attach the mass hangers to the red safety string. Place 10 steel washers on one mass hanger. This will be m_1. Place eight steel washers and six plastic washers on the other mass hanger. This will be m_2. Place the string over the dynamic pulley.
3. Pull m_2 down to the stand base. Place a sponge or some other small cushion on the base to protect it from the falling m_1. Let go of m_2 and observe the motion of the Atwood's machine.

a. Which mass moves downward and why?

b. What would happen if m_1 and m_2 were equal masses? Why?

c. Do the masses accelerate when they move? Explain.

d. How does the acceleration of m_1 compare to the acceleration of m_2?

e. The system's net force equals the weight difference of the mass hangers. Write a simple formula that will allow you to use the mass difference and *g*, the acceleration due to gravity, to calculate the weight difference of the mass hangers.

3 Collecting data

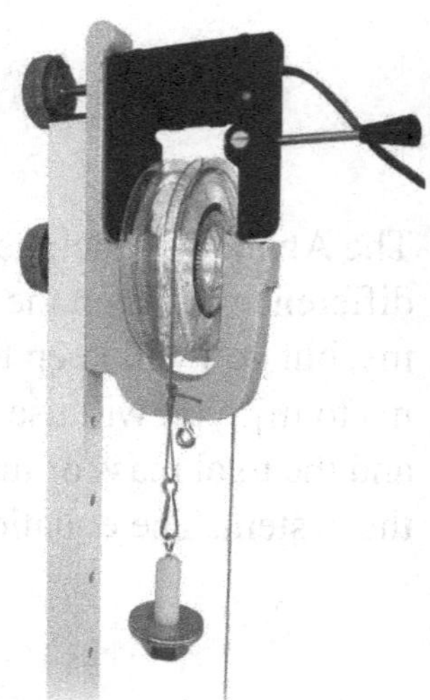

1. Find the total mass of m_1 and m_2 and record in Table 1.
2. Attach a photogate to the double pulley as shown at right. Plug the photogate into the DataCollector (input A). The striped pattern on the pulley will break the light beam of the photogate as the pulley rotates.
3. Turn on the DataCollector. At the home window, select Data Collection mode.
4. At the Go window, tap on the Setup option at the bottom of the screen.
5. In the Setup window, choose Standard mode.
6. Set samples to 500 and set the data rate to 5 Hz.
7. For input, select the SP_A setting, then select speed in cm/s. Set the other inputs to none, working from bottom to top.
8. Pull m_2 to the base. Tap Go at the bottom of the Setup window.
9. When the experiment has started, release m_2. When the hanger falls onto the cushion, press the button on the top left of the DataCollector to stop the experiment.

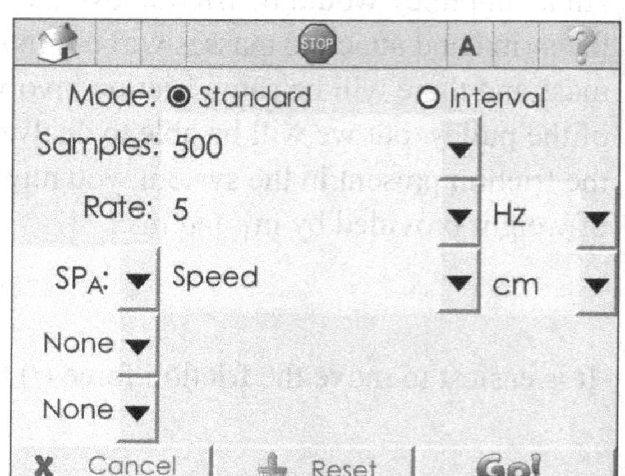

10. Select the Graph tab at the bottom of the screen, and study the graph of the speed data. The speed should be increasing at a constant rate. Determine the acceleration by calculating the slope of the speed vs. time graph. Convert to units to m/s^2. Record the acceleration in Table 1. Calculate the net force (see your answer to 2e) and record it in Table 1.
11. Transfer one of the plastic washers from m_2 to m_1.
12. Press the button on the DataCollector to resume data collection. Repeat steps 7–10 until you have transferred all of the plastic washers to m_1.

Table 1: Acceleration and force data

Acceleration (m/s^2)	Calculated net force (N)

4 Analyzing the data

a. Make a net force vs. acceleration graph (force on the *y*-axis and acceleration on the *x*-axis). Draw a best-fit line through the data points.

b. What kind of relationship does the graph show? Is this consistent with Newton's second law? Explain.

c. Determine the slope of your line. What is the significance of the slope in your experiment?

d. Compare your slope and the known total mass of the system. What is the percent difference? What could account for any difference?

e. Determine the *y*-intercept of your line. The equation for a line is $y = mx + b$ (m is the slope and b is the *y*-intercept). Substitute your variables in for y, m, and x. Compare this equation to the one presented in Part 1.

$$F_{net} = (m_1 + m_2)a + f$$

f. Based on your answer to the previous question (4e), what does the *y*-intercept represent? Does this value make sense? Explain.

5.3 Newton's Third Law: Action and Reaction

What happens when equal and opposite forces act on a pair of Energy Cars?

When you apply a force to throw a ball, you also feel the force of the ball against your hand. That is because all forces come in pairs called *action* forces and *reaction* forces. Newton's third law of motion states that there can never be a single force (action) without its opposite (reaction) partner. Action and reaction forces always act in opposite directions on two different objects. You can set up two Energy Cars to study Newton's third law.

In this investigation, you will:

- link two Energy Cars to create an action/reaction force pair.
- use different numbers of marbles in each car to see how motion is affected.
- relate the cars' motion to Newton's third law.

Materials List

- SmartTrack
- Energy Cars (one blue and one orange)
- Energy Car link
- Steel marbles
- Rubber band
- Track feet (2)
- Bubble level
- Velocity sensor
- DataCollector
- Safety goggles

1 Setting up an action/reaction force pair

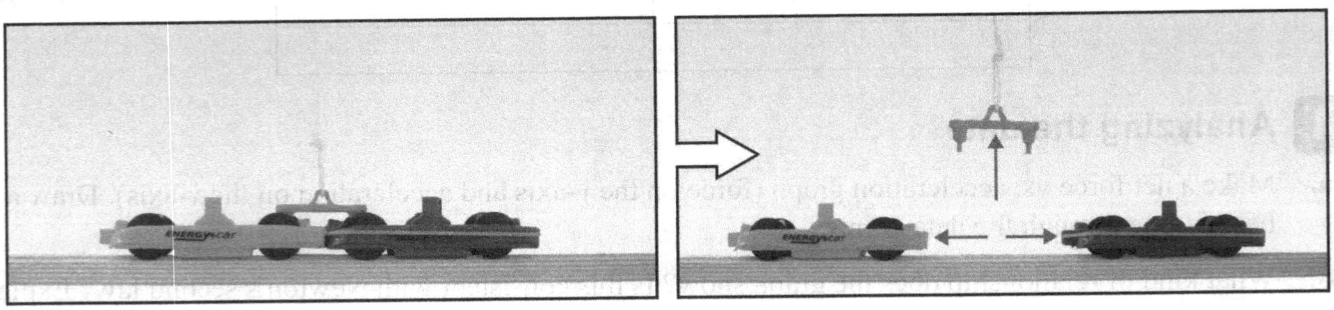

1. Adjust the SmartTrack so it sits level on the table. Check it with the bubble level and adjust the feet as necessary to make the track level.
2. Attach the velocity sensor to the end of the SmartTrack.
3. Place one steel marble in each car, and wrap one car with a rubber band.
4. Place the two cars "nose to notch" in the middle of the track.
5. Squeeze the cars together and attach them with the car link.
6. Position the attached car pair in the middle of the track so the blue car is closest to the velocity sensor. Make sure all four wheels of both cars are on the track.
7. With a very quick upward motion, pull the link straight up and away from the cars. ***CAUTION: Wear eyeglasses or safety glasses to avoid injury.***

a. Describe how the cars move when you remove the car link.

b. How does Newton's third law of motion explain the motion of the cars when you remove the car link?

c. You will use different numbers of marbles in each car to see how that affects the cars' motion. Write a hypothesis to address the question, "What happens when equal and opposite forces act on objects that have different masses?" Your hypothesis should follow this format: "If equal and opposite forces act on objects of different masses, then _______."

2 Collecting data

1. Place one additional marble in the orange car for a total of two marbles. Remove the marble from the blue car. Make sure one car is still wrapped with a rubber band.
2. Place the two cars "nose to notch" in the middle of the track.
3. Squeeze the cars together and attach them with the car link.
4. Position the attached car pair in the middle of the track so the blue car is closest to the velocity sensor. Make sure all four wheels of both cars are on the track.
5. Plug the velocity sensor into input 1 on the DataCollector.
6. Turn the DataCollector on. At the home window, select Data Collection mode.
7. At the Go window, choose Setup at the bottom of the screen.
8. At the Setup window, choose Standard mode, 200 samples, and 0.02 Hz. This will allow the DataCollector to collect 50 samples of data from the velocity sensor each second.
9. Press Go on the DataCollector.
10. With a very quick upward motion, pull the link straight up and away from the cars. ***CAUTION: Wear eyeglasses or safety glasses to avoid injury.***
11. When the cars stop moving, press the button on the DataCollector enclosure to stop the experiment.
12. Switch from Meter to Table and Graph views to study your data. Go to Setup to note the name of the experiment in case you want to go back and look at the data later. Record the maximum velocity of the blue car in Table 1. You can only record the blue car's data, because it was the car the motion detector could "see."
13. Attach the cars with the energy link again, and position them so the *orange* car is closest to the velocity sensor. Do not change the placement of the marbles in the cars. Set up a new experiment on the DataCollector and repeat steps 9–13.
14. Change the marble configurations as indicated in Table 1 and repeat the experiment. Continue until you have completed Table 1. For each trial, you will have to collect two sets of data—one with the blue car facing the velocity sensor, and one with the orange car facing the velocity sensor.

Table 1: Action/reaction pair data

Trial	Marble pairings for connected cars		Maximum velocity (cm/s)		Experiment file name	
	Blue	Orange	Blue	Orange	Blue	Orange
1	0 marbles	2 marbles				
2	0 marbles	0 marbles				
3	2 marbles	0 marbles				
4	2 marbles	2 marbles				

3 Analyzing the data

Note: Adding two steel marbles doubles the mass of the car.

a. How does the velocity of each car compare when masses are equal?

b. How does the velocity of each car compare when one car has twice the mass of the other car?

c. Explain how your velocity data supports the idea that equal and opposite action and reaction forces acted on the once-linked car pairs.

d. If the action-reaction forces are equal in strength when the cars separate, why does one car move at a different velocity from the other car when the masses are unequal, as in trials 1 and 3?

e. Which of Newton's laws of motion best explains the answer to the previous question (3d)?

f. Compare and contrast the velocity vs. time graphs for the blue car from trial 1 and trial 3.

g. Compare and contrast the velocity vs. time graphs for the orange car from trial 2 and trial 3.

h. What is one common characteristic from all of the velocity vs. time graphs for each car at every trial? Explain why you think this is so.

4 Why don't equal and opposite forces cancel each other out?

It is easy to get confused about action-reaction forces. People often ask, "Why don't they cancel each other out?" The reason is that the action and reaction forces act on *different* objects. Prove that action-reaction forces never act on the same object by doing the following.

a. Create a sketch showing your calculator sitting on top of your textbook, which is sitting on top of your desk, which is standing on the tile floor.

b. Identify the forces that serve as the action-reaction forces and draw them in your sketch.

c. Now draw a free-body diagram that shows each object by itself (the calculator, the textbook, the table, the tile floor, and Earth) and use arrows to represent the forces acting on each particular object.

d. Look at the forces on any one object. Does that one object have *both* forces from any single action-reaction pair acting on it? Is this true for all of the objects? (This is the key to Newton's third law: The action-reaction forces are equal in size and opposite in direction, but since they act on different objects, they do not cancel each other out.)

6.1 Mass Versus Weight

How are mass and weight related on Earth?

People often use the words *mass* and *weight* to mean the same thing. In physics, mass and weight are related but they are *not* the same thing. Mass describes the amount of matter in an object. Mass is a fundamental property of matter. An object has the same mass everywhere in the universe, as long as it has the same amount of matter (under most conditions). Weight is not a fundamental property of matter. Weight is a measure of the pulling force of gravity on a mass. Since the strength of gravity varies throughout the universe, the weight of an object can vary.

In this investigation, you will:

- collect mass and weight data for different numbers of steel washers.
- make a graph of weight versus mass, find the slope, and identify its significance.

Materials List

- Spring scale (0-5 N)
- Loop of string
- Washers (15)
- Electronic scale or triple beam balance
- Graph paper

1 Collecting weight and mass data

Weight comes from the action of gravity on an object's mass. How strong is the pulling force of gravity on Earth? You will collect mass and weight data for steel washers, graph the results, and analyze the graph to find out how much force acts on each kilogram of mass on Earth.

1. Attach three steel washers to a loop of string.
2. Use a calibrated spring scale to measure the weight of the washers in newtons (N). Record the weight in Table 1.
3. Use an electronic scale or triple beam balance to measure the mass in grams (g). Convert each mass to kilograms and record the values in Table 1.
4. Repeat for 6, 9, 12, and 15 washers.

Calibration for weighing an object

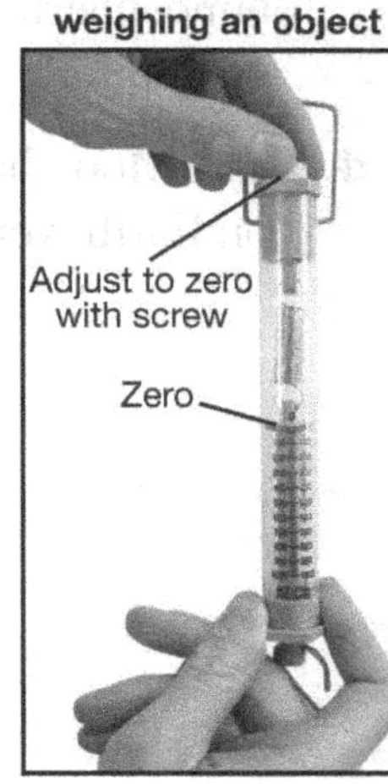

Table 1: Weight and mass data

Number of washers	Weight (N)	Mass (g)	Mass (kg)
3			
6			
9			
12			
15			

2 Analyzing the data

What do the results tell you about the relationship between weight in newtons and mass in kilograms? Create a graph as described below to answer this question.

a. Make a weight vs. mass graph of your data.

b. Describe the graph. What does it tell you about the relationship between mass and weight?

c. Calculate the slope of your graph. What is the significance of the slope for this graph?

d. Write an equation that relates weight in newtons, mass in kilograms, and the strength of gravity (*g*). It should be in the form "weight = ____________."

e. If an object has a mass of 10 kilograms, how much does it weigh in newtons?

3 Arguing from evidence

a. Explain how you could estimate the weight and mass of seven of your steel washers.

b. Find the actual weight and mass for seven of your steel washers. What is your percent error? Explain where in your procedure the error is most likely to occur.

c. You found the value for the strength of gravity (*g*) on Earth to be 9.8 N/kg. At the top of Jupiter's atmosphere, the value for *g* is 23 N/kg. How much would a person who weighs 450 N on Earth weigh on Jupiter?

d. The Moon has about one-sixth the gravity that Earth has. How much would a person who weighs 450 N on Earth weigh on the Moon? Draw a free-body diagram to illustrate your answer.

6.2 Friction

What happens to the force of sliding friction as you add mass to a sled?

Friction is a resistive force that opposes the motion that produces it. You encounter friction all the time. Without friction, cars wouldn't move, you couldn't write with a pencil, and you would have difficulty walking or staying on a chair. In this investigation you will explore the sliding friction between a sled and the SmartTrack.

In this investigation, you will:

- determine the force of friction present when you launch sleds of different mass on the SmartTrack.
- calculate the coefficient of sliding friction.

Materials List

- SmartTrack
- Energy Car sled
- Steel marbles
- Rubber band
- Track feet (2)
- Bubble level
- Velocity sensor
- DataCollector
- Electronic scale or triple beam balance

1 Developing a model for sliding friction

1. Attach a foot to each end of the SmartTrack so it sits level on the table. Check it with the bubble level and adjust the feet as necessary to make the track level.
2. Put a rubber band onto the arms of the launcher in an *x* pattern by giving the rubber band a half-twist, then secure it in place between the retaining washers with the securing screws. Attach the launcher to the track facing away from the stopper end and adjust it so the rubber band is at the 100-cm mark, as shown in the picture above.
3. Place a sled with no marbles on the track and launch it.

a. What force causes the sled to move? What force opposes the sled's motion?

b. What happens to the sled's velocity as it travels along the track?

c. What would happen to the sled's motion if you added marbles to it? Why?

We want to study the force of friction that acts on the sled. It would be convenient if you could measure the force of sliding friction by dragging the sled along the track with a spring scale, but the friction force is too small to measure this way. Instead, you are going to use the velocity sensor to collect acceleration data for the car's motion, and use Newton's second law to calculate the force of friction.

Think through this scenario: You launch the sled and it moves a certain distance along the track before coming to a stop. The net force that acts on the sled equals the force of friction. This friction force decelerates the sled. According to Newton's second law, the force equals the mass times the acceleration (or deceleration, as in this case). So, if you know the sled's mass and deceleration (from the velocity sensor), you can calculate the force of friction that acts on the sled.

$$f = m \times a$$

$$\text{force of friction} = \text{mass of sled} \times \text{deceleration}$$

2 Planning your investigation

a. How will adding mass to the sled affect the force of sliding friction? Develop a hypothesis to address this question. Your hypothesis should follow this format: "If the sled's mass affects the force of sliding friction, then ________."

b. *Predict*: When you carry out the experiment and see the velocity vs. time graph for the sled's motion, what will the graph look like?

c. *Predict*: When you carry out the experiment and see the acceleration vs. time graph for the sled's motion, what will the graph look like?

Design an experiment to test your hypothesis (2a). You will use the launcher to get the sled moving. Use the velocity sensor to measure the sled's acceleration, and use an electronic scale or triple beam balance to find the sled's mass. Vary the number of marbles. Here are some things to consider:

- Choose a launch force that will give a reasonable travel distance with all sled masses, and keep the launch force the same throughout the experiment.
- The sled decelerates, which is negative acceleration.
- Consider whether the force of friction is negative or positive.

d. Write out your experiment's procedure in numbered steps.

3 Analyzing the data

a. Create a data table for all your measurements and calculations. Stay true to the units (kg, m/s^2, N).

b. Be sure to view velocity vs. time and acceleration vs. time graphs for each trial of the experiment. How do they compare to your predictions in 2b and 2c?

c. Was your hypothesis confirmed? Briefly summarize your results in a few sentences. Be sure to refer back to your hypothesis.

d. Use physics to describe why you got the results you did in this experiment.

e. The coefficient of friction (μ) is a ratio of the strength of sliding friction between two surfaces compared to the force holding the surfaces together, called the normal force. In the case of the sled on the flat track, the normal force equals the weight of the sled. Write an equation for calculating the coefficient of friction between the sled and track.

f. Calculate the coefficient of sliding friction with all the friction force and weight data you collected. Compile the results in a data table, and find the average coefficient of sliding friction for your plastic-on-wood situation.

g. Study the table of representative coefficients of sliding friction below and describe how your calculated coefficient of sliding friction for plastic-on-wood compares.

Surfaces	μ (sliding)
rubber on concrete	0.65
wood on wood	0.20
ice on ice	0.03
glass on glass	0.40
steel on steel	0.57

6.3 Equilibrium of Forces and Hooke's Law

How do you predict the force on a spring?

Springs come in two basic types. An extension spring is designed to be stretched, or *extended*. A compression spring is designed to be squeezed, or *compressed*. Hooke's law describes the relationship between force, the spring constant, and deformation.

In this investigation, you will:

- conduct experiments to determine the spring constants for an extension spring and a compression spring.
- create and test a graphical model for spring data.

Materials List

- Stand with ruler card
- 2 Extension springs
- Compression spring
- Mass hanger
- Mass holder
- 12 Washers
- Electronic scale (or triple beam balance)

1 Investigating with an extension spring

The goal of the first part of the experiment is to determine how much an extension spring extends per newton of applied force. You will measure the amount the spring extends as the applied force is changed. The data will allow you to determine the spring constant, which measures the strength of the spring.

1. Set up the Hooke's Law apparatus as shown at right.
2. Hang the blue-tabbed extension spring from the top of the stand, and attach the mass hanger with five washers on it.
3. Adjust the measurement card so the 0-cm mark lines up with the bottom of the mass hanger.
4. How much does the spring extend when you add more washers? Use Table 1 to help guide your experiment as you investigate the strength of the spring.

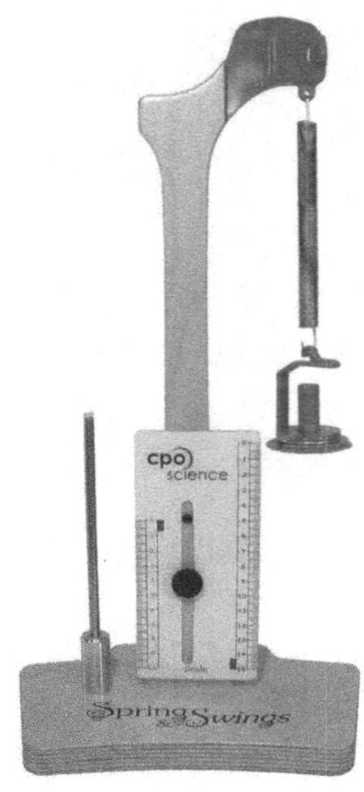

Table 1: Force and extension of a spring

Number of washers	Mass of holder and washers (kg)	Calculated force of holder and washers (N)	Spring's extension (cm)
5			0
7			
9			
11			

2 Analyzing the data

a. Make a graph of force versus extension for the spring.

b. The force from a spring can be described by a formula known as Hooke's law. The spring constant (k) is a measure of the strength of the spring. For example, a spring with $k = 1$ N/cm produces 1 newton of force for every centimeter of extension.

> ***HOOKE'S LAW***
> (springs)
>
> $$F = -kx$$
>
> **Force** (N): F; ***Spring constant*** (N/cm): k; ***Deformation*** (cm) (extension or compression): x

Use your graph of force versus extension to determine the spring constant for the spring in the experiment. Express your result in N/cm.

3 Testing the model

a. Use your graph to predict how much the spring should extend for a hanger with 15 washers.

b. Set up the experiment and test whether the graph gives the correct prediction.

c. How close did your prediction come to the actual extension of the spring? Calculate your percent error.

d. If you use the Hooke's law equation and the spring constant you found in 2b to predict the spring extension, you get a predicted extension that is greater than expected. Why is the answer off by that factor? Explain why you have to make a calculation adjustment to your experimental data to accommodate the Hooke's law equation prediction.

e. You have a second extension spring in the kit—this one has a white tab on it. Do you think this spring has a lower k value than the blue-tabbed spring you just studied, or a higher k value? Explain your choice.

f. Repeat Parts 1 and 2 of this investigation with the white-tabbed spring. How do the results compare to your prediction? Explain.

4 Investigating with a compression spring

This part of the investigation uses a compression spring. Compression springs obey Hooke's law also, but are designed to be compressed rather than extended. To measure the spring constant, you need to measure how much the spring compresses when different amounts of force are applied.

1. Place the compression spring on the spindle, with the mass holder and three washers on top, as shown at right.
2. Adjust the measurement card so the 0-cm mark lines up with the bottom of the mass holder.
3. How much does the spring compress when you add more washers? Use Table 2 to help guide your experiment as you investigate the strength of the compression spring.

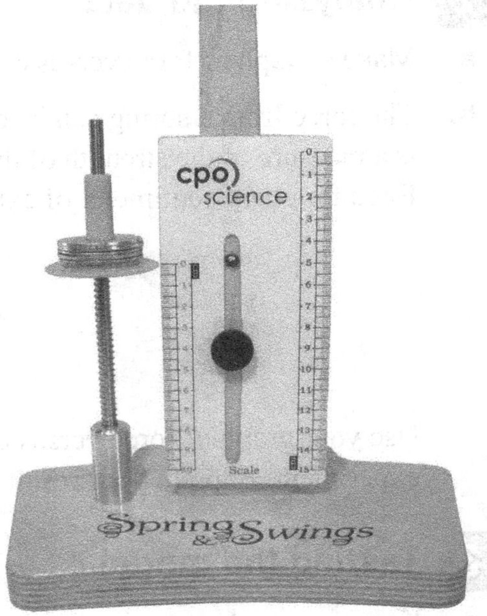

Table 2: Force and compression of a spring

Number of washers	Mass of holder and washers (kg)	Calculated force of holder and washers (N)	Spring's compression (cm)
3			0
6			
9			
12			

5 Analyzing the data

a. Make a graph of force versus compression for the spring.

b. The force from a compression spring can also be described by Hooke's law. As with extension springs, the spring constant (k) is a measure of the strength of the spring. For example, a spring with $k = 1$ N/cm produces 1 newton of force for every centimeter of compression.

HOOKE'S LAW (springs)

$$F = -kx$$

Force (N) F; **Spring constant** (N/cm) k; **Deformation** (cm) (extension or compression) x

Use your graph of force versus compression to determine the spring constant for the spring in the experiment. Express your result in N/cm.

6 Testing the model

a. Use your graph to predict how much the spring should compress for a holder with 15 washers.

b. Set up the experiment and test whether the graph gives the correct prediction.

c. How close did your prediction come to the actual compression of the spring? Calculate your percent error.

d. If you use the Hooke's law equation and the spring constant from 5b to predict the spring compression, you get a predicted compression that is higher than expected. Why is the answer off by that factor? Explain why you have to make a calculation adjustment to your experimental data to accommodate the Hooke's law equation prediction.

e. Could this spring be used to measure the mass of one washer? If so, explain a procedure for how to make the measurement. If not, explain why not.

f. How much force would it take to cause a compression of 1 millimeter?

7.1 Vectors and Direction

How do you give directions in physics?

A *position vector* is a quantity that includes information about both distance and direction. Vectors are the most descriptive way to give directions because they tell exactly how far and in exactly what direction you need to move.

In this investigation, you will:

- learn how to read a compass.
- draw a polar coordinate vector map and follow it.
- draw an *x-y* coordinate vector map, find the resultant, and follow it.
- convert *x-y* coordinates to polar coordinates.

Materials List

- Compass
- Graph paper
- Protractor
- Metric ruler

1 Reading a compass

A compass is a device that includes a magnetic needle that points north. Around the needle is a rotating housing that has degrees marked clockwise from zero to 360. Zero degrees is straight north, 180 degrees is straight south, east is 90 degrees, and west is 270 degrees. A compass *bearing* means direction in degrees relative to north. For example, a bearing of 45 degrees is a direction northeast of where you are.

The clear plastic base of the compass has an arrow you can use to find the compass bearing to a landmark (see photo at right). For example, suppose you wish to know the direction to a certain tree. Point the arrow on the base of the compass at the tree. Hold the compass firmly, keeping the arrow on the base pointing at the tree, and rotate the adjustable housing until the red (north) end of the needle is sitting within the north position on the housing. You can now see the tree's direction by reading the degree value from the adjustable housing scale. In the example, the tree is at a bearing of about 40 degrees. The tree is in a northeast direction.

Reading a compass

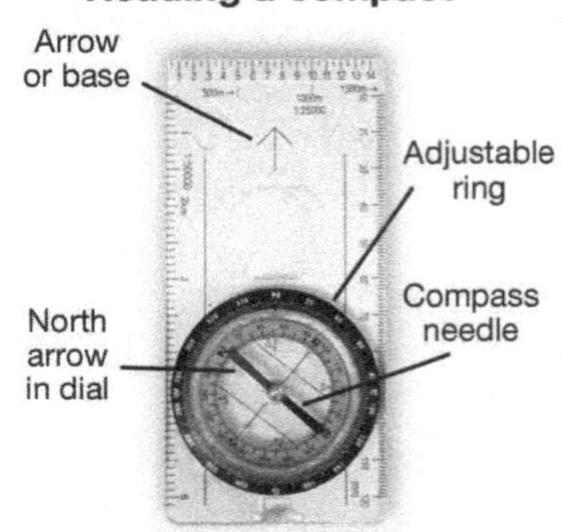

Point the arrow on the base toward the tree. Then, turn the compass housing until the north end of the needle (red) lines up with the north setting.

Read the compass bearing on the adjustable housing (the example above shows a bearing of 40 degrees).

The arrow on the base of the compass is also used to tell which direction to take when you know the compass bearing. For example, suppose you wish to walk on a compass bearing of 40 degrees. Rotate the adjustable housing until 40 degrees on the ring is aligned with the arrow on the base of the compass. Now take the compass, hold it level, and rotate the whole compass until the north tip of the needle is aligned with 0 degrees (the north setting in the adjustable housing). The arrow on the base now points in the direction you need to walk. This technique works best when the compass is held near your body, and you rotate the compass by rotating yourself, so you always face the direction of the arrow on the base of the compass.

2 Making a map

Your teacher will give you a sequence of two vectors. The vectors are in polar coordinates, making them easy to travel to with a compass.

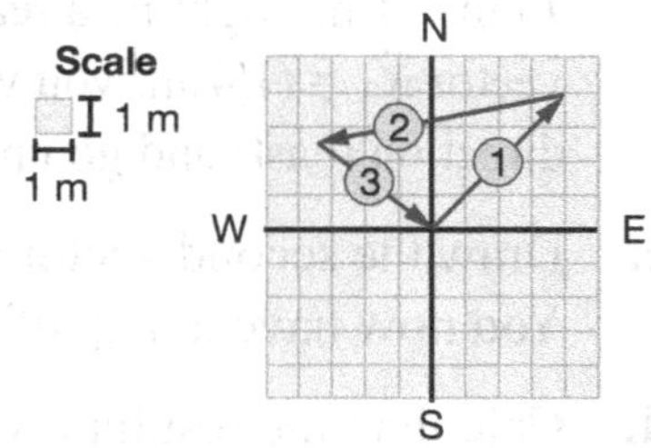

a. Take a sheet of graph paper and draw *x*- and *y*-axes pointing east-west (*x*) and north-south (*y*).

b. Draw the first vector your teacher gives you on the graph, which should be to scale. For example, if you had graph paper with a 1-centimeter grid, an appropriate scale would be 1 cm = 1 m.

c. Draw the second vector your teacher gives you, starting from the end of the first vector. The graph paper now shows a map of two legs of a walking trip.

d. Take the graph paper and draw a third vector from the end of the second vector back to the start—this is the vector back home. Use a protractor and a ruler to measure the angle of the vector and determine its length in meters. Remember, you want compass bearings. The easiest way to measure the angle is to draw a north/south line that runs through the junction of vectors 1 and 2, and measure the angle clockwise from the positive *y*-axis you just constructed, which is north on the compass.

3 Calibrating your pace

If you were exploring uncharted territory, you could keep track of how far you traveled by counting steps. For this technique to be accurate, you must know your own step length. Walk 10 steps at an average pace and use a meter stick to measure how much distance you covered. An average step for most people is between 1/2 and 3/4 meter. Record the step length for each group member in Table 1.

Table 1: Step lengths

Name	Distance of 10 paces (m)	Avg. length of 1 pace (m)

4 Walking the vectors

1. Take each of the three vectors and convert the distance in meters to the number of steps for each person in the group. Use Table 2 to record the calculations.
2. Take a compass and go out and walk the first vector.
3. Walk the second vector starting where the first vector ended.
4. Walk the third vector and see how close you come to where you started.

Table 2: Vectors in units of steps

Vector	Distance in meters	Distance in steps

5 Vectors in *x-y* coordinates

a. Take a sheet of graph paper and draw *x*- and *y*-axes pointing east-west (*x*) and north-south (*y*).

b. Draw the first vector on the graph. The graph should be to scale; 1 cm = 1 m might be a reasonable scale. For example, to draw the vector $\vec{x}_1 = (1,4)$m, you would go over 1 cm in the positive *x* direction (east) and go up 4 cm in the positive *y* direction (north).

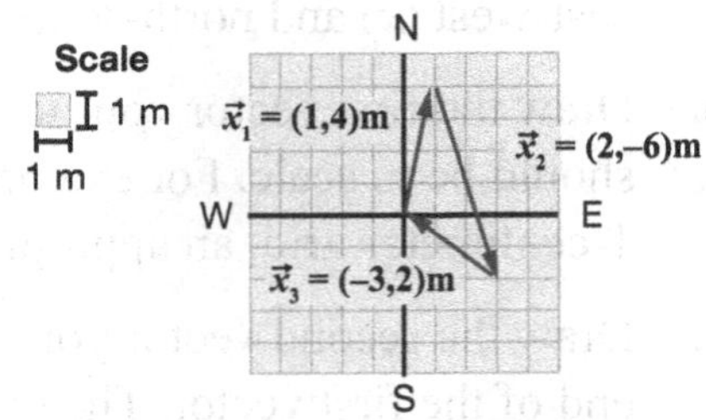

c. Draw the second vector *starting from the end of the first vector.* You now have a map of two legs of a walking trip.

d. Calculate the resultant vector by adding up the *x*- and *y*-components of the two vectors you already have. For example, if $\vec{x}_1 = (1,4)$m and $\vec{x}_2 = (2,-6)$m, then $\vec{x}_1 + \vec{x}_2 = (3,-2)$m.

e. The *resultant vector* is the vector you could have walked straight from the origin (home) to the final destination. To get back home, you need to walk the resultant vector in the opposite direction. Mathematically, that means multiplying the components by –1. In the example, the vector back home would be $\vec{x}_3 = (-3,2)$m. You can see on the map that this is correct.

6 Converting to compass coordinates

A compass uses polar coordinates that specify a distance and an angle. Take each of the three vectors and convert them to polar coordinates by sketching the triangle shown in the diagram. You might have to add 90, 180, or 270 degrees to the angle θ of the triangle to get a compass bearing. Record the converted vectors in Table 3.

(*Hint*: Try using the Pythagorean Theorem: $a^2 + b^2 = c^2$.)

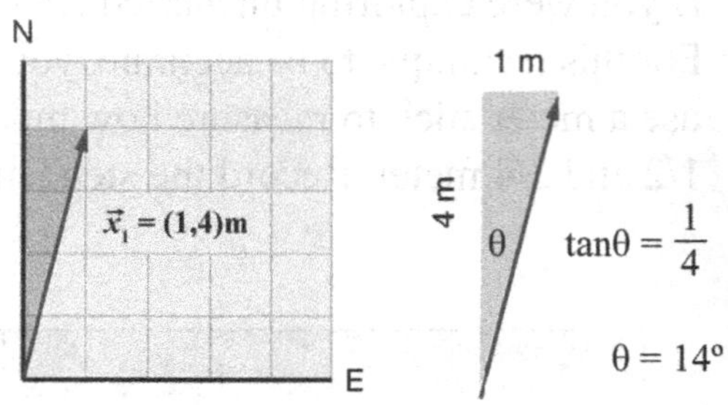

Table 3: Converting to polar coordinates

Vector in *x-y* coordinates	Angle from north (degrees)	Distance (m)	Distance (steps)
1,4			
2,–6			
3,–2			

7 Walking the vectors

1. Take each of the three vectors and convert the distance in meters to the number of steps for each person in the group.
2. Take a compass and go out and walk the first vector.
3. Walk the second vector starting where the first vector ended.
4. Walk the third vector and see how close you come to where you started.

7.2 Projectile Motion

How can you predict the range of a launched marble?

The speed and direction of a projectile's velocity vector changes as it moves. You recognize this when you see the typical curved path, or trajectory, of a projectile. How can you set up a model to predict where a launched projectile will land?

In this investigation, you will:

- use a mathematical or an experimental model to predict where a marble will land, given its launch angle and initial velocity.

Materials List

- Marble Launcher
- DataCollector and photogate
- Plastic marble
- Tape measure
- White paper and carbon paper
- Masking Tape
- Safety glasses

1 Analyzing the motion of the marble in two dimensions

How can you predict the range, or horizontal distance, of a launched marble? Since gravity pulls *down* and not sideways, the motion of the marble must be separated into components. It makes sense to pick one component (y) in the vertical direction aligned with gravity. The other component (x) is then chosen to be in the horizontal direction, perpendicular to the force of gravity. The diagram below shows the velocity of the marble (v) at three points in its trajectory, resolved into x and y components, v_x and v_y.

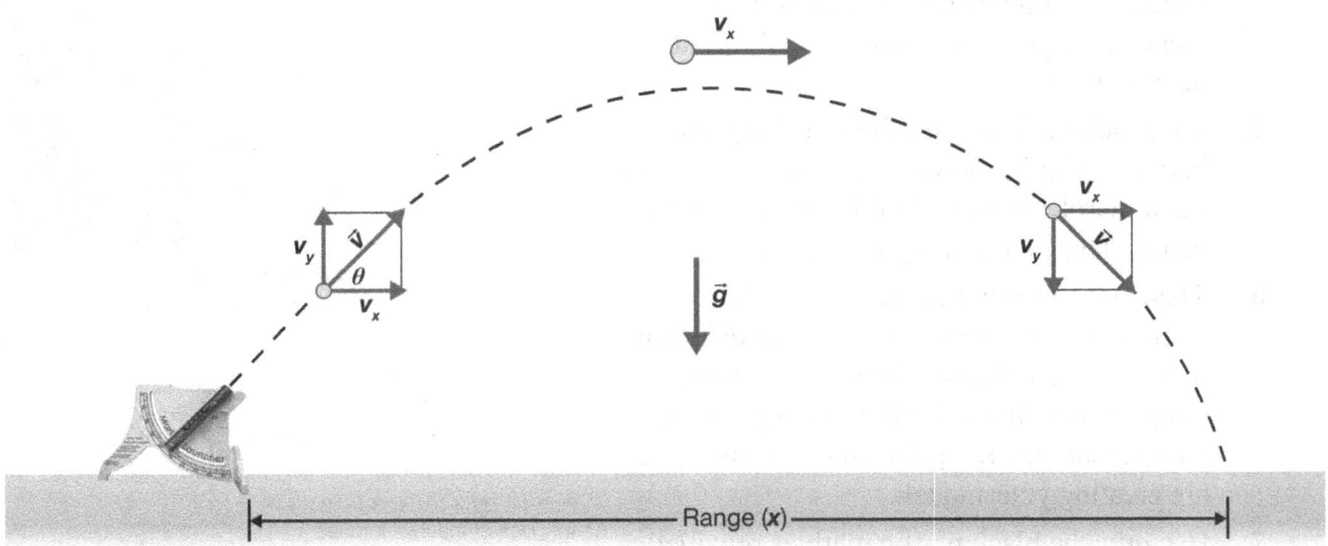

a. Use the diagram to explain why projectiles travel in a curved path or trajectory.

2 Where will the marble land?

Your goal is to work with your group to design a model that will predict where the marble will land from any given angle and initial velocity. You may choose to create a mathematical model or an experimental model, and you will be given hints on how to create both types of models in Parts 3 and 4. At the end of the investigation, your group will be given a launch angle, and you will launch the marble from the third spring setting. You must predict where the marble will land by placing a piece of paper on the floor. If your marble lands on the paper, you have succeeded.

Safety

- **Never launch marbles at people.**
- **Wear safety glasses or other eye protection when launching marbles.**
- **Launch only the black plastic marbles that come with the marble launcher.**

Before your group starts working on a model to predict the range of the marble, it is a good idea to get familiar with how the marble launcher works, and try some practice launches.

1. Always wear eye protection when using the marble launcher.
2. Connect a single photogate to the "A" DataCollector input.
3. At the DataCollector home window, choose Timer mode. Choose Interval by tapping the Interval function button at the bottom of the screen.
4. Attach one photogate to the marble launcher so the marble breaks the light beam as it comes out of the barrel. The launcher can launch at angles from 0 (horizontal) to 90 degrees (vertical).
5. To launch marbles, pull the launching lever back and slip it sideways into one of the slots. Put a marble in the end of the barrel and the marble launcher is ready to launch.
6. There are five notches that change the compression on the spring and give different launch speeds. When it is time to test your model, you will use the third spring setting to launch marbles. Keep this in mind when you are creating your model.
7. Use *only* the black plastic marbles provided.
8. Do a few practice launches to get familiar with how the launcher works and how to measure initial velocity and range.
9. When you are ready, study Parts 3 and 4 on the next pages to decide which type of model you would like to create to predict the range of the marble.
10. Choose your model method, work through the model, and get ready to test your prediction for the marble's landing spot in Part 5.

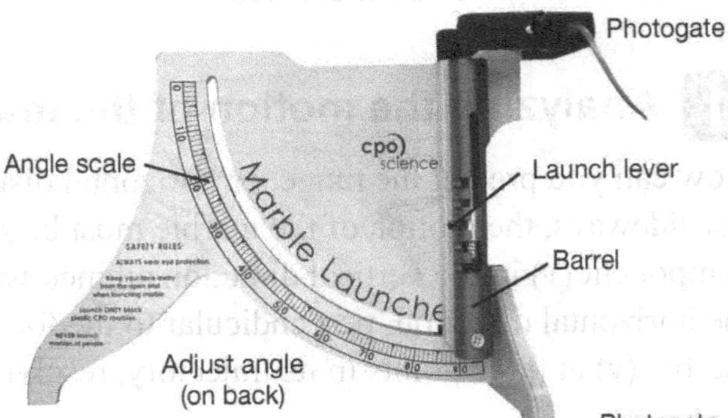

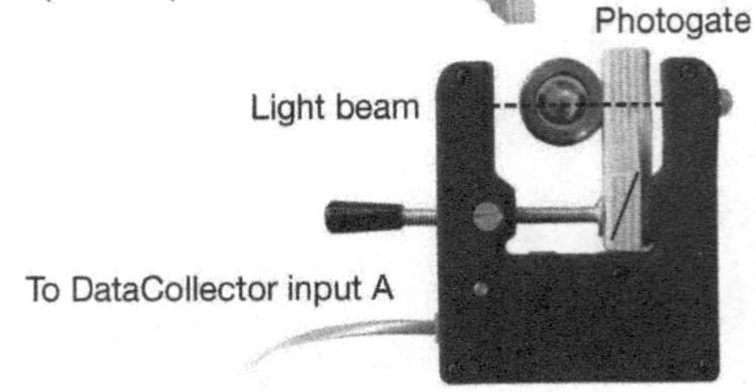

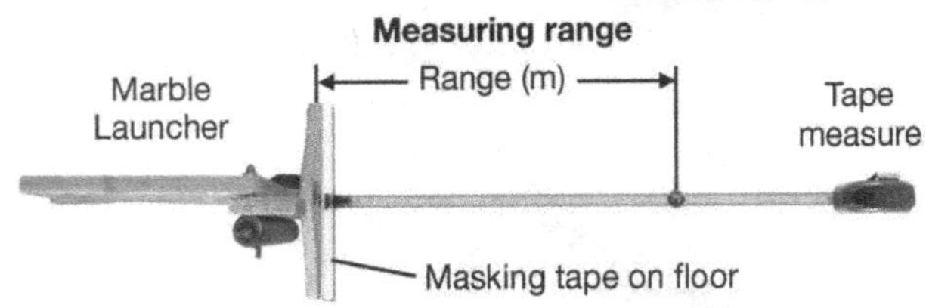

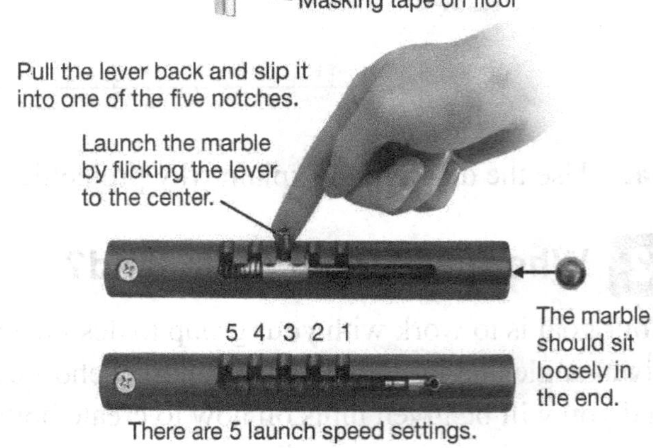

3 Hints for creating a mathematical model

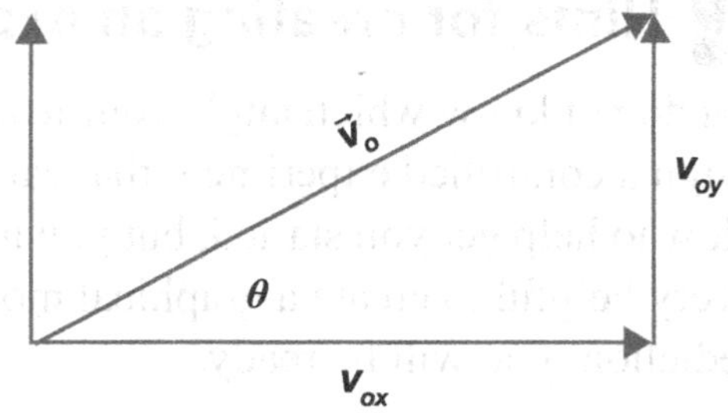

Analyzing the motion of a projectile always requires the initial velocity to be resolved into its horizontal and vertical components, v_{ox} and v_{oy}. The components are $v_{ox} = v_o\cos\theta$ and $v_{oy} = v_o\sin\theta$, with the launch angle θ measured from the horizontal. The two velocity components are independent of each other. The horizontal velocity remains constant during the entire flight, and the vertical velocity changes due to the influence of gravity.

The range of a projectile depends on the horizontal velocity component and the time of flight. A projectile's *flight time* depends only on the initial vertical velocity. The fact that a projectile's vertical velocity is zero at its highest point is useful when finding the flight time. The time to the top is calculated and then doubled to find the total air time. Range can then be found by multiplying the horizontal velocity and the total air time.

	Equation	Notes
Basic idea of how to find range, x	$x = v_{ox}t$	Range can be found by multiplying the horizontal velocity component, v_{ox}, by the time of flight. You do not have a way of directly measuring these variables, so you will have to substitute variables you can define or measure.
Substituting for v_{ox}	$v_{ox} = ?$	Use the right triangle in the sketch above to find a relationship between v_{ox}, θ, and v_o. Velocity, v_o, can be measured with the photogate and DataCollector, and θ is easily measured on the marble launcher.
Substituting for t	$t = \frac{v_{oy}}{g}$	Flight time depends on vertical velocity, v_{oy}. This simple motion equation relates velocity, acceleration, and time. This time is only the time to the top, so *double it* to represent the whole flight time. Gravity acts to accelerate the marble in the vertical direction.
Substituting for v_{oy}	$v_{oy} = ?$	Use the right triangle in the sketch above to find a relationship between v_{oy}, θ, and v_o. Velocity, v_o, can be measured with the photogate and DataCollector, and θ is easily measured on the marble launcher.
Put it all together	$x = ?$	Now you have all the parts you need to create a mathematical range model that relates variables you know or can measure.

Once you find a range equation based on initial velocity and launch angle, you will have to find the average initial velocity for the marble at the third launch spring position, so you will be ready to plug in an initial velocity when your teacher assigns an angle to your group.

4 Hints for creating an experimental model

You do not know which angle your teacher will choose when your model is tested. Work with your group to design a controlled experiment that can be conducted within the time allowed. A sample data table is given below to help get you started, but you must formulate a procedure and determine how much data to collect. It is very helpful to create a graphical model from your experimental data, so when it is time to make your prediction, you will be ready.

Table 1: Sample data table (partial) for creating an experimental model

Spring setting	Launch angle (degrees)	Range (m)	Distance (m)	Photogate time (s)	Initial velocity (m/s)
3	*20*		*0.019*		
3	*30*		*0.019*		
3	*40*		*0.019*		

5 Testing your model

1. Your teacher will give you an angle setting for your marble launcher.
2. You will use the third spring launch setting to launch your marble.
3. Set the given angle on your marble launcher and place a piece of white paper on the floor where you predict the marble will land. Adjust the paper so the middle is your predicted landing spot. Tape the paper on the floor. Put a piece of carbon paper, carbon-side down, on top of the white paper. The marble will leave a mark on the paper where it hits.
4. Launch your marble and see how close your prediction was to the actual landing spot of the marble.

6 Evaluating your model

a. Describe the model you used to predict the landing spot of the marble. Include any quantitative data you collected.

b. How close was your prediction to the actual landing spot? Be as quantitative as possible in your answer.

c. Discuss possible limitations of your particular model.

7.3 Forces in Two Dimensions

How do forces balance in two dimensions?

If an object is in equilibrium, all of the forces acting on it are balanced, and the net force is zero. If the forces act in two dimensions, then all of the forces in the *x*-direction and *y*-direction balance separately. Pulling on a ring in different directions with three different force scales is an interesting way to explore forces in two dimensions.

In this investigation, you will:

- resolve force vectors into their components.
- calculate the net force on an object.

Materials List

- Tri-force ring
- 3 String loops
- 5-N spring scale
- 10-N spring scale
- 20-N spring scale
- Scientific calculator

1 Setting up the spring scales

1. Hold each spring scale vertically from its hook and check to see that it is properly calibrated. If it needs to be adjusted, turn the nut at the top of the scale until it reads zero when you hang the scale from its hook.
2. Hook the 5-N scale on one of the string loops attached to the tri-force ring.
3. Repeat step 2 for the 10- and 20-N scales.

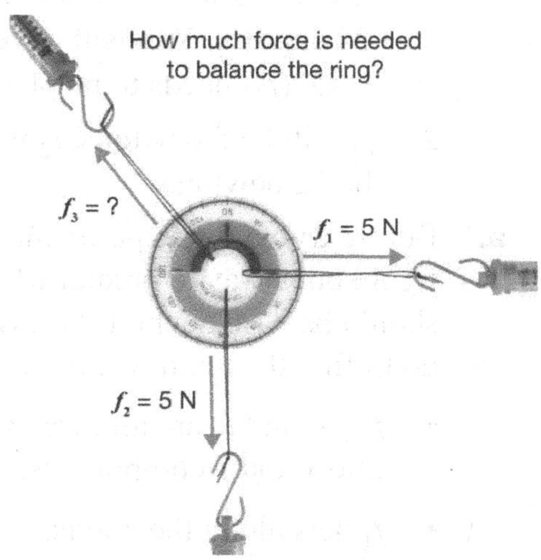

2 Developing a model for the force needed for equilibrium

1. Study the photo of the tri-force ring setup at right. The ring has three spring scales attached at different angles. A 5-N spring scale is pulled from the 0-degree mark with 5 newtons of force. A 10-N spring scale is pulled from the 90-degree mark, also with 5 newtons of force. The 20-N spring scale is attached at the 130 degree mark, but the force is unknown. How much force do you need to apply at f_3 so the ring is in equilibrium?

a. Before trying the experiment, work with your group to figure out a way to predict what the third force (f_3) should be. Use the hints below and the illustration at right for help. *Show the work* you do to find the unknown force. The solution requires several steps.

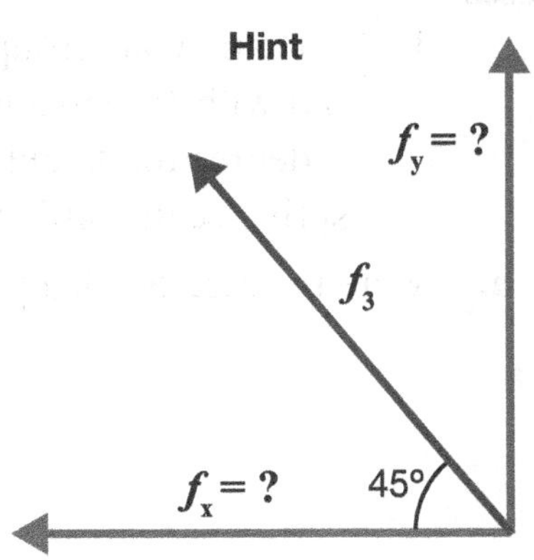

- f_1, f_2, and f_3 are all force vectors that can be resolved into *x* and *y* components.
- f_1 acts along the x-axis and f_2 acts along the y-axis.
- The force ring is in equilibrium. Therefore, the sum of the *x* components of all three forces and the sum of the *y* components of all three forces must equal zero.
- Trigonometry will help you find *x* and *y* components of a right-angle triangle (see sketch above).

3 Finding the force needed for equilibrium

1. Work with your group to set up the force ring as described in parts 1 and 2. Someone will pull on the 5-N scale with 5 newtons of force, someone will pull on the 10-N scale with 5 newtons of force, and the third person will pull on the 20-N scale with as much force as needed to balance the other forces. Remember to set each scale at the angles described in part 2.

a. How much force do you need to make with the 20-N spring scale to place the ring in equilibrium? How does this compare with your prediction? Discuss.

4 Developing a model for the direction needed for equilibrium

1. This time, you will be given all three forces, and you will have to determine an unknown force direction. As you can see from the photo at right, the 5-N scale (f_1) is at the 0-degree mark. The 10-N scale (f_2) is at 75 degrees. You will have to predict where the 20-N scale (f_3) needs to be placed.
2. f_1 will be 5 newtons, f_2 will be 10 newtons, and f_3 will be 12 newtons.

Where should the scale be placed so 12 N of force creates equilibrium?

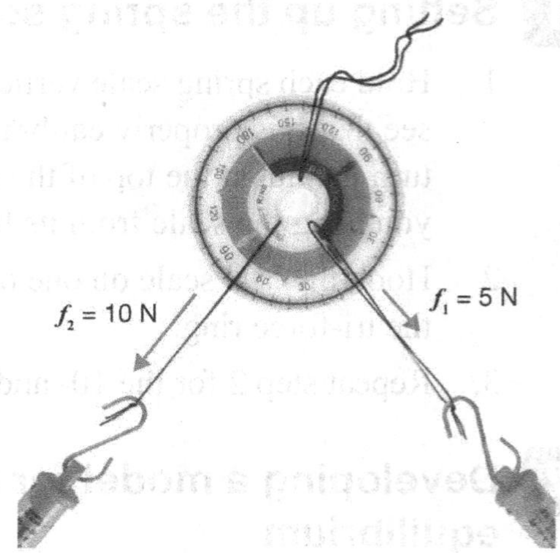

a. Before trying the experiment, work with your group to figure out a way to predict what the direction (angles) for f_3 should be. Use the hints below for help. *Show the work* you do to find the unknown force.

- f_1, f_2, and f_3 are all force vectors that can be resolved into x and y components.
- f_1 acts along the x-axis.
- The force ring is in equilibrium. Therefore, the sum of the x components of all three forces and the sum of the y components of all three forces must equal zero.
- Trigonometry will help you find x and y components of a right-angle triangle.

5 Finding the direction needed for equilibrium

1. You and your group members must pull the 5-N scale (f_1) with a force of 5 newtons, the 10-N scale (f_2) with 10 newtons, and the 20-N scale (f_3) with 12 newtons. Keep the 5-N scale on the 0-degree mark and the 10-N scale on the 75-degree mark, and adjust the position of the 20-N spring scale until the ring is in equilibrium. It takes some time to do this accurately.

a. Were the results what you expected? How does your data show that the force ring was in equilibrium?

8.1 Motion in Circles

How do you describe circular motion?

British scientist George Atwood (1746–1807) used two masses on a light string running over a pulley to measure acceleration due to gravity. You experimented with the Atwood's machine in Chapter 5 to study Newton's second law. In this investigation, you will use the Atwood's machine to study the relationship between linear and angular velocity.

In this investigation, you will:

- set up an Atwood's machine and find the maximum velocity of the moving mass hangers.
- predict the maximum angular velocity of the pulley wheel.
- predict the angle turned by the pulley wheel while the mass hangers move.

Materials List

- Physics Stand
- Double pulley
- Red safety string
- 2 Mass hangers
- Metal washers
- Measuring tape
- Photogate
- DataCollector

1 Linear and angular velocity

To describe the circular motion of the Atwood's machine pulley, we need to develop concepts similar to the ones we use to describe the linear motion of the mass hangers. The *angular velocity* (ω) is the rate at which an object's angle changes as it rotates. Units for angular velocity could be degrees per second. For example, an object could rotate 90 degrees every second. If ω = 90 deg/s, the object completes one full turn in four seconds.

ANGULAR VELOCITY

$$\omega = \frac{\theta}{t}$$

Angular velocity (rad/s) = ***Angle turned*** (rad) / ***Time taken*** (s)

In physics, it is more convenient to work with angles in units of radians. One radian is approximately 57.3 degrees. If the angular velocity is expressed in radians per second, a simple relationship exists between the angular velocity and the linear speed. This relationship comes from the fact that a point on the edge of a circle moves one circumference in one rotation.

LINEAR AND ANGULAR VELOCITY

$$v = \omega r$$

Linear velocity (m/s) = ***Angular velocity*** (rad/s) × ***Radius*** (m)

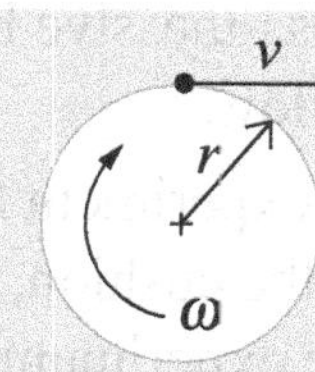

2 Investigating with the Atwood's machine

The Atwood's machine is well-suited for this investigation because linear motion (mass hangers) and rotational motion (pulley wheel) happen at the same time. You will find the maximum linear velocity of the mass hangers, and relate it to the angular velocity of the pulley wheel.

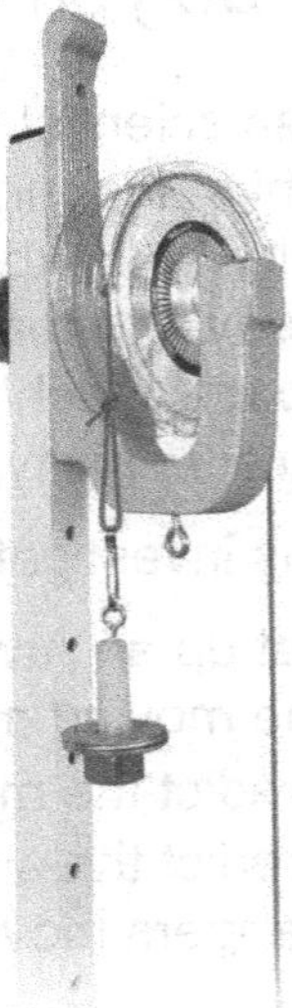

1. Attach the double pulley to the top of the physics stand. You will only use the striped pulley.
2. Attach the mass hangers to the red safety string. Place four metal washers on one mass hanger (m_1) and three metal washers on the other mass hanger (m_2). Place the string over the striped pulley.
3. Pull m_2 down to the stand base. Place a sponge or some other small cushion on the base to protect it from the falling m_1 hanger. Let go of m_2 and observe the Atwood's machine as m_1 falls.

a. As m_1 accelerates downward, m_2 accelerates upward. When will the velocity of the mass hangers be the greatest?

3 Measuring maximum linear velocity

1. Attach a photogate to the double pulley as shown at right. Plug the photogate into the DataCollector (input A). The striped pattern on the pulley will break the light beam of the photogate as the pulley rotates.
2. Turn on the DataCollector. At the Home window, select Data Collection mode.
3. At the Go window, tap on the Setup option at the bottom of the screen.
4. In the Setup window, choose standard mode.
5. For photogate A (PG_A), select velocity in cm/s. Set the PG_B option to none.
6. Pull the less-massive hanger to the base. Tap go at the bottom of the setup window.
7. When the experiment has started, release the hanger. When the hanger falls onto the cushion, press the button on the DataCollector enclosure to stop the experiment.
8. Select Table at the bottom of the screen to view the data table. Scroll down until you find the maximum velocity, and record it in Table 1.
9. Repeat the experiment three times, record the velocities in Table 1, and find the average.

8.1 Motion in Circles

How do you describe circular motion?

British scientist George Atwood (1746–1807) used two masses on a light string running over a pulley to measure acceleration due to gravity. You experimented with the Atwood's machine in Chapter 5 to study Newton's second law. In this investigation, you will use the Atwood's machine to study the relationship between linear and angular velocity.

In this investigation, you will:

- set up an Atwood's machine and find the maximum velocity of the moving mass hangers.
- predict the maximum angular velocity of the pulley wheel.
- predict the angle turned by the pulley wheel while the mass hangers move.

Materials List

- Physics Stand
- Double pulley
- Red safety string
- 2 Mass hangers
- Metal washers
- Measuring tape
- Photogate
- DataCollector

1 Linear and angular velocity

To describe the circular motion of the Atwood's machine pulley, we need to develop concepts similar to the ones we use to describe the linear motion of the mass hangers. The *angular velocity* (ω) is the rate at which an object's angle changes as it rotates. Units for angular velocity could be degrees per second. For example, an object could rotate 90 degrees every second. If ω = 90 deg/s, the object completes one full turn in four seconds.

ANGULAR VELOCITY

$$\omega = \frac{\theta}{t}$$

Angle turned (rad) — θ
Angular velocity (rad/s) — ω
Time taken (s) — t

In physics, it is more convenient to work with angles in units of radians. One radian is approximately 57.3 degrees. If the angular velocity is expressed in radians per second, a simple relationship exists between the angular velocity and the linear speed. This relationship comes from the fact that a point on the edge of a circle moves one circumference in one rotation.

LINEAR AND ANGULAR VELOCITY

$$v = \omega r$$

Linear velocity (m/s) — v
Angular velocity (rad/s) — ω
Radius (m) — r

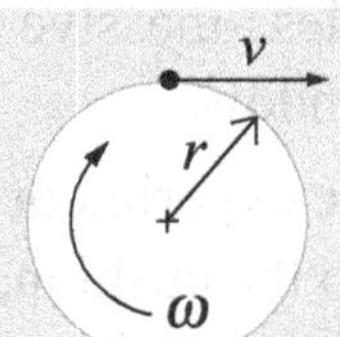

2 Investigating with the Atwood's machine

The Atwood's machine is well-suited for this investigation because linear motion (mass hangers) and rotational motion (pulley wheel) happen at the same time. You will find the maximum linear velocity of the mass hangers, and relate it to the angular velocity of the pulley wheel.

1. Attach the double pulley to the top of the physics stand. You will only use the striped pulley.
2. Attach the mass hangers to the red safety string. Place four metal washers on one mass hanger (m_1) and three metal washers on the other mass hanger (m_2). Place the string over the striped pulley.
3. Pull m_2 down to the stand base. Place a sponge or some other small cushion on the base to protect it from the falling m_1 hanger. Let go of m_2 and observe the Atwood's machine as m_1 falls.

a. As m_1 accelerates downward, m_2 accelerates upward. When will the velocity of the mass hangers be the greatest?

3 Measuring maximum linear velocity

1. Attach a photogate to the double pulley as shown at right. Plug the photogate into the DataCollector (input A). The striped pattern on the pulley will break the light beam of the photogate as the pulley rotates.
2. Turn on the DataCollector. At the Home window, select Data Collection mode.
3. At the Go window, tap on the Setup option at the bottom of the screen.
4. In the Setup window, choose standard mode.
5. For photogate A (PG_A), select velocity in cm/s. Set the PG_B option to none.
6. Pull the less-massive hanger to the base. Tap go at the bottom of the setup window.
7. When the experiment has started, release the hanger. When the hanger falls onto the cushion, press the button on the DataCollector enclosure to stop the experiment.
8. Select Table at the bottom of the screen to view the data table. Scroll down until you find the maximum velocity, and record it in Table 1.
9. Repeat the experiment three times, record the velocities in Table 1, and find the average.

Table 1: Linear velocity

Trial	Maximum velocity (cm/s)
1	
2	
3	
average	

4 Developing a model for the maximum angular velocity

a. As the mass hangers move, the pulley wheel rotates. If you know the linear velocity maximum for the hangers, what other information would you need to relate this linear velocity to the maximum angular velocity of the pulley wheel?

b. Write down the equation that you would use to predict the angular velocity of the pulley wheel, in rad/s, if you know the linear velocity and the variable you identified in the previous question (4a).

c. Use the equation from 4b to predict the maximum angular velocity (ω) of the pulley in rad/s. Show your work. (The pulley wheel radius is 5.00 cm.)

5 Testing your model

1. Set up your Atwood's machine as in Part 3.
2. Go back to the DataCollector home window and select Data Collection mode.
3. At the Go window, tap on the Setup option at the bottom of the screen.
4. In the Setup window, choose standard mode.
5. For photogate A (SP_A), select speed for angular velocity in rad/s.
6. Pull m_2 to the base. Tap Go at the bottom of the setup window.
7. When the experiment has started, release the hanger. When the hanger falls onto the cushion, press the button on the DataCollector enclosure to stop the experiment.
8. Select Table to view the data table. Scroll down until you find the maximum velocity, and record it in Table 2.
9. Repeat the experiment three times, record the angular velocities in Table 2, and find the average.

Table 2: Angular velocity

Trial	Maximum angular velocity (rad/s)
1	
2	
3	
average	

a. How close was your angular velocity prediction to the average maximum angular velocity you measured? Find the percent error.

b. Convert your average maximum angular velocity (rad/s) to revolutions per second (rps). (*Hint*: There are 2π radians for each revolution.)

c. Convert the revolutions per second to revolutions per minute. A typical computer hard disk drive can spin at about 5,000 to 15,000 rpm. How does your pulley wheel compare? How could you increase the pulley wheel rpm for your Atwood's machine?

6 Using a model to measure total radians

How many radians does the pulley turn through while the mass hangers move, from start to stop?

1. Pull m_2 down to the base and let it go. Watch the pulley wheel turn as m_1 falls.

a. How many turns do you think the pulley makes as the mass hanger falls? State your best guess.

b. How many radians does this prediction equal? There are 2π (about 6.28) radians in a full turn.

c. Work with your group to devise a way to use the photogate and DataCollector to test your total radians prediction. (*Hint*: There are 10 stripes on the pulley for every radian of angle.) In Timer mode, you can select the Count function. Photogate A (Na) will count the stripes (beam breaks). Describe your procedure.

d. Conduct your group's procedure. How many total radians does the pulley turn? How does this measured value compare to your guess? If your guess was very close, was it a lucky guess, or did you base your prediction on something helpful? If your guess was quite different from the measured angle turned, why was your guess so far off the mark?

e. Would the pulley's turn angle change if you increased the mass difference between the mass hangers? Explain.

7 Arguing from evidence

a. An automobile's speedometer only works correctly if the tire radius matches the manufacturer's specification. Why does the speedometer performance depend on tire radius? Do some research on how a speedometer works to help you answer this question.

b. Suppose you remove the tires from a new car and replace them with larger tires. Is it possible that you might exceed a speed limit without knowing it? Why or why not?

8.2 Centripetal Force

Why does a roller coaster stay on the track when upside down on a loop?

Have you ever wondered how roller coasters stay on the track as they go around loops? Understanding centripetal force is essential to designing roller coasters that stay on the track as they move through loops.

In this investigation, you will:

- study a marble's forces and motion as it rolls along a looped track.
- derive a formula that predicts the minimum speed a marble needs to have to make it around a loop on a track.
- test the minimum speed theory by conducting an experiment.

Materials List

- Loop Track
- Physics Stand
- Steel marble
- Plastic marble
- DataCollector
- Photogate

1 Introduction

The loop track has a downhill section where the marble picks up speed. If the marble stays on the track, it drops into the catcher without hitting the wooden lip at the opening of the slot. If the marble leaves the track, the catcher may not catch it at all, or the marble may catch but bounce against the wood at the opening of the slot. You can tell if the marble stayed on the track all the way around the loop by the sound it makes when it lands in the catcher.

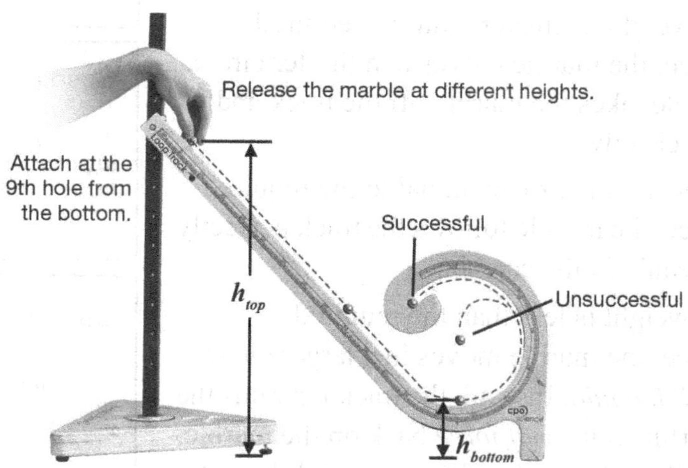

1. Attach the loop track to the stand using the ninth hole from the bottom. This sets the sloped part of the track to an angle of 45 degrees.
2. Drop the steel marble from different heights and observe whether it makes it around the track. Be careful to release the marble without pushing it.
3. Try dropping the plastic marble which has a much lower mass. Is the performance of the plastic marble different from the performance of the steel marble?

a. What do you observe as you dropped the marble from different heights? Think about the speed of the marble and whether it makes it around the loop successfully.

b. Write down a hypothesis about the conditions that are necessary for a marble to make it successfully around the loop.

2 Developing a model for centripetal force

The force that causes an object to move in a circle is called a *centripetal force*. Any type of physical force can be a centripetal force if it results in circular motion. Some common examples of centripetal forces come from string tension, normal forces, friction, and weight. A centripetal force is always directed toward the center of the circle that the object's motion follows.

The centripetal force required to cause an object to move in a circle of a given radius (r) is given by the formula on the right. If the centripetal force is smaller, the object moves in a circle of larger radius. If the centripetal force is larger, the object moves in a circle of smaller radius. This is because a stronger centripetal force is able to bend the motion into a sharper curve.

CENTRIPETAL FORCE

$$F_c = \frac{mv^2}{r}$$

Centripetal force (N); Mass (kg); Linear speed (m/s); Radius of path (m)

Consider the forces acting on the marble at the top of the loop. The marble's weight (mg) pulls straight down. Think about three different cases that apply to the motion of the marble. The centripetal force required to make the marble follow the circular track of radius (r) with speed (v) is $mv^2 \div r$.

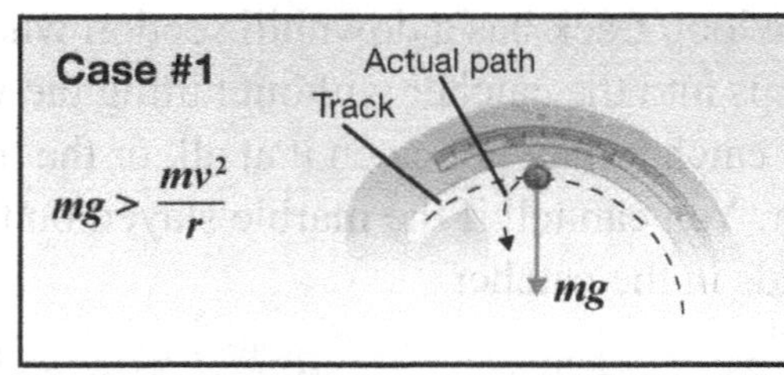

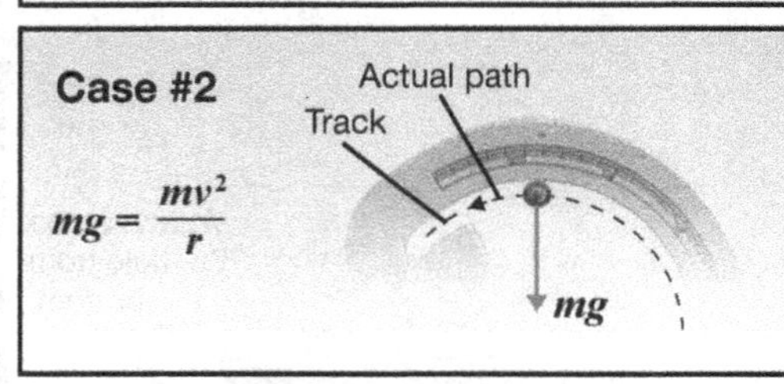

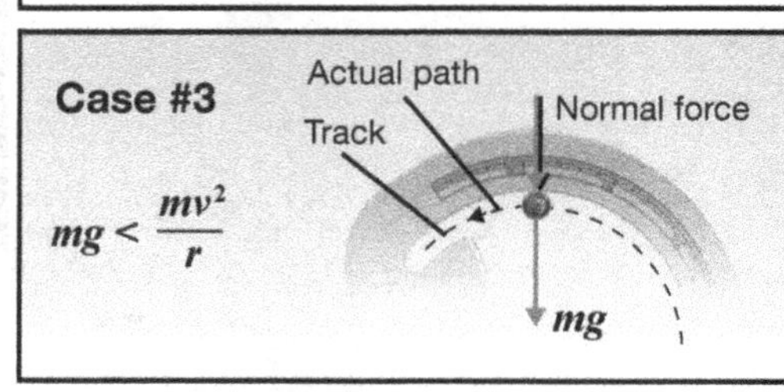

1. Case 1: If the weight is greater than the required centripetal force, the marble moves in a tighter circle. The tighter circle takes the marble off the track and it does not catch cleanly.
2. Case 2: If the weight is exactly equal to the required centripetal force, the marble follows the track perfectly and catches cleanly in the catcher.
3. Case 3: If the weight is less than the required centripetal force, the marble moves in a larger circle than the track *if it could.* Instead, the track restrains the marble by exerting a *normal force* back on the marble, forcing it to follow the circle of the track tightly and the marble catches cleanly in the catcher.

The condition for the marble to follow the track and catch cleanly is that *the weight must be less than or equal to the centripetal force required to make the marble move in the circle of the track.*

a. Using the formulas for weight and centripetal force, write a formula expressing the minimum speed (v) needed to keep the marble on the track.

b. Use the formula to calculate the minimum speed required for the marble to make it around the loop and stay on the track. The radius of the track is 10 centimeters.

c. Draw a free-body diagram for the marble when it is at the top of the loop. Label the two forces acting on the marble and use vectors to show their directions.

d. Explain why it would *not be* correct to draw a third arrow on your free-body diagram and label it "centripetal force."

3 Testing your model

1. Place a photogate at the top of the loop. The photogate allows you to measure the speed of the marble at the very top of the loop.
2. Release the steel marble from different heights and record the time it takes to pass through the photogate and whether the marble stayed on the track. Take data for a range of heights that give the marble speeds both greater than and less than the minimum speed required to make it around the loop.
3. Calculate the speed of the marble at the top of the loop for each trial. Use the diameter of the marble (0.019 m) for the distance in the speed calculations. Also, calculate the weight of the marble and the centripetal force required to keep the marble moving in a circle with a radius of 0.100 meters. The mass of the steel marble is 0.028 kilograms. Record the data Table 1.
4. Repeat steps 2 and 3 using the plastic marble. The mass of the plastic marble is 0.004 kg.

Table 1: Experimental loop-the-loop data

Mass (kg)	Weight (N)	Photogate time (s)	Speed (m/s)	Centripetal force at 0.10 m radius (N)	Did the marble stay on the track (yes or no)

4 Analyzing the data

a. Compare the weight to the centripetal force for your experimental data on both marbles. How should they be related? Do your experimental results confirm this?

b. If a group in your class found the centripetal force to be significantly larger than the marble's weight in trial 1 for one of the marbles, what would you suspect they did wrong?

c. How would you use the centripetal force and the weight to calculate the normal force on the marble?

d. Imagine that your job is to design a roller coaster containing a hill followed by a circular loop with a radius of 20 meters. Use what you learned in this investigation to estimate the minimum speed required for the roller coaster to make it around the loop. Do you have to take the mass of the passengers riding the roller coaster into account? Explain.

e. Real roller coasters are always designed to enter a loop after going down a hill that has more than the minimum height. Explain why.

f. You may have noticed that many roller coaster loops are not circular but teardrop-shaped (see photo at right). This is called a *clothoid* loop, and it has a smaller radius at the top than at the sides. Use what you have learned in this investigation to explain the advantages of a clothoid loop over a circular loop.

8.3 Universal Gravitation and Orbital Motion

How strong is gravity in other places in the universe?

All objects that have mass attract each other through gravity. The strength of the gravitational force that acts between two objects depends on the mass of those objects and the distance between their centers. This idea is known as *Newton's law of universal gravitation*.

Materials List

- Scientific calculator

In this investigation, you will:

- use the law of universal gravitation to calculate the gravitational force of attraction between objects.
- calculate the gravitational field strength (g) on the surface of different planets using the gravitational constant (G) which is the same everywhere in the universe.

LAW OF UNIVERSAL GRAVITATION

$$F = G\frac{m_1 m_2}{r^2}$$

Force (N): F

Mass 1, Mass 2 (kg): m_1, m_2

Gravitational constant (6.67×10^{-11} N•m²/kg²): G

Distance between masses (m): r

1 The gravitational force between everyday objects

How gravitationally "attracted" are you to another person? Use Newton's law of universal gravitation to calculate the gravitational force of attraction between the following pairs of objects. For each situation, estimate the masses of the objects and the distances between them.

a. your pencil and your calculator, having been placed next to each other on your desk

b. your left shoe and your right shoe as you stand comfortably

c. you and another person of similar mass at a distance of two meters

d. two cars parked next to each other

2 The gravitational force of attraction between people and planets

Gravitational forces tend to be important only when one of the objects has a very large mass, like a planet. Note that the mass of any of the eight planets in our solar system is hugely greater than the mass of any ordinary object. For example, Earth's mass is 5.97×10^{24} kilograms. Expressed without the help of scientific notation, that is 5,970,000,000,000,000,000,000,000 kg, or almost six trillion trillion kilograms. That is a massive object! Table 1 lists the masses and radii of the other planets in the solar system.

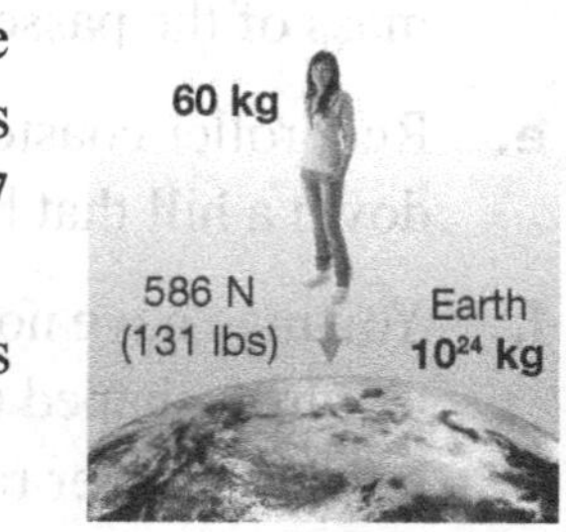

Table 1: Mass and radius of the planets

Planet	Mass (kg)	Radius (m)
Mercury	3.18×10^{23}	2.43×10^{6}
Venus	4.88×10^{24}	6.06×10^{6}
Earth	5.97×10^{24}	6.37×10^{6}
Mars	6.42×10^{23}	3.37×10^{6}
Jupiter	1.90×10^{27}	6.99×10^{7}
Saturn	5.68×10^{26}	5.85×10^{7}
Uranus	8.68×10^{25}	2.33×10^{7}
Neptune	1.03×10^{26}	2.21×10^{7}

3 Calculate your weight on different planets

When we talk about the gravitational force of attraction between an object and a planet, we call that force *weight*. Use the law of universal gravitation and the information in Table 1 to calculate your weight at several different locations in the solar system. Remember, the r in the formula stands for the distance between the *centers* of the two objects being considered.

a. at the surface of Earth

b. at a distance of 1,000 kilometers above the surface of Earth

c. at the surface of Jupiter

d. at a distance of 5,000 kilometers above the surface of Saturn

4 The value of g

An object's weight can be calculated using the equation $F_w = mg$, where m represents the mass of the object and g represents the strength of gravity. At the surface of Earth, $g = 9.8$ N/kg, meaning that Earth exerts a gravitational force of 9.8 newtons per kilogram of mass. The law of universal gravitation can be used to calculate g on other planets.

a. Set the weight (m_1g) equal to the gravitational force ($Gm_1m_2 \div r^2$) and solve the equation to get a formula for g in the form $g =$ ______.

b. Calculate the value of g at the surface of each of the planets (including Earth), using the formula you found in 4a above.

c. On the surface of which planet would you weigh the most? On the surface of which would you weigh the least? How do you know?

d. What happens to the strength of gravity as you get farther from the surface of a planet? Sketch a graph that illustrates this relationship.

e. *Advanced Problem*: A bowling ball has a mass of about 5 kilograms. Suppose you shrank the ball until it had a surface gravity equal to Earth's ($g = 9.8$ N/kg). What would the radius of the ball be? A single atom has a radius of about 10^{-10} meters. How does the size of the shrunken bowling ball compare with an atom? As strange as it seems, neutron stars exceed this incredible density!

9.1 Torque

How does force create rotation?

Force is needed to change an object's rotational motion just as it is needed to change an object's linear motion. For rotational motion, however, you need to know more than the strength of the force. You also need to know where the force is applied relative to the center of rotation.

The torque produced by a force

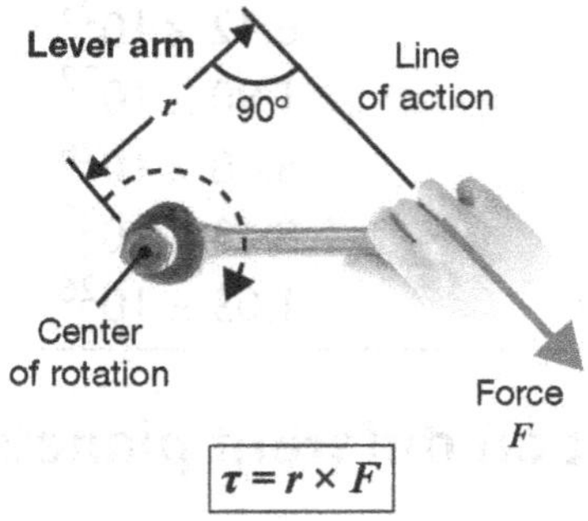

$\tau = r \times F$

Materials List

- Physics stand
- Spinner
- Red safety string
- 2 mass hangers
- Steel washers
- Electronic scale or triple beam balance

A *torque* created by a force is equal to the force multiplied by the length of the *lever arm*. The lever arm is the perpendicular distance between the line of action of the force and the center of rotation. For example, a force (F) applied a distance (r) from the center of rotation (see diagram above) exerts a torque of $\tau = r \times F$. You can see in the diagram that this torque tends to cause clockwise rotation.

In this investigation, you will:

- apply balanced and unbalanced torque to a rotating object.
- create rotational equilibrium by balancing torques.

1 Setting up the investigation

1. Attach the spinner to the top hole in the stand using a threaded knob.
2. Attach five washers to one mass hanger. Make a 1-meter length of string with a small loop on one end. Attach the hanger to the other end and hook the loop over one of the notches in the pulley wheel.
3. Wind the string up onto the pulley. When the hanger is released, the pulley spins because the hanging mass creates a torque equal to its own weight multiplied by the radius of the pulley.

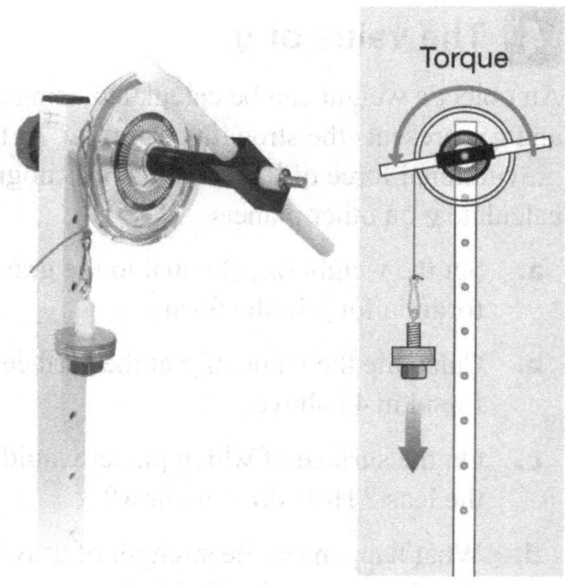

a. Measure and record the mass of the hanger in kg and its calculated weight in N.

b. The radius of the pulley is 5.00 cm. Calculate the torque created by the weight of the hanger and washers. Give your answer in units of N-m.

2 Equilibrium and torque

The net torque is the total of all torques acting on an object about a given center of rotation. For linear forces, a system is in equilibrium when the net force is equal to zero. For rotation, a similar equilibrium exists when the net torque is zero. To add torques, we usually assume a sign convention in which counterclockwise torque is considered positive and clockwise torque is considered negative.

1. Add some washers to one side of the spinner arm using nuts to hold them in place. Adjust the position of the washers (in or out) until the spinner arm is in equilibrium with the arm horizontally.

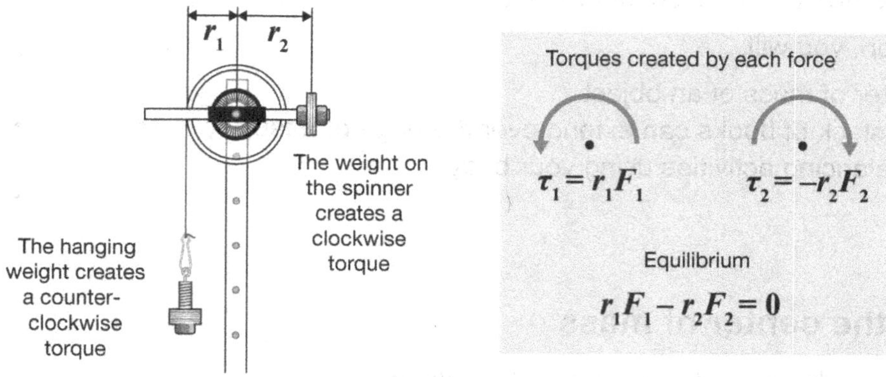

a. Calculate the torque created by the nuts and washers you added to the threaded rod. To do the calculation, use the distance measured from the spindle to the approximate center of the washers.

b. If you move the washers farther from the center, does the torque increase or decrease? For this question, consider only the magnitude of the torque and not the direction.

3 Torque when the force and distance are not perpendicular

1. Modify the configuration of the last step by moving the washers out from the center so the spinner rotates when you let go. This system is not in rotational equilibrium.
2. If you move the spinner so it makes an angle with the horizontal, you can find a position where the system is in rotational equilibrium. Find the position of the spinner that creates equilibrium.

Can this be true? Explain why.

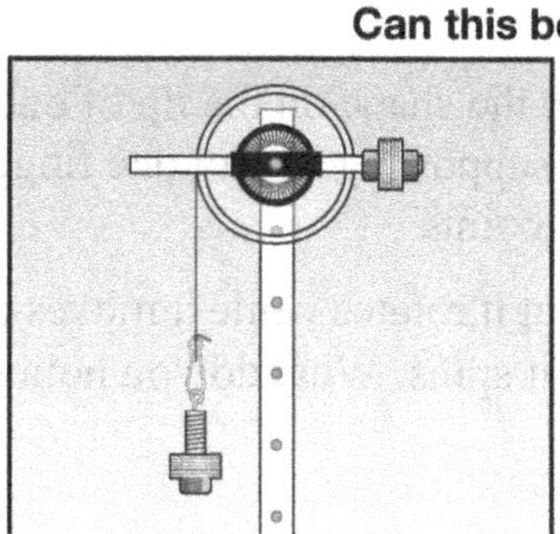
This system is NOT in rotational equilibrium.

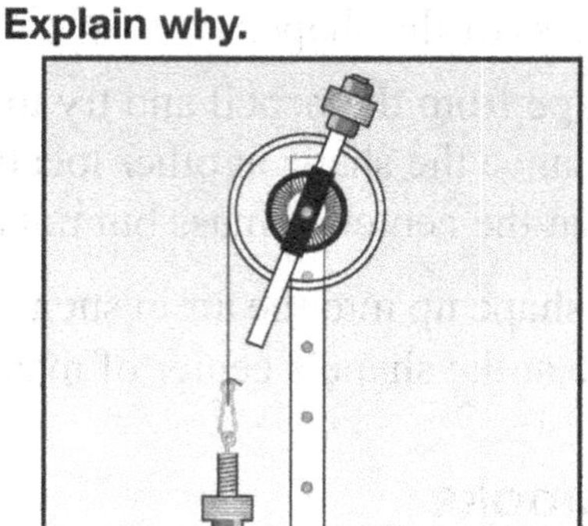
This system IS in rotational equilibrium.

a. Draw a free-body diagram showing the torques acting on the system by the hanger on the string and the mass on the spinner arm. Make your drawing as accurate as you can, especially the angle of the spinner arm relative to horizontal.

b. Explain why the system is *not* in equilibrium when the spinner is horizontal but *is* in equilibrium when the spinner is at a certain angle. Your explanation should use the concept of lever arm and include a diagram showing the lever arm for both torques acting on the system.

9.2 Center of Mass

How do objects balance?

The *center of mass* is the average location of an object's mass. The center of mass is important because objects balance around it. If you hang an object from a string, the object will rotate so its center of mass is directly below the point of support. Closely related to the center of mass is the *center of gravity*. The center of gravity is the average location of the force of gravity exerted on an object (its weight).

In this investigation, you will:

- locate the center of mass of an object.
- explain why a stack of books can extend over the edge of a table.
- perform two balancing activities using your body.

Materials List

- Cardboard
- Scissors
- Hole punch
- String
- Washers
- 5 Identical books
- Classroom chair
- Metric ruler
- Graph paper

1 Locating the center of mass

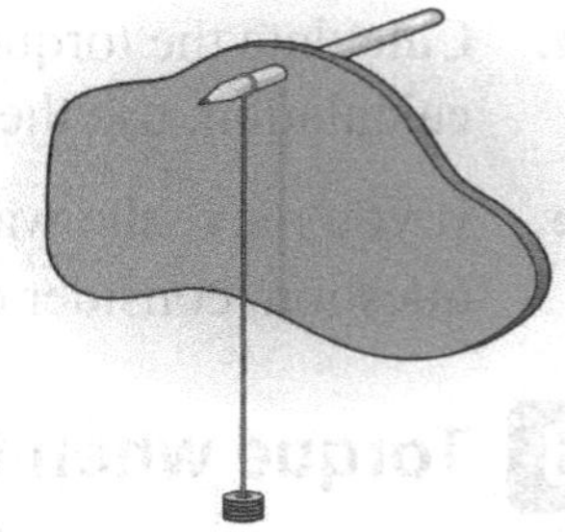

1. Cut an irregular shape from a piece of cardboard.
2. Make a plumb line by tying a mass or some washers to the bottom of a piece of string. The string should be at least twice as long as the largest dimension of the shape you cut out.
3. Punch a small hole in your shape near the outside edge. Suspend the shape and the plumb line from the tip of a pencil as shown in the diagram. The shape should be able to swing freely on the pencil.
4. Carefully trace the path the string makes on the shape.
5. Repeat steps 3 and 4 using holes punched at different locations at the edge of the shape. Repeat again for a total of three lines.
6. The three lines should intersect at one point. Make a dark circle at this location. This point is the center of mass of the shape.

a. Remove the shape from the pencil and try to balance the shape on the tip of one finger at its center of mass. Try to balance the shape at other locations by supporting it with one fingertip. Explain why the shape balances at the center of mass but not at other points.

b. Gently toss the shape up into the air in such a way that it rotates while it moves up and then back down. Watch the circle at the shape's center of mass while it spins. What do you notice?

2 Balancing books

1. Place a book (or, alternatively, a domino or a block) on a table so that it extends over the edge as far as possible without falling. Measure the length of the book and its overhang distance. Record your measurements in Table 1.

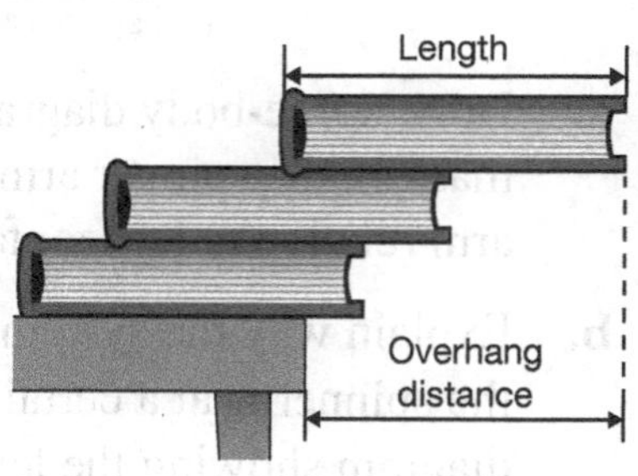

2. What fraction of the length of the book is the overhang distance? Place a second identical book (or other identical object) on top of the first. Adjust the positions of the two books to get the maximum overhang distance. Measure the distance and record. One group member should be ready at all times to catch the books in case they fall off the table's edge.
3. If the books are placed so that they have the greatest amount of overhang possible, where is the center of mass of the two-book combination in relation to the edge of the table? Explain.
4. Repeat step 3 with three, four, and five books. Try to achieve the greatest overhang distance for each number of books and measure each distance.

Table 1: Balancing books

Length of one book (cm)	Number of books	Overhang distance (cm)
	1	
	2	
	3	
	4	
	5	

a. Explain the method you used to determine how to position the books. Use the term *center of mass* in your explanation.

b. Make a graph of overhang distance versus number of books. Plot the distance on the *y*-axis and the number of books on the *x*-axis.

c. Explain the relationship between the number of books and the overhang distance. Is it linear?

d. How many books had to be stacked to reach an overhang distance of one book length?

e. How many books do you think it would take to reach an overhang distance of two book lengths?

3 Your body's center of mass

a. Stand with your back and heels against a wall. Slowly bend down and try to touch your toes. Explain what happens and why.

b. Position a chair so that it is facing you. Stand with your toes two feet away from the front edge of the chair. Bend over at the waist, touch the sides of the chair, and bend your body up to a standing position. Now bend over in the same way but grab the sides of the chair this time. Try to bend upward while holding the chair. Explain what happens and why.

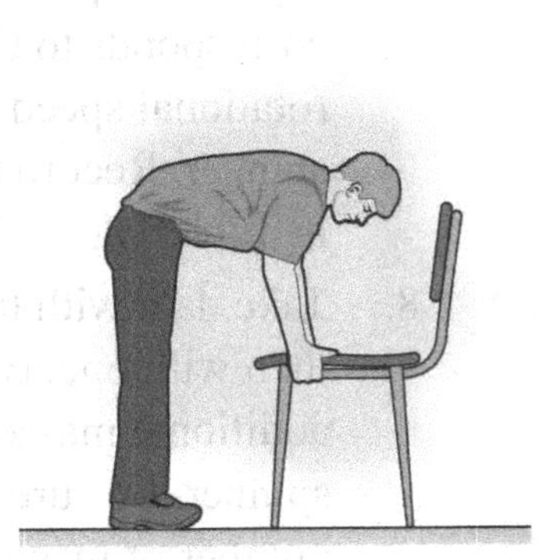

c. Why might these activities be easier for some students than for others?

9.3 Rotational Inertia

What property describes a body's resistance to rotational motion?

Inertia describes the resistance of an object to changes in its linear motion. Rotational inertia is the analogous property for rotational motion. In this investigation, you will:

- measure the rotational speed of a rotating object.
- determine how the distribution of mass affects rotational inertia.

Materials List

- DataCollector
- Physics Stand
- Spinner
- Pulley
- Photogate
- Hanging masses
- String

1 Setting up the spinner

1. Attach the spinner with a knob as shown. Attach a photogate just below the spinner so the stripes on the pulley break the light beam.
2. Make a loop at one end of a length of string and tie the string to the weight hanger so the length of the assembly is about 50 centimters.
3. Wind the string on the pulley until the weight just touches the pulley.
4. Let the spinner release and it will spin.
5. Set the DataCollector to measure frequency and collect 50 samples at a rate of 20 Hz.
6. Hit Go and release the spinner. The DataCollector will take 2.5 seconds' worth of data, then stop.
7. Once the DataCollector takes the samples, switch to table view. Scroll back through the data to find the highest frequency recorded. This corresponds to the maximum rotational speed reached by the spinner. Record this and all other data in Table 1.
8. Take data with the spinner arm empty, then with one, two, and finally three additional masses on each arm of the spinner. Be sure to have the same number of masses on each arm, as in the diagram to the right.

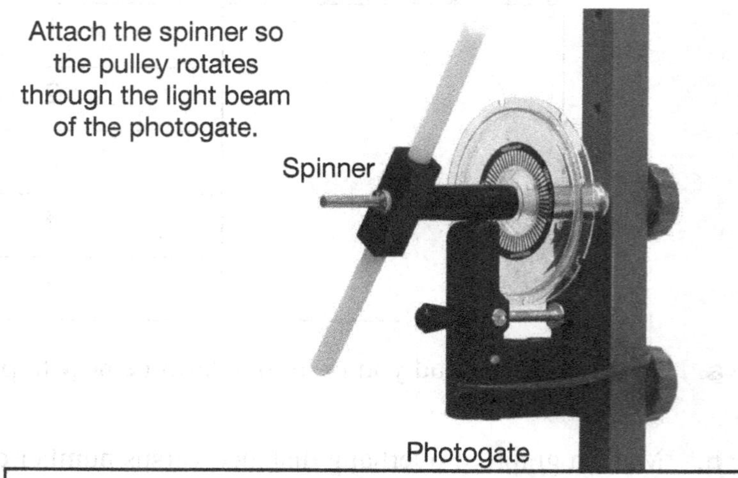

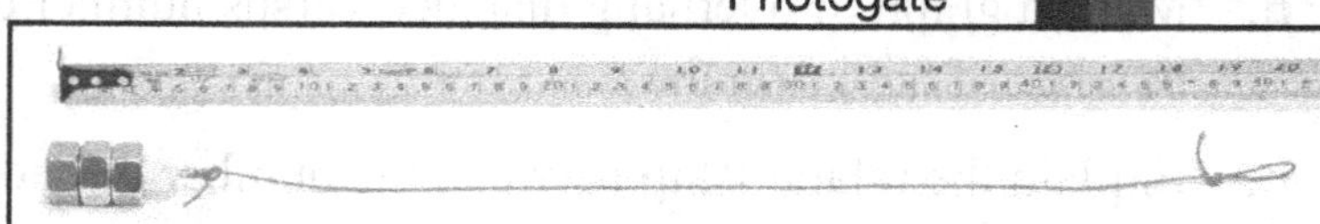
Attach the hanging weight to a 50 cm string with a loop on the end.

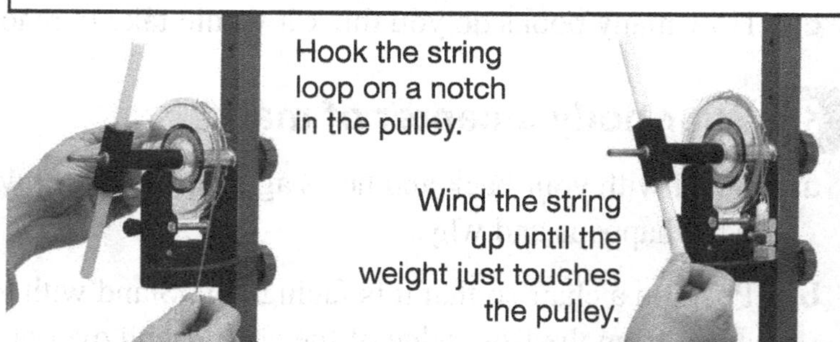

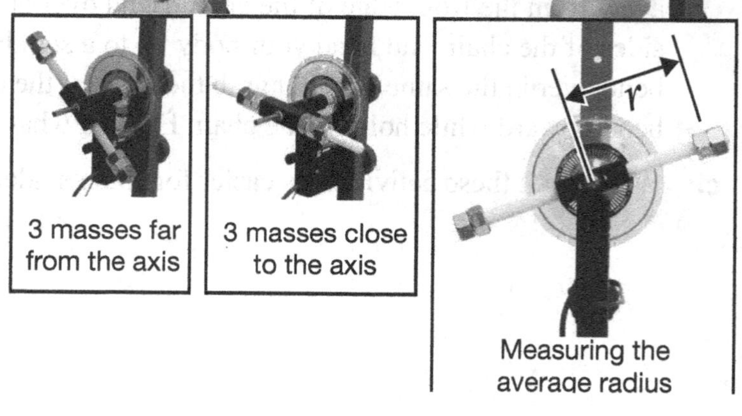

3 masses far from the axis

3 masses close to the axis

Measuring the average radius

9. Take data with different masses at different distances from the axis of rotation. Measure the average radius to the center of the added mass.

Table 1: Frequency and angular velocity data

Number of masses on spinner	Total added mass (g)	Average radius (cm)	Frequency (Hz)	Calculated mr^2 (g-cm^2)	Angular velocity (rad/s)

2 Rotational inertia

a. For each trial, calculate the total mass added to the spinner (in grams), then calculate the contribution to the total moment of inertia by multiplying the total added mass by the average radius squared.

b. Calculate the angular velocities in radians per second. The pulley has a stripe pattern that breaks the light beam 10 times for each radian of rotation. That means the angular velocity is the frequency ÷ 10.

c. In this experiment, the torque is fixed because the same hanging mass is applied at the same radius. Create a graph showing how the angular velocity changes as the moment of inertia is increased.

d. You do not know the moment of inertia of the spinner without any added mass because the mass is distributed over an irregular shape. The masses of the components are given in the diagram. The spindle does not rotate, so it contributes zero moment of inertia. Estimate the contribution to the total moment of inertia from the other components.

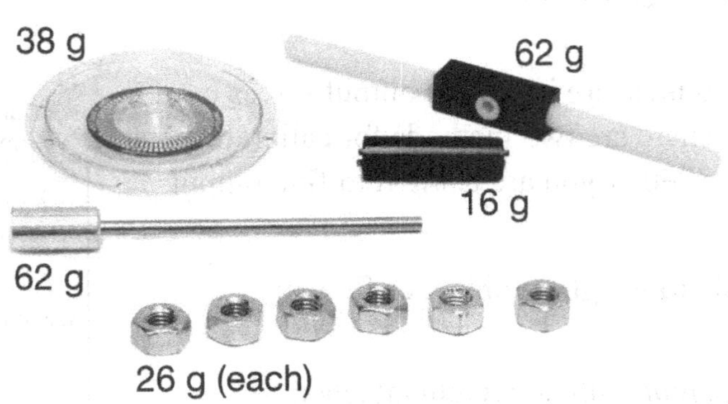

e. Which situation has more rotational inertia?
a) the masses near the axis
b) the masses far from the axis

f. How do your experimental results support this conclusion?

g. Estimate the average error in your measurements. How do the differences you observe compare to the average error? Does this make your conclusion stronger or weaker?

10.1 Machines and Mechanical Advantage

How do simple machines work?

A machine is something created by human technology that makes a task easier. Of course, the word *easier* means different things for different tasks. A simple machine is a machine that accomplishes its task with no source of energy except for forces that are applied directly to the machine as the task is being done. This investigation explores a simple machine that multiplies force.

In this investigation, you will:

- build a simple machine that multiplies force.
- identify input and output forces for a pulley setup.
- determine the mechanical advantage of different pulley setups.

Materials List

- Physics Stand
- Dynamic Pulleys
- Yellow pulley string
- 3 Weights
- Spring scales
- Red safety string

1 Identifying input and output forces

The photo at right shows a pulley system like one you will set up. *Input force* is the force required to operate a simple machine. *Output force* is equal to the weight of the load that is raised by the rope and pulleys setup.

a. Describe where the input force is on the pulley system at right.

b. Describe where the output force is on the pulley system at right.

2 Calibrate the spring scale

You will use a spring scale to measure input and output forces used in the dynamic pulleys setup. There are two methods for calibrating the spring scale, depending on whether you are using it to find output force or input force.

Before using the spring scale to weigh an object or find an output force:

1. Hold the scale *upright* with no weight attached.
2. Use the thumbscrew at the top of the scale to adjust the plunger so it sits right at the zero mark.

Before using the spring scale to pull down on a rope and pulley string to find input force:

1. Hold the scale *upside down* by the hook.
2. Use the thumbscrew to adjust the plunger so it sits right at the zero mark.

Calibration for weighing an object

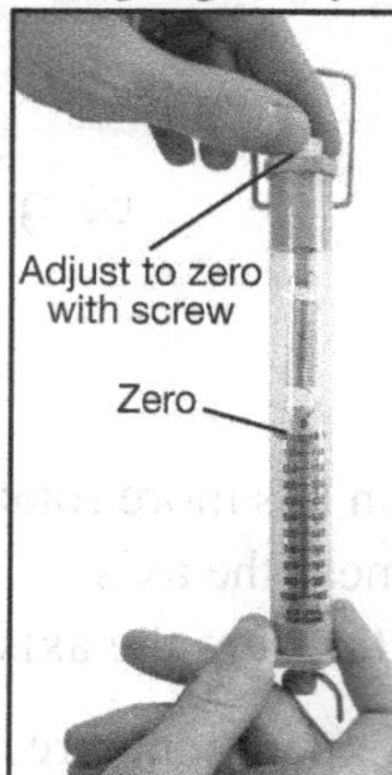

Calibration for pulling down on string

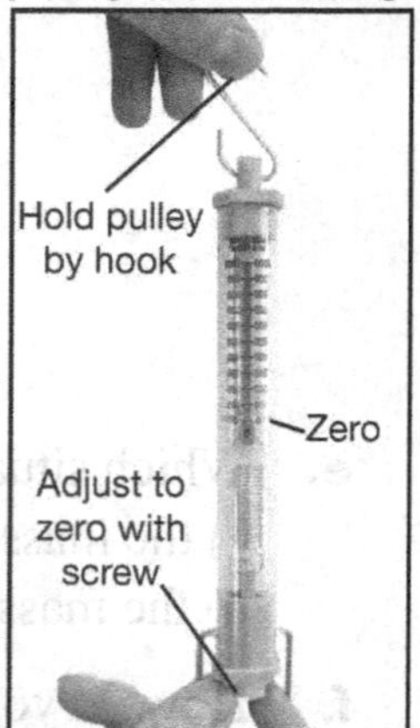

3 Setting up a simple machine

To begin exploring simple machines, you will measure how much force it takes to directly lift a weighted block. Then you will use a single supporting strand on a pulley to lift the same "load." Will it take more or less force to lift the load using the pulley? Try the experiment to find out.

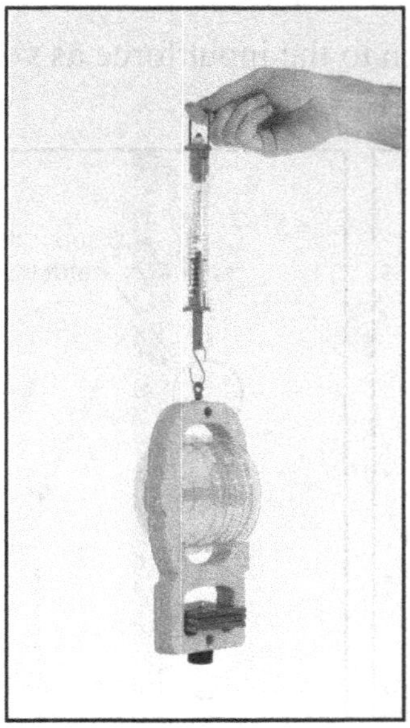

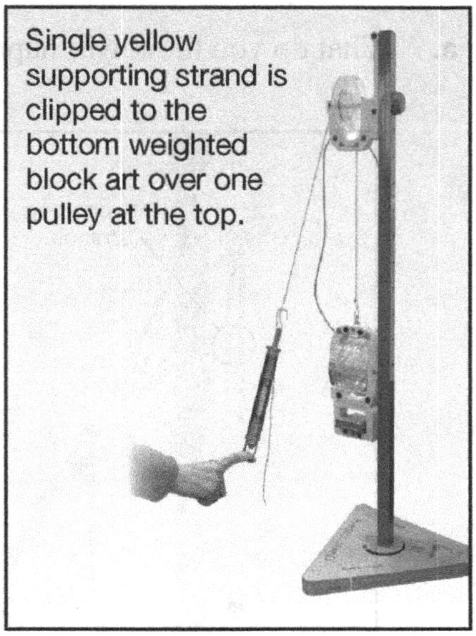

1. Make your weighted block by placing three weights on the triple pulley block. This will be the "load" you must lift throughout the investigation.
2. How much force does it take to lift the load without any help from a simple machine? Use the 20-N spring scale to measure the weight, in newtons, of the load. (Don't forget to calibrate the spring scale as shown on the first page of the investigation.) Record the weight of the load in Table 1. This is the *output force,* since lifting the load is the output task of the machine.
3. Before using the top and bottom blocks together as a simple machine, attach the red safety string, as described on the dynamic pulley instruction sheet that was packed with your kit.
4. Set up a simple machine to lift the load. Attach the top block to the second hole from the top on the Physics Stand. Make sure the eye hook is oriented downward.
5. Get the 20-N spring scale ready by calibrating it as shown on the first page of the investigation. Be sure to use the second procedure described.
6. Clip the yellow pulley string to the load and loop it over one of the pulleys in the top block.
7. *Safety tip*: Pull the string downward. Don't pull the string horizontally, or you could tip the stand over. Have someone be a "spotter" for the equipment.
8. Move the cord lock up to a convenient location and slip the hook of the spring scale around the cord lock so it is safely anchored. Pull *down* on the spring scale to raise the load. Measure the force required to hold up the load by applying a downward force on the string. This is the *input force,* since it is the force applied to the machine to accomplish the output task (lifting the block). Record the force in Table 1.

Table 1: Input and output force

Number of pulley wheels	Input force (N)	Output force (N)
1		

a. Compare the input and output force for the pulley setup. Is one force larger than the other, or are they the same? Was this what you expected? Explain.

4 The block and tackle machine

A block and tackle is a machine built with multiple pulleys (blocks) and a single rope (tackle). You will create a block and tackle machine using a string and up to four pulley wheels. As in Part 3, the input force is applied to the string using a spring scale. The output force is the loaded triple pulley's weight. The output force will stay the same throughout the experiment.

a. What do you think will happen to the input force as you add more pulleys to the setup?

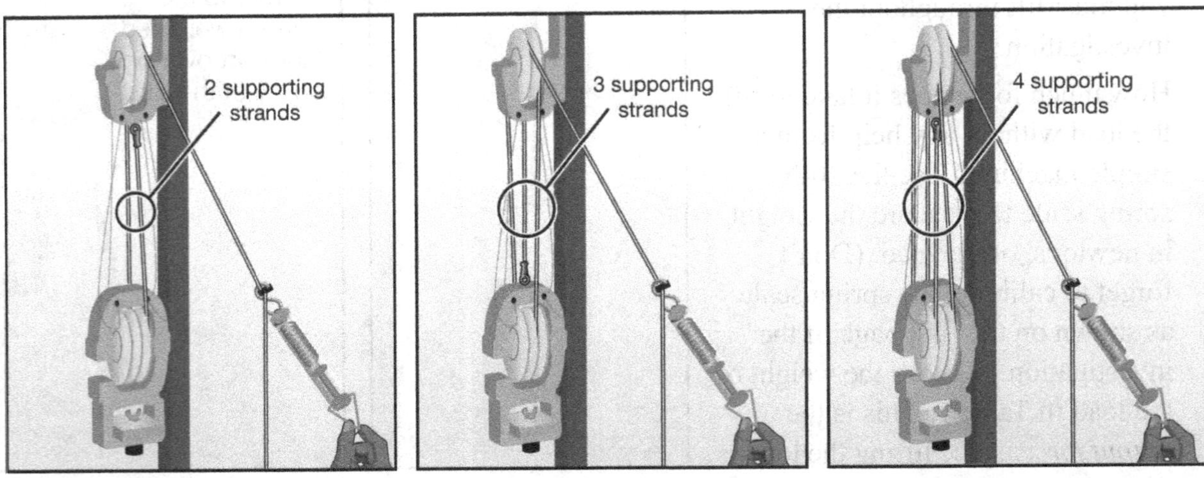

1. Record the output force you measured in Part 3 (the weight of the triple pulley block with three weights attached) in each row of Table 2. Copy the data from Table 1 into the first row of Table 2.
2. Set up a two-pulley system. Clip the yellow pulley string to the double pulley block, then loop it around one of the triple pulleys and finally up around one of the double pulleys (see photo above). Notice that when you lift the pulley block by pulling down on the string, there are two strands of the string that directly support it (the string you pull down on is not supporting the weight and doesn't count as a supporting strand).
3. Move the cord lock up to a convenient location and slip the hook of the spring scale around the cord lock so it is safely anchored. Pull down on the spring scale to raise the triple pulley block. Record the input force in Table 2.
4. Configure the strings and pulleys to create three and four strands supporting the pulley block (see photos above). Always set up the system so you pull *down* on the string to lift the load. Record input forces in Table 2. Switch spring scales when necessary.

Table 2: Input and output forces

Number of supporting strands	Input force (N)	Output force (N)
1		
2		
3		
4		

b. As you arrange the yellow string so it passes over more pulleys, what happens to the force needed to lift the triple pulley block?

c. Write a math equation that would allow you to calculate the input force needed to lift the load if all you knew was the number of supporting strands and the weight of the load. The variables in the equation should be input force, output force, and number of supporting strands.

d. If you apply your math equation to each trial in Table 2, your theoretical input force might not always be the same as each measurement. Explain why this could happen.

e. What input force would you expect to measure if you created a block and tackle machine that used all five wheels and five supporting strands?

f. Figure out a way to rig up a block and tackle machine that uses all five wheels and five supporting strands. For this challenge, you might have to pull *up* on the string to lift the load. If you pull up, remember to calibrate the spring scale using the first method described on the first page. When you figure out how to rig up this block and tackle machine, measure the input force. How does the input force compare to your prediction in question e? Explain why the input force is *not* one-fifth the output force.

5 Mechanical advantage

A system of ropes and pulleys can make work easier by reducing the amount of force you need to apply. In a block and tackle machine, a small input force can create a large output force. *Mechanical advantage* is a term that describes the ratio of the output force to the input force. In a machine with a mechanical advantage greater than 1, the output force is larger than the input force.

To find the mechanical advantage of a simple machine, divide the output force by the input force. Mechanical advantage does not have units. Because mechanical advantage is a ratio of force divided by force, the units cancel.

a. Copy your data from Table 2 into Table 3. Calculate the mechanical advantage for each arrangement of the ropes and pulleys. Record your calculations in Table 3.

Table 3: Mechanical advantage

Number of supporting strands	Output force (N)	Input force (N)	Mechanical advantage
1			
2			
3			
4			

b. What is the relationship between the mechanical advantage and the configuration of the block and tackle machine?

c. Explain in a few sentences why the mechanical advantage of a block and tackle machine should be what you proposed in b. This is a challenging question. *Hint:* The string's tension is the same along the entire length of string.

d. Describe how friction affects the mechanical advantage of a machine like the block and tackle. Would friction increase the mechanical advantage, decrease it, or have no effect?

e. What is the mechanical advantage for the five-pulley wheel system you set up in Part 4, question e? Is this what you would expect? Explain.

6 A block and tackle pulling contest

Almost all the pulley systems you built for this investigation served the same purpose: to lift something *up* by pulling *down* on a rope. Try analyzing a different type of setup. In the illustration below, two groups are having a pulling contest. The pulleys are set up in a horizontal system. The team on the left is pulling on a strand of rope that comes off of the pulley farthest from the tree. The team on the right is pulling a strand of rope that is attached to the pulley's block. It may help to turn the picture sideways with the tree at the top before answering the questions that follow.

a. Is the moving pulley the one that is closest to the tree or the one that is farthest from the tree?

b. Which group could use the smaller force to move the opposing team? Explain.

10.2 Work

What is the trade-off for multiplying forces in a machine?

In the previous investigation, you learned how a simple machine can be arranged to provide a mechanical advantage. Ropes and pulleys (and other simple machines) create large output forces from small input forces. In this investigation, you will explore the nature of work and energy and come to an interesting conclusion that is true for all machines.

In this investigation, you will:

- build a simple machine that multiplies force.
- measure and compare input and output forces and distances for different pulley setups.
- calculate and compare work input and work output.
- explore the relationship between work and energy.

Materials List

- Physics Stand
- Dynamic Pulleys
- Yellow pulley string
- 3 Weights
- Spring scales
- DataCollector
- Photogate
- Measuring tape

1 Investigating the block and tackle machine

You learned from the previous investigation that the input force decreases as you add more supporting strands to a block and tackle machine. What is the trade-off? What variable in the system increases to compensate for the decrease in input force? Do this experiment to find out.

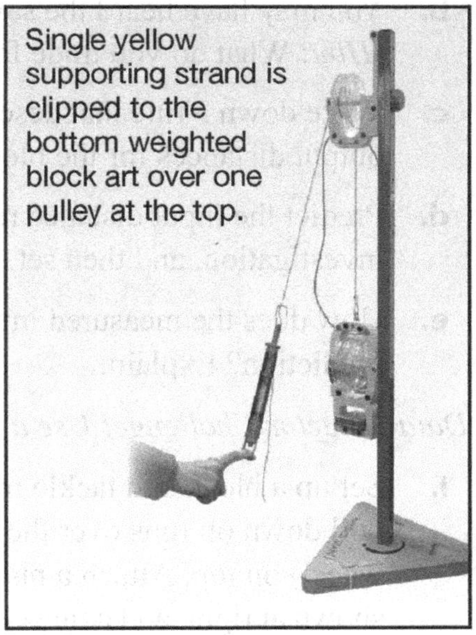

1. Set up a one-pulley wheel block and tackle machine like you did in the beginning of the previous investigation. Choose a distance that you will lift the lower pulley block during each trial of the experiment. This is the *output distance*. Your output distance should be at least 30 centimeters. Use the same output distance for each trial of the experiment. Record that distance in Table 1.
2. There are two cord locks you will use to mark the string-pulling distance. Start with both cord locks next to each other, just touching the upper pulley where the string leaves it.
3. Pull the string to lift the lower pulley block your chosen distance. Measure how the length of the string you had to pull to raise the pulley block (input distance). To easily find this *input distance*, hold the string with the lower pulley block at the chosen height, slide one cord lock back up to the pulley wheel and measure between the two cord locks.
4. Using the spring scale, measure the force needed to lift the triple pulley block. This is the *input force*. Record the input force.
5. Disconnect the lower pulley block and measure its weight with a spring scale. Record the weight as the output force in Table 1. The output force is the same for each trial of the experiment.
6. Rearrange the pulleys and string so that you achieve mechanical advantages of two, three, and four. For each arrangement, be sure you pull down to lift the load up, and find the input force and input distance.

Table 1: Force and distance data

Mechanical advantage	Output force (N)	Output distance (m)	Input force (N)	Input distance (m)
	(same for each trial)	(same for each trial)		
1				
2				
3				
4				

2 Analyzing your data

a. As the mechanical advantage increases, what happens to the length of the string you have to pull to raise the triple pulley block? (Remember, the length of string you pull is called the *input distance*.)

b. You may have heard the saying "Nothing is free." Explain why this is true of the ropes and pulleys. (*Hint*: What do you trade for using less input force to lift the lower pulley block?)

c. Write down a rule that describes the relationship between mechanical advantage and the input and output distances for the block and tackle machine.

d. Predict the input distance required for the special five-pulley wheel setup you created in the previous investigation, and then set it up and measure the input distance.

e. How does the measured input distance for the special pulley setup compare to your prediction? Explain.

DataCollector Challenge: Use a photogate to measure the input distance!

f. Set up a block and tackle machine. Be sure the strand you pull down on runs over the dynamic pulley wheel (striped wheel) on top. Attach a photogate to the top of the stand as shown at right, and connect it to the DataCollector photogate A input. Turn on the DataCollector and select Data Collection mode and then choose the Setup option. From Standard mode, use the pull-down menu next to photogate A (SP_A) to choose displacement in meters (m). The photogate is triggered by the stripes on the moving dynamic pulley wheel, and the information can be interpreted by the data collector as a displacement. Press the Go button and smoothly pull the string to lift the lower pulley block your chosen distance. What is the displacement? How close is it to the input distance you measured before? Try this method of measuring input distance with several different pulley setups. Summarize your findings.

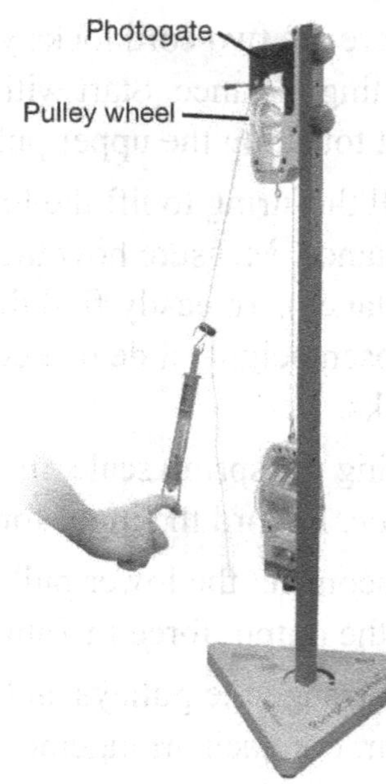

3 What is work?

The word *work* is used in many different ways. For example, you work on science problems, you find that your toaster doesn't work, or you feel that taking out the trash is too much work. In science, however, *work* has one specific meaning: it describes a quantity that is calculated by multiplying force and distance. Your data show that there is a relationship between the force needed to lift the block and the amount of string you had to pull to lift the block. As input force *decreased*, the distance over which you needed to apply the force *increased.*

The work done by a force is equal to the magnitude of the force multiplied by the distance moved in the direction of the force.

$$\textbf{work} = \textbf{force (N)} \times \textbf{distance (m)}$$

Just as you compared input and output force in the previous investigation, you can compare input and output work. Input work describes the amount of work done by you as you pulled the string. Output work describes the work done on the triple pulley block to lift it.

Work is measured in units of newton-meters (N-m). Newton-meters are also called *joules* (J) after British scientist Sir James Joule who discovered the importance of measuring work in a series of experiments performed between 1843 and 1847. One joule is equal to one newton-meter.

4 Calculating work

1. The work done on the triple pulley block is called the *output work.* This work is equal to the output force (weight of the loaded pulley block) multiplied by the output distance (the height the block was lifted). Calculate the output work for each mechanical advantage and record it in Table 2.
2. Next, use your data to calculate the work you did as you pulled on the string to lift the block. This is the *input work.* Multiply the input force by the input distance to find input work.

Table 2: Output and input work

Mechanical advantage	Output work (J)	Input work (J)
1		
2		
3		
4		

a. For each mechanical advantage, how do output and input work compare?

b. Is output work ever greater than input work? Can you explain this?

c. Explain any differences between input and output work in your data.

5 The relationship between work and energy

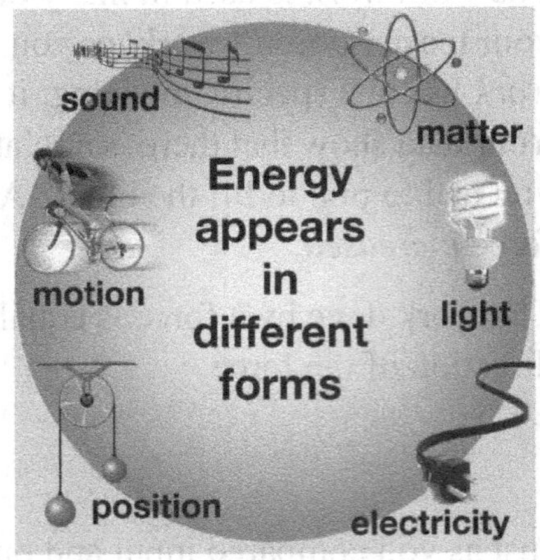

You may already be familiar with the unit *joule* as a measurement of *energy*. Work and energy are related but not the same. Work is one form of energy, but energy can also take other forms. Scientists sometimes define energy as *the ability to do work*. This means that anything with energy, can produce a force that is capable of acting over a distance. The force can come from many different sources. Here are some examples:

- A moving ball has energy because it can exert forces on whatever tries to slow it down or stop it.
- Wind has energy because it can exert forces on any object in its path.
- Electricity has energy because it can turn a motor to exert forces.
- Gasoline has energy because it can be burned in an engine to exert force to move a car.

Energy and work are both measured in joules because energy is *stored work*. Any object with energy can use its energy to do work, meaning it exerts a force that acts over a distance.

The ropes and pulleys experiment illustrates a rule that is true for *all* simple machines: *You can never get more work out of a simple machine than you put into it.* In fact, in most machines, the output work is *less* than the input work. Forces such as friction use some of the input work before it reaches the output of the machine.

a. Give a detailed description of the places in your block and tackle setups where friction plays the greatest role. Include answers to these questions: Does friction's role increase, decrease, or stay the same as you use more pulley wheels in your setups? Does friction affect the spring scale's performance?

b. A bicycle is a machine that does a very good job of turning input work (work done by human muscles) into output work (forward motion). However, about one out of every 20 joules of input work is lost to friction rather than producing output work. Think about how a bicycle works. Name at least three sources of friction that use some of the input work.

10.3 Energy and Conservation of Energy

How is motion on a loop track related to energy?

As a marble rolls along a loop track, what happens to its velocity? What happens to its energy? Where does the marble move fastest, where does it have the most potential energy, and where does it have the most kinetic energy? The rolling marble's potential-to-kinetic energy trade-offs are a direct result of the law of conservation of energy.

In this investigation, you will:

- design an experiment to test a hypothesis about how the marble's energy changes as it moves along a loop track.
- use the law of conservation of energy to predict the minimum release height the marble needs to successfully complete the loop, and the marble's velocity at the top of the loop when released from the track peg.

Materials List

- Physics Stand
- Loop Track
- DataCollector
- Photogate
- Plastic and steel marbles
- Electronic scale or triple beam balance

1 The marble's energy

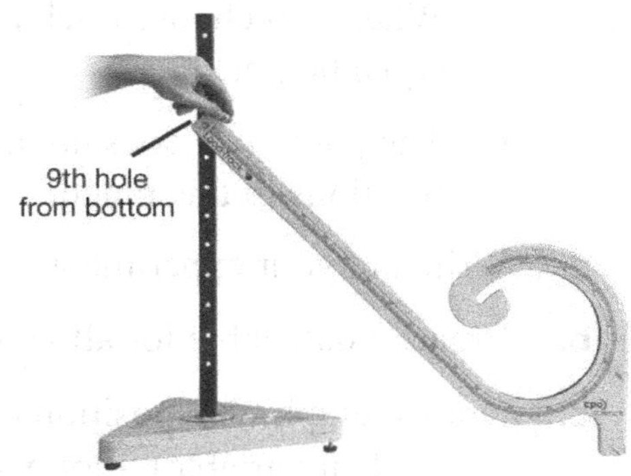

Attach the loop track to the ninth hole from the bottom of the stand. Place the plastic marble against the starting peg. Release the marble and observe its motion. The marble's potential energy at different positions on the track depends on its height. The marble's kinetic energy at different positions depends on its velocity. Does the height of the marble change as it moves along the track? Does the marble's velocity change? Think about these questions as you form your hypothesis about the marble's energy trade-offs.

a. How do potential and kinetic energy change as the marble rolls along the track? Develop a hypothesis by completing this statement: "If the marble's energy trade-offs are related to track position, then . . ."

b. The potential energy the marble has because of its height off the table is exactly the same as the work you must put in to move the marble up to that position in the first place. You know that work = force × distance. In this case, the force is the weight of the marble. Write out the equation that describes the marble's potential energy at any track position.

The rotational kinetic energy (E_{Kr}) of a rigid, rolling object can be calculated using the angular velocity (ω) and the moment of inertia (I) of the object.

$$E_{Kr} = \frac{1}{2} I \omega^2$$

c. You can easily measure the marble's velocity and mass, but you can't directly measure its angular velocity or moment of inertia with the equipment you have. You learned how to calculate the moment of inertia of a solid sphere in section 9.3. You learned how linear velocity and angular velocity are related in section 8.1. Use this information to derive a kinetic energy equation for the rolling marble that only includes the variables of mass and linear velocity. When you derive the equation for the marble's rotational kinetic energy, don't forget to add it to the linear kinetic energy ($1/2mv^2$) to get the marble's total energy. Write out the equation and ask your teacher to check both the potential energy equation and the kinetic energy equation for the marble's motion.

2 Designing an investigation

Design an investigation to test your hypothesis. Collect data so you can calculate the marble's kinetic and potential energy at enough track positions to clearly confirm the hypothesis. Here are some things to consider:

- Be sure the photogate is flat against the bottom of the track. If the photogate is not attached properly, the light beam will not cross the center of the marble and the velocity you calculate will be inaccurate.
- There are two small bumps on each photogate that mark the position of the light beam. The bumps are also in line with the center of the marble as it passes through the photogate. If you measure the height of the marble, measure from the tabletop to one of the bumps.
- The diameter of the marble is 0.019 m.
- When you choose track positions for photogate placement, be sure to include two positions of equal height.
- The photogate does not attach to the track properly between 0.73 and 0.81 m, so do not use positions in this range.

a. Write out your experiment's procedure in numbered steps.

b. Create a data table for all your measurements and calculations. Stay true to the units (m, kg, m/s).

c. Draw a graph with position on the *x*-axis (independent variable) and kinetic energy on the *y*-axis (dependent variable). Plot potential energy and total energy on the same graph (the graph will have three lines).

3 Arguing from evidence

a. Write a brief description of what your graph shows about the marble's energy trade-offs. Include answers to these questions: Where is the potential energy largest and smallest? Where is the kinetic energy largest and smallest? How are the potential and kinetic energy related to the total energy?

b. Does your experimental evidence confirm your hypothesis? Explain.

c. According to the law of conservation of energy, the total energy of the marble stays constant while the potential and kinetic energy trade-offs happen along the track. Do your data support this? Explain.

4 Developing a model for the minimum release height

In Investigation 8.2, Centripetal Force, you found that the marble must move at a velocity of at least 1 m/s at the top of the loop if it is going to land successfully in the catcher. Using this minimum velocity, kinetic energy, and potential energy, you can predict the minimum release height the marble needs to land successfully in the catcher.

a. Work with your group to decide which calculations you need to predict the minimum release height. Describe your approach.

b. Do the necessary calculations and show your work. Explain how the law of conservation of energy was important to your prediction.

c. Release the marble from the predicted height. Did it make it around the loop? Will it make it around if you release the marble from lower than this point? Explain.

d. Roller coaster cars are always designed to enter a loop after going down a hill that has more than the minimum height. Explain why.

5 Predicting the maximum velocity at the top of the loop

Energy conservation can also be used to predict the marble's velocity at the top of the loop if it is dropped from a specific height. What should the marble's velocity be if you release it from the starting peg at the top of the track?

a. Work with your group to decide what calculations you should make to predict the marble's velocity at the top of the loop when released from the starting peg. Describe your approach.

b. Do the necessary calculations and show your work. Explain how the law of conservation of energy was important to your prediction.

c. Use a photogate to test your prediction. Calculate the percent difference between your predicted velocity and the measured velocity. Give at least two reasons why the predicted and measured velocities might be different.

d. Does mass make a difference? Will a steel marble of the same diameter have a velocity that is greater than, less than, or equal to the plastic marble's velocity at the top of the loop when released from the starting peg? Explain the reasoning behind your prediction.

e. Test your prediction. How does the steel marble's velocity at the top of the loop compare to the plastic marble's velocity? Explain.

11.1 Efficiency

How efficient is the SmartTrack?

The conversion of potential energy to kinetic energy obeys the *law of conservation of energy.* This law states that energy cannot be created or destroyed, but can be converted from one form to another. As the Energy Car rolls down the track, its potential energy is converted into kinetic energy.

The law of conservation of energy does *not* require that 100 percent of the potential energy become kinetic energy. Other forms of energy conversion are also allowed. For example, friction converts some potential energy to heat and wear. The *efficiency* of a process is the ratio of the energy you get out of the process divided by the energy you start with. In this investigation, you will explore whether the conversion from potential to kinetic energy for a car rolling down a track is 100 percent efficient.

In this investigation, you will:

- calculate the kinetic and potential energy of the Energy Car at the top and bottom of the SmartTrack.
- compare results from cars of different masses.
- determine the efficiency of the SmartTrack.

Materials List

- SmartTrack
- Physics Stand
- Energy Car
- Velocity sensor
- DataCollector
- Electronic scale or triple beam balance

1 Setting up the SmartTrack

1. Attach the SmartTrack to the Physics Stand at the 10th hole from the bottom.
2. Attach the velocity sensor to the top of the track and plug it into the DataCollector.

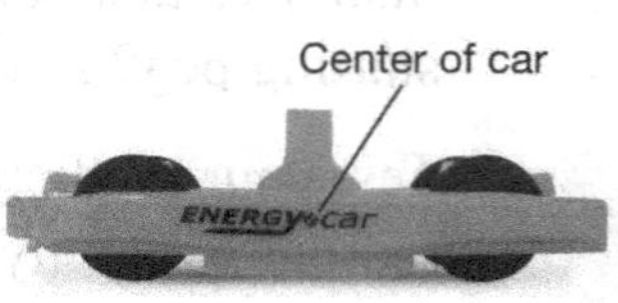

2 Collecting data

1. Turn on the DataCollector. At the home window, select Data Collection mode.
2. At the Go window, choose Setup at the bottom of the screen.
3. At the setup window, choose standard mode, 200 samples, and 50 Hz. This will allow the DataCollector to collect 50 samples of data from the velocity sensor each second.

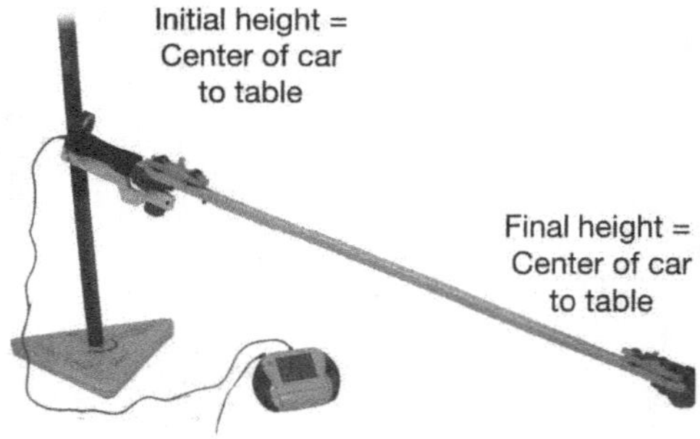

4. Place the car (with no marbles) so it is against the starting peg at the top of the track. Tap go at the bottom of the setup screen on the DataCollector, and release the car.
5. When the car reaches the bottom of the track, press the button on the DataCollector enclosure to stop the experiment. Switch from Meter to Table and Graph view to study your data.
6. Record the car's maximum velocity in Table 1.
7. Do two more trials of the experiment, record the velocities, and calculate the average.

8. Place the car at the top of the track so it is held in place by the starting peg. Secure the car at the top of the ramp with the peg and measure the vertical distance from the center of the car to the tabletop (see photo on previous page). Record the height in meters in Table 1.
9. When the car reaches maximum velocity it is at the bottom of the track. Find the height of the car at the bottom of the track by measuring the vertical distance from the tabletop to the center of the car (see photo on previous page). Record the height in meters in Table 1.
10. Measure the mass of the car in kilograms and record in Table 1.

Table 1: Velocity, height, and mass data for car with no marbles

Trial	Velocity m/s	Initial height (m)	Final height (m)	Mass of car (kg)
1				
2				
3				
Average				

a. If you repeat this experiment with a car that holds two marbles, which values from Table 1 will change significantly?

11. Place two marbles in the car and repeat steps 1–10. Record your data in Table 2.

Table 2: Velocity, height, and mass data for car with two marbles

Trial	Velocity (m/s)	Initial height (m)	Final height (m)	Mass of car (kg)
1				
2				
3				
Average				

3 Potential and kinetic energy

The total energy of the car on the track is a combination of potential and kinetic energies. Potential energy, also known as stored energy, depends on the *height* of the car. Kinetic energy is related to motion and depends on the *velocity* of the car.

1. What is the velocity of the car at the instant you release it at the top of the track? Record this in Table 3. What is its kinetic energy? Record this in Table 3.
2. Copy the average velocities you found at the bottom of the track for each car in Table 3.

3. Calculate the potential and kinetic energy of the car at the top and bottom of the track. Record your results in Table 3.
4. The total energy at each position on the track is calculated by adding the potential and kinetic energies. Find the total energy at each position and record the results in Table 3.

Table 3: Potential, kinetic, and total energy of cars

Number of marbles in car	Location on track	Velocity (m/s)	Kinetic energy (J)	Potential energy (J)	Total energy (J)
0 marbles	top				
	bottom				
2 marbles	top				
	bottom				

4 Efficiency

You might not have realized it, but the SmartTrack is a simple machine that converts energy from one form to another. The energy put into any machine is called *input energy*. You provide the input energy to the car as potential energy when you lift the car up to the top of the track. The useful energy supplied by a machine is called *output energy*. Imagine that you use the track as an egg-breaking machine. You place an egg at the bottom and roll a car down the track. The output energy of the machine is the car's kinetic energy at the bottom of the track. The greater the kinetic energy, the greater the car's ability to do work on the egg and smash it.

The ratio of the output energy to the input energy for a machine is called its *efficiency*. An efficiency of 100 percent means that all of the input energy is converted to useful output energy. No machine is 100 percent efficient. Automobiles, for example, are typically only about 13 percent efficient.

EFFICIENCY (energy efficiency)

$$\text{Efficiency } \varepsilon = \frac{E_o}{E_i} \quad \begin{matrix}\textbf{Energy output} \text{ (J)} \\ \textbf{Energy input} \text{ (J)}\end{matrix}$$

a. Compare the total energy of each car at the top of the track to its total energy at the bottom. Did each car's total energy increase, decrease, or remain unchanged?

b. Calculate the efficiency for each car.

c. The law of conservation of energy states that energy cannot be created or destroyed. If the efficiencies were less than 100 percent, does this mean the law of conservation of energy is not true? Was energy lost? Where did the energy go? (*Hint*: The motion of the car includes rotational as well as linear.)

d. If the efficiency for the car with two marbles was 100 percent, how fast would it be moving at the bottom of the track?

11.2 Energy and Power

How powerful are you?

The power output of motors, cars, and other machines is often listed on the machine or in its owner's manual. What about the power output of the human body?

In this investigation, you will:

- calculate the work and power output while lifting an object.
- describe the difference between energy and power.
- compare your power output with that of a common electric light bulb.

Materials List

- DataCollector or stopwatch
- Meter stick
- Object for lifting such as a small barbell, loaded backpack, brick, or heavy book
- Scale to determine the object's weight or mass
- Piece of masking tape

1 Lifting power

1. During the first part of this investigation, you will calculate the work you do and your power output while you lift an object. Listen to directions from your instructor about the type of object you will lift. Determine the weight of the object in newtons and record it in Table 1.
2. Find a location along a wall where your group will lift the object. Measure 1.5 meters up from the floor and mark this height on the wall with a piece of tape.
3. Now each person in your group will lift the object from the floor until the bottom of the object is at a height of 1.5 meters. Measure the time (in seconds) that it takes for each person to lift the object. You may choose to lift it quickly or slowly. Record each person's name and time in Table 1.
4. Calculate the work done in lifting the object. The force you should use is the weight of the object in newtons. Record the work done in Table 1.
5. Calculate each person's power in lifting the object. Record in Table 1.

Table 1: Weight lifting data

Name	Weight (N)	Distance (m)	Time (s)	Work (J)	Power (W)
		1.5			

2 Analyzing and interpreting data

a. Why is the object's weight used in the calculation of work done?

b. How did the amount of work done by each person compare? Why?

c. What determined the power of each person? Explain.

d. A typical light bulb has a power of 75 watts. How does this compare to your power output while lifting the object?

e. Discuss two ways you could increase the amount of work you do in this activity.

f. Discuss two ways you could increase your power output in this activity.

3 Stair-climbing work and power

1. This part of the investigation is done with your entire class. Select several people who want to walk or run up a flight of stairs while others record their times. Each person who volunteers must know his or her weight in newtons. (Weight in pounds can be converted to newtons using the conversion factor *1 pound = 4.45 newtons.*) Record the names and weights of the volunteers in Table 2.

Table 2: Stair-climbing data

Name	Weight (N)	Vertical distance (m)	Time (s)			Avg. time (s)	Work (J)	Power (W)

2. Locate a suitable stairwell and measure the vertical distance from one flight to the next. An easy way to do this is to measure the height of one stair and count the number of stairs students are climbing. Record the measurement in Table 2.

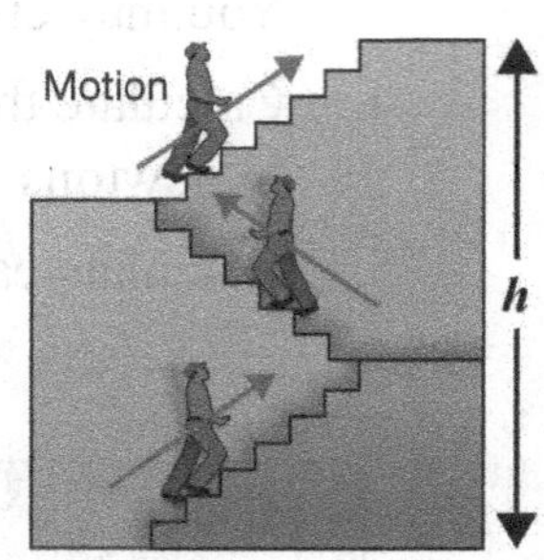

3. Select three people to measure the time it takes for the three volunteers to climb the stairs. All three timers will measure the time at once; each stair climber will make only one trip up the stairs.
4. Each volunteer may climb the stairs slowly or quickly. Record the three times for each person.
5. Calculate the average time for each climber and record it in the table.
6. Calculate the work done and power output for each climber. Record the results in the table.

4 Constructing explanations

a. How did the work done by the different people compare? In the first part of the investigation, every group member did the same amount of work. Why was this not the case with the stairs?

b. Which person had the greatest power? Which two factors determined each person's power?

c. Calculate the average work done and the average power for all of the stair climbers.

d. How do the average amounts of work and power for the stair climbing compare with your own work and power when lifting the object in the first part of the investigation?

e. The human body is able to do work because energy is absorbed when we eat food. In physics, we usually measure work and energy in joules, but food energy is usually measured in calories. One calorie is equal to 4,186 joules. Select one of the people who climbed the stairs and calculate his or her work done in calories. These calories come from food that is "burned" by the person's body. You may be surprised by how small the number is.

f. Imagine that two people having equal weights climb the same flight of stairs. One runs and the other walks. Do they do the same number of calories of work (in the physics sense)? Do they have the same power output? Explain.

11.3 Energy Flow in Systems

Where did the energy go?

The flow of energy drives all processes both in human technology and in natural systems, including living things. This investigation challenges you to trace the flow of energy through its different forms in processes that you may experience.

Materials List

- Pencils and paper

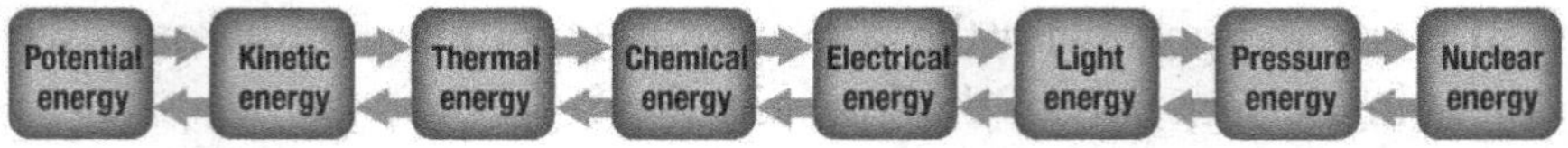

Energy can never be created or destroyed, only changed from one form to another.

In this investigation, you will:

- follow sequences in which energy changes form in technical, natural, and biological systems.
- sketch an energy flow diagram of technical, natural, and biological systems.

1 What is an energy flow diagram?

An energy flow diagram is a type of flow chart that shows how energy moves and what forms it takes. For example, in a pendulum the energy flows back and forth between potential energy and kinetic energy. An energy flow diagram for a pendulum is shown below.

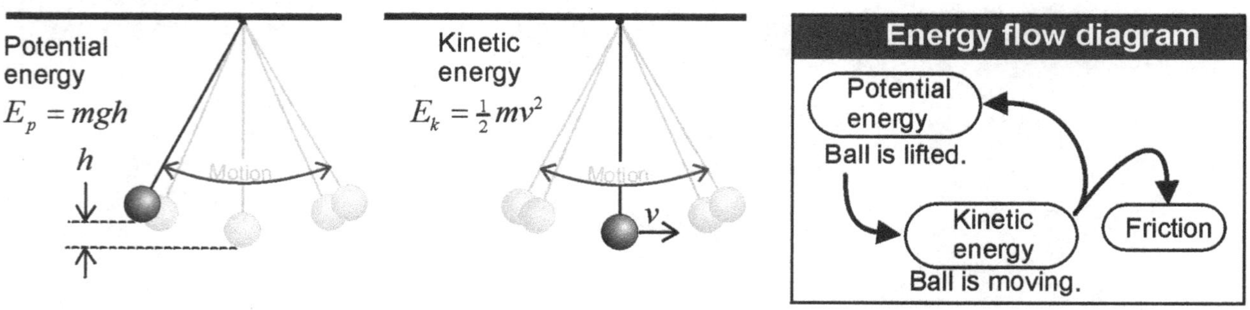

2 Drawing energy flow diagrams

Draw an energy flow diagram for each scenario described below. Your diagram must have enough detail to show at least three forms that energy takes. A short description (a few words) should accompany each different form of energy on your diagram. You may make your diagram more complex by including more forms of energy.

a. A car drives down a hill and then up and over the top of another hill.

b. A seagull in flight swoops down and snatches a fish to eat out of the water.

c. A calculator operates from a solar cell.

d. A camper uses a wood fire to heat up a pot of water for tea. The pot has a whistle that lets him know when the water boils.

e. A bicyclist rides at night, switching on her bike's generator so that the headlight comes on. The harder she pedals, the brighter the headlight glows.

12.1 Momentum

What are some useful properties of momentum?

This investigation is about *momentum*, a property of moving matter. According to Newton's first law, moving objects tend to continue moving with the same speed and direction. Like mass, momentum measures the resistance of objects to changes in speed or direction. Unlike mass, momentum is a *vector,* meaning it has direction as well as magnitude. You can think of momentum as a property of moving mass that tends to maintain an object's direction and speed.

In this investigation, you will:

- explore collisions and show how they obey the law of conservation of momentum.

Materials List

- SmartTrack
- Velocity sensor
- 2 Energy Cars
- Photogate
- 4 Steel marbles
- 1 Rubber band
- Electronic scale or triple beam balance

1 Making a collision

1. Set up the SmartTrack with a rubber band on one end and a clay ball on the other end. Use the bubble level to set the track level.
2. Place one steel marble in each car.
3. Wrap the thick rubber band around the moving car (the blue one). Place both cars on the track so their noses are pointed toward the rubber band launcher.
4. Place the target car (the orange one) near the center of the track. Use the screw to launch the moving car using the same deflection of the rubber band each time. This means the same force is applied to each launch. You will use this car to create the collision.
5. To make a collision, release the moving car from the rubber band launcher. It will speed down the track and hit the target car. This is an efficient way to produce collisions on the track.

a. Does the *moving* car bounce back after the collision?

b. Does the *moving* car keep going forward after the collision?

c. Does the *moving* car stop at the collision?

d. How does the *target* car behave?

2 Using your model

a. Describe in words the motion of the two cars before and after the collision.

b. The target car must exert a force on the moving car to stop it. How strong is this force relative to the force the moving car exerts on the target car to get it moving? How could you use the velocity sensor and photogate to provide evidence for your answer?

3 Gathering evidence

1. Try the experiment again, but now use the velocity sensor and photogate to find out what happens to the cars' velocities. Record your data in the first row of Table 1.

Table I: Collision data

Collision Trial	Velocity of moving car right before collision (m/s)	Time for target car to pass through photogate after collision (s)	Velocity of target car after collision (m/s)
1			
2			
3			
4			
5			

a. Newton's third law tells us that when the moving car exerts a force on the target car, the target car exerts an equal and opposite force on the moving car. Does your data provide evidence for this? Explain.

b. You can compare the velocities of the cars before and after the collisions. How can comparing these velocities show that there are equal and opposite forces at work?

4 Changing the masses

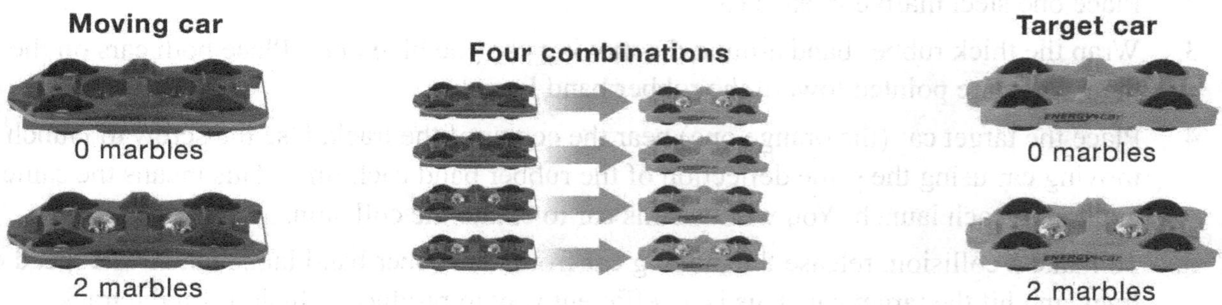

1. Try the experiment with the four mass combinations shown above. Record your data in Table 1 in rows 2 through 5 to keep track of what happens to each car's velocity before and after each collision.
2. Measure and record the mass of each car used in your experiment. Use these mass values to calculate the momentum of the system before and after each collision.

5 Arguing from evidence

a. Compare and contrast the momentum of the system before and after each collision. Do your data support the law of conservation of momentum? Explain.

b. Explain how your observations support the idea that there are action and reaction forces.

c. If the action and reaction forces are equal in strength, why do the two cars move at different speeds after some collisions?

12.2 Force and Momentum

How does force affect momentum?

Momentum changes when a net force is applied. The converse is also true: when momentum changes, forces must be present. If momentum changes quickly, large forces are involved.

In this investigation, you will:

- create and observe elastic and inelastic collisions.
- deduce the relative size of forces from changes in momentum.

Materials List

- Physics Stand
- String
- Standard-sized brick
- Rubber superball
- Clay ball of similar size
- Paper clip
- Meter stick

1 Force, momentum, and the second law

The relationship between force and motion follows directly from the second law. Acceleration is the change in speed divided by the change in time. In this type of logic, it is convenient to use the symbol Δ (the Greek letter *delta*), which translates to "the change in." When you see the Δ symbol, replace it in your mind with the phrase "the change in." Then acceleration can be written as $\Delta\vec{v}/\Delta t$, which translates to "the change in speed divided by the change in time."

NEWTON'S SECOND LAW $\vec{F} = m\dfrac{\Delta\vec{v}}{\Delta t}$ *Translation* Force = mass × $\dfrac{\text{change in velocity}}{\text{change in time}}$

Mass multiplied by the change in velocity ($m\Delta\vec{v}$) is the same as the change in momentum ($\Delta\vec{p}$). We can rewrite the second law in a form that shows force is equal to the rate of change in momentum. Large forces produce a proportionally large change in momentum.

NEWTON'S SECOND LAW (momentum form)

$$\vec{F} = \frac{\Delta\vec{p}}{\Delta t}$$

Force (N); ***Change in momentum*** (kg·m/s); ***Change in time*** (s)

Consider two balls falling on the floor. One ball is made of rubber and bounces off the floor in an *elastic* collision. The other ball is made of clay and does not bounce when it hits the floor.

a. Which collision creates the greater change in momentum? Remember, you need to consider the sign of the velocity before and after the collision when you calculate the change in momentum.

2 Observing elastic and inelastic collisions

For this experiment, you need a rubber ball about 5 centimeters (2 inches) in diameter and an equal amount of soft clay. You also need a standard sized brick, around 3 5/8 in. × 2 1/4 in. × 8 in.

1. Measure and record the mass of the rubber ball in Table 1.
2. Make a clay ball that has the same mass as the rubber ball.
3. Attach 1 meter of string to each ball. Attach the string to the rubber ball by poking half a paperclip into the rubber to make a loop. Attach the string to the clay ball by burying it in the clay so that it passes through the center.
4. Tie a small loop in the free end of both strings. Make the distance from the loop to the ball about 70 centimeters. Use the same distance for both rubber and clay balls.
5. Slip the physics stand wrench in a hole near the top of the stand. Either ball may now be swung so it hits the brick by looping the string over the steel shaft.
6. Set the brick on the base of the stand next to the pole. Hang the rubber ball from the spindle and swing it so that it collides with the brick when you release it.
7. Find the *lowest* height from which you can release the rubber ball and have it still knock over the brick every time. Use a meter stick to record and measure the height. It will take many trials to locate the minimum height at which the brick is always knocked over.

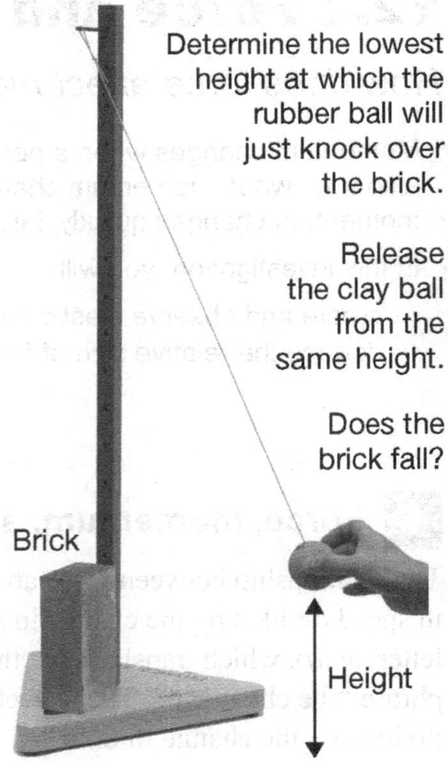

8. Replace the rubber ball with the clay ball. Release the clay ball from the same height you determined in step 7. Observe what happens when the clay ball hits the brick. The mass is the same, but is the result the same?

Table 1: Brick collision data

Type of ball	Mass of ball (g)	Minimum height to knock over brick (cm)
Rubber		
Clay		

3 Constructing explanations

a. Compare the velocities you observed of the rubber ball and the clay ball just before they collided with the brick. Are the velocities different or are they about the same?

b. Compare the momentum of the rubber ball and the clay ball just before they collided with the brick. Does the rubber ball have more, less, or about the same momentum as the clay ball?

c. Did the clay ball knock over the brick? Report your observations using the fact that force is equal to the rate of change of momentum.

d. Which ball experienced a greater change in momentum during the collision?

12.3 Angular Momentum

How does momentum apply to rotational motion?

You can think of Newton's first law in terms of momentum. The momentum of an object does not change unless force is applied. A similar situation applies to rotating motion, except that the quantity that tends to stay the same is the angular momentum.

In this investigation, you will:

- create a rotating system and observe the action of a torque.
- deduce the relative size of forces from changes in momentum.

Materials List

- 1 Meter of string
- 3 Metal washers
- Plastic drinking straw with a flexible joint
- DataCollector

1 Thinking about angular momentum

Consider a small mass rotating at the end of a string. The mass follows a circle as it moves. Since the system of the mass and string is rotating, it has angular momentum. Like linear momentum, angular momentum obeys a conservation law. If a change is made to the system without producing any torque, the system reacts in such a way as to keep the total angular momentum constant.

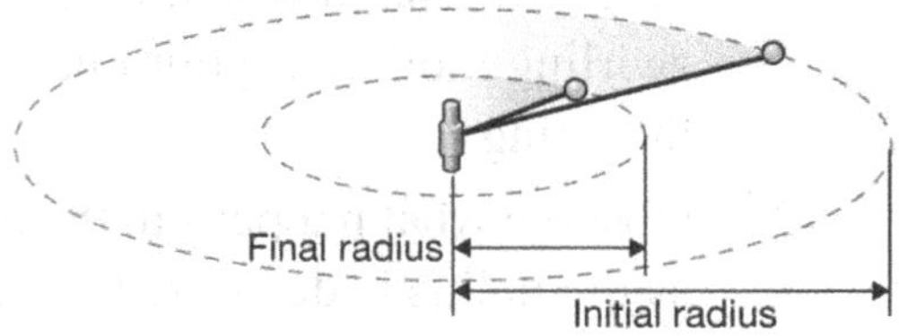

a. List three variables that determine the angular momentum of the rotating mass on a string.

b. Discuss with your group what would happen if the radius of the circle were suddenly made smaller. Assume that the mass stays constant. How can the angular momentum stay the same when the radius decreases? Write a hypothesis describing what you believe will happen in an experiment in which you are observing a mass revolving at the end of a string and the string is shortened during the motion.

2 Setting up the investigation

1. One person in each group should take the flexible straw and thread a string through the straw. Take one washer and put the straw and string through its center. The string should be about 1 meter long.
2. Securely tie a washer on either end of the string so the distance between washers is about 0.8 meter. The setup should look like the diagram at right.

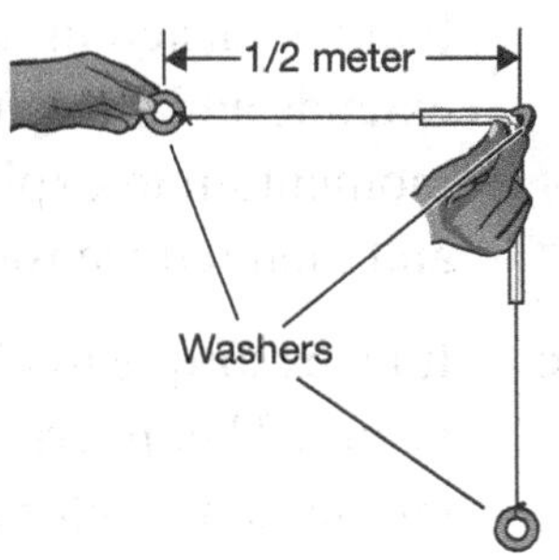

3. Bend the straw so it makes it a 90-degree angle as shown in the diagram.
4. Hold one washer in one hand and use your other hand to hold the longer, straight section of the straw. Arrange your hands so the bend in the straw is about the halfway point on the string and supported by the middle washer. Using a gentle movement of the hand holding the straw, you can whirl the washer around in a slow circular motion. The experiment requires slow, steady circular motion.

Safety Notes: Do not whirl the washer too quickly! Spinning the washer around at a high speed can be dangerous. Make sure no one is too close when you are starting or spinning your washer.

3 Observing the conservation of angular momentum

By pulling the string through the straw, it is possible to change the radius of the motion while the mass is still moving. If the string is pulled carefully downward, the force applied is along a radius and therefore does not exert a torque about the center of rotation. If the torque is zero, angular momentum should be conserved. Pulling the string down evenly while maintaining steady circular motion is tricky and requires practice.

1. Once you have established a steady rhythm, others in the group should measure how long it takes to complete 10 full circles. The DataCollector in stopwatch mode may be used for this purpose. The data allows you to repeat the experiment under the same conditions if necessary.
2. With your lower hand, slowly pull down on the hanging string. This will pull in the rotating washer, thus decreasing its radius of motion. Try to keep the swirling motion of your upper hand steady while you pull the string.
3. Observe what happens to the motion of the rotating washer as the radius is decreased. You may wish to try the experiment several times to perfect your technique.

4 Constructing explanations

a. How do the conservation of angular momentum and the concept of moment of inertia help to explain your observations from Part 3.

b. In many diving competitions, the diver jumps from the board or platform, turns one or two complete somersaults, and straightens out to land in the water with arms and head straight down and feet straight up in the air. While completing the somersault, the diver's body is tucked in tightly. Before entering the water, the diver's body extends and straightens out. Use the conservation of angular momentum to explain how a diver can spin rapidly in a somersault and then reduce rotational speed to enter the water vertically.

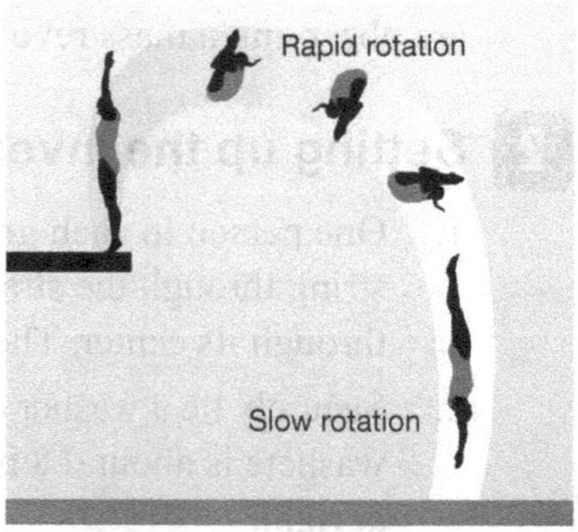

c. It takes torque to change the direction of the angular momentum vector. The inverse is also true. If the direction of the angular momentum vector changes, reaction torques act back on the rotating object to resist the change. For example, it is harder to balance a bicycle that is standing still than a bicycle that is rolling. Use the conservation of angular momentum to explain why this is true.

Angular momentum behaves like a vector. Unlike the linear momentum vector, however, the angular momentum vector does not point in the direction of an object's velocity. Instead, the angular momentum vector points along the axis of rotation. You can use your right hand to determine its direction. When the fingers of your right hand curl in the direction of rotation, your thumb points in the direction of the angular momentum vector. The law of conservation of angular momentum applies to the direction of the angular momentum vector as well as its magnitude.

13.1 Harmonic Motion

How do we describe the back-and-forth motion of a pendulum?

Two common types of motion are linear motion and harmonic motion. One kind of motion goes from one place to another like a person walking from home to school. This is *linear motion*. We use words such as distance, time, speed, and acceleration to describe linear motion. The second kind of motion repeats itself over and over like a child going back and forth on a swing. This kind of motion is called *harmonic motion*. The word *harmonic* comes from the word *harmony* meaning "multiples of." Any system that exhibits harmonic motion is called an *oscillator*.

In this investigation, you will:

- experiment with a pendulum and see what you can do to change its period.
- set up your pendulum to measure a 30-second time interval.

Materials List

From Springs and Swings:

- Small stand
- Mass hanger
- Washers
- Pendulum string
- Beam breaker

Other materials:

- DataCollector
- Photogate
- Measuring tape

1 A pendulum

A pendulum is an oscillator made from a mass on a string. The mass is free to swing back and forth.

- A *cycle* is one complete back-and-forth motion.
- The *period* is the time it takes to complete one full cycle. For a pendulum, this is the time it takes for the pendulum to swing from left to right *and back again.*
- The *amplitude* describes the size of the cycle. The amplitude of a pendulum is the amount the pendulum swings away from equilibrium.

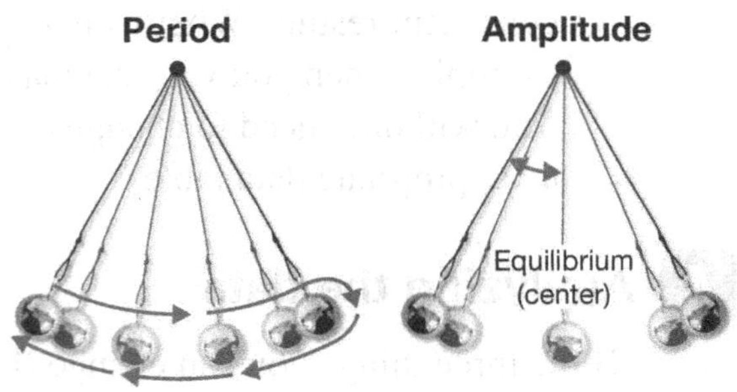

The period is the time to complete one cycle.

The amplitude is the maximum amount the system moves away from equilibrium.

2 Setting up the pendulum

1. Slide the pendulum string into the bracket at the top of the stand as shown at right.
2. Attach the mass hanger to the loop end of the pendulum string. You now have a freely swinging pendulum.
3. Attach the beam breaker to the string just below the bracket (see photo at right).
4. Attach a photogate to the top of the stand as shown at right.
5. Plug the photogate into the DataCollector (photogate A input). Turn on the DataCollector and choose Timer mode from the home screen.

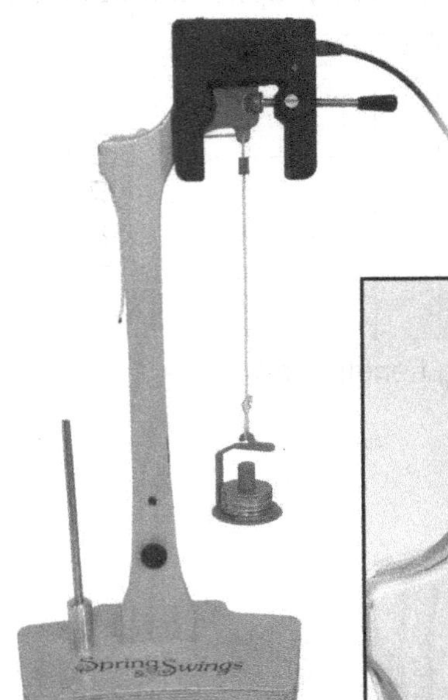

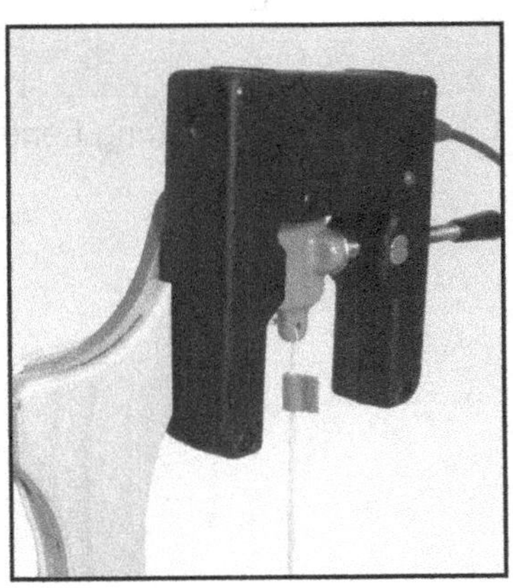

6. In Timer mode, choose the Period function (p). Period is the time for one cycle. The pendulum breaks the beam once as it swings through its complete cycle. Thus, the DataCollector's period readout corresponds directly to the time it takes the pendulum to complete one full cycle.

3 The three pendulum variables

In this experiment, the period of the pendulum is the dependent variable. There are three independent variables: the pendulum mass, the amplitude of the swing, and the length of the pendulum string.

1. Put a few washers on the mass hanger, adjust the string length to about 15 cm, and swing the pendulum through the photogate. Notice the period measured by the photogate.
2. The length of the string can be changed by removing it from the slotted bracket and placing it back in. You can change the mass by varying the number of washers on the mass hanger. The amplitude can be changed by varying the starting angle of the pendulum (low, medium, and high angle).
3. Design an experiment to determine which of the three variables has the greatest effect on the period of the pendulum. Your experiment should provide enough data to show that one of the three variables has a much greater effect than the other two. Be sure to use a technique that gives you consistent results. When you are changing a variable, it is a good idea to change it by a lot. For example, when you vary the mass, try experimenting with no washers, 5 washers, and 10 washers. You will only need to change each variable three times to see the effect. Be sure to design and use an appropriate data table.

4 Analyzing the data

a. Of the three things you can change (length, mass, and amplitude), which has the greatest effect on the pendulum, and why? In your answer, you should consider how gravity accelerates objects of different mass.

b. Split up your data. Make a separate graph showing how each variable affects the period. Then, study the effect of each variable. To make comparison easier, make sure all the graphs have the same scale on the *y*-axis (period). The graphs should be labeled as shown below.

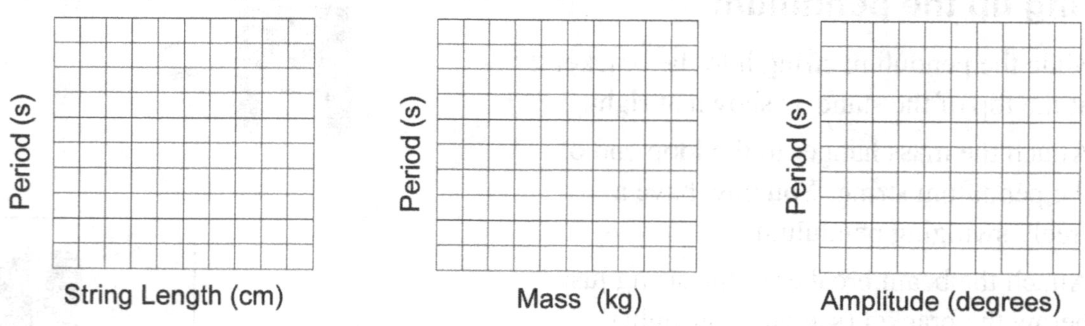

5 Planning and carrying out an investigation

Pendulum clocks were once among the most common ways to keep time. You can still buy beautifully-made, contemporary pendulum clocks. To make a pendulum clock accurate, the period must be set so that a certain number of periods equals a convenient measure of time. For example, you could design a clock with a pendulum that has a period of one second. The gears in the clock mechanism would then have to turn the second hand $1/60^{th}$ of a turn per swing of the pendulum.

a. Using your data, design and construct a pendulum that you can use to accurately measure a time interval of 30 seconds. Test your pendulum clock against the DataCollector stopwatch mode.

b. Mark a horizontal line on your string length vs. period graph at the period you chose for your pendulum.

c. How many cycles did your pendulum complete in 30 seconds?

d. If mass does not affect the period, why is it important that the pendulum in a clock be heavy?

e. Calculate the percent error in your prediction of time from your pendulum clock.

13.2 Why Things Oscillate

What kinds of systems oscillate?

Motion results when we disturb a system in equilibrium, and sometimes the result is harmonic motion. For example, consider a cart balanced on top of a hill. If you push the cart even a little, it quickly rolls downhill and does not return. This is motion, but not harmonic motion. A cart in a valley is a good example of a system that *does* show harmonic motion. If you move the cart partly up the hill and release it, harmonic motion results as the cart rolls back and forth in the bottom of the valley.

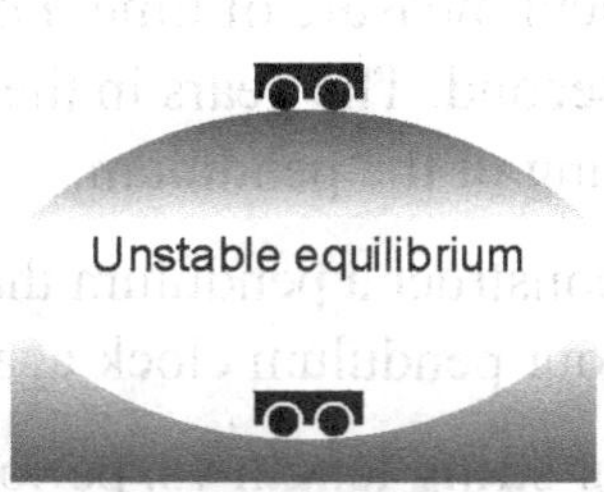

In this investigation, you will:

- build a mechanical oscillator.
- find the oscillator's period and natural frequency.
- change the oscillator's natural frequency.

Materials List

From Springs and Swings:

- Small stand
- Mass hanger
- Beam breaker
- Washers
- Extension springs (blue and white tab springs)

Other materials:

- DataCollector
- Photogate

1 Stable and unstable systems

The top of a hill is an example of *unstable* equilibrium. In unstable systems, forces act to pull the system away from equilibrium when disturbed. The bottom of a valley is an example of *stable* equilibrium. In a stable system, forces always act to restore the system to equilibrium when it is disturbed. We find harmonic motion in stable systems.

a. Describe your example of a stable system in one or two sentences. What happens when you push it a little away from equilibrium? Write one sentence that describes the motion.

b. Describe your example of an unstable system in one or two sentences. What happens when you push the unstable system a little away from equilibrium? Write one sentence that describes the motion.

2 Restoring forces

With mechanical systems like a pendulum, harmonic motion comes from the action of forces and inertia. A *restoring force* is any force that always tries to pull a system back toward equilibrium. For example, gravity always pulls the pendulum toward the center, no matter which side it is on. Because it has inertia, the pendulum passes right through the middle and keeps going. The restoring force then slows it and accelerates the pendulum back toward equilibrium. However, the pendulum overshoots again and goes too far. The cycle (see below) repeats over and over to create harmonic motion.

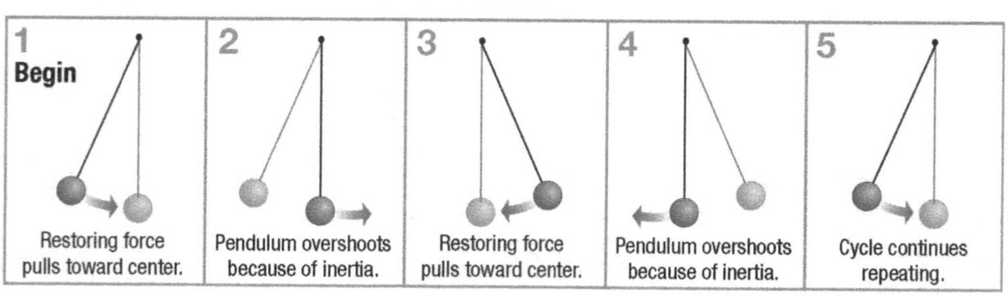

3 Building an oscillator

To make a mechanical oscillator, you need to provide some kind of restoring force connected to a mass that provides inertia. Rubber bands, strings, and springs can all provide restoring forces. Steel marbles, wood blocks, or even a rubber band have mass to supply inertia. The oscillation can go back and forth, up and down, twist, circle, or create any other motion that can be repetitive.

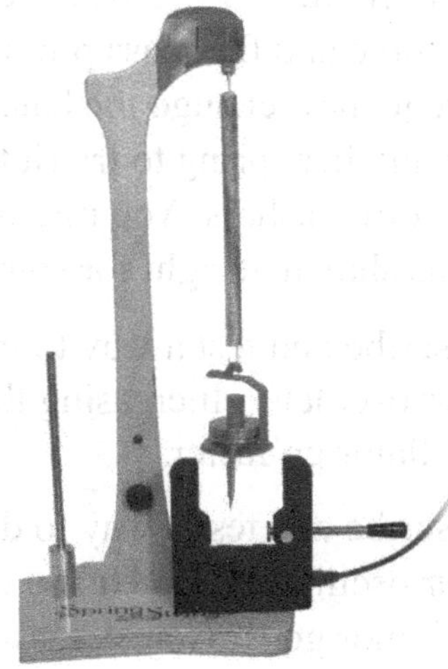

1. Set up a spring system that oscillates up and down (see photo at right). Use the extension spring that has the *blue* tab.
2. Pull the mass hanger down a very small amount and gently let go to create smooth oscillations.

a. Draw a sketch of your system and identify what creates the restoring force, and also the relocation of the mass that provides inertia.

4 The natural frequency

When you pluck a stretched string, it vibrates. If you pluck the same string 10 times in a row, it vibrates at the same frequency every time. The frequency at which objects tend to move in harmonic motion is called the *natural frequency.* Everything that can oscillate has a natural frequency, and most oscillators have more than one.

1. Place seven washers on the mass hanger, and place the photogate on the small stand as shown in the photo above.
2. Plug the photogate into the A input on the DataCollector. Select Timer mode, then the period (p) function. For slow oscillators like this spring/mass system, it is more accurate to measure period and then *calculate* the natural frequency than to measure frequency. Recall from the previous investigation that the period is the time it takes the oscillator to complete one cycle.

PERIOD AND FREQUENCY RELATIONSHIP

Frequency (Hz) $f = \frac{1}{T}$ Period (s)

3. Gently pull down on the spring (you should be using the one with the blue tab) to begin a controlled, smooth oscillation. The period in seconds will vary at the last three decimal places. Practice measuring the period several times, and choose a period with two significant digits. Avoid creating any swinging motion with the oscillator. Practice until you can get smooth up-and-down oscillations with very little movement in any other direction.

a. Calculate the natural frequency by dividing 1 by the period. Show your work.

5 Changing the natural frequency

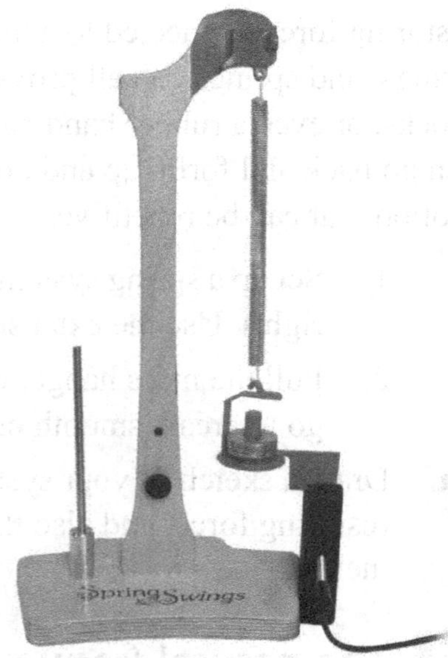

The natural frequency is a balance between the strength of the restoring force and the mass providing the inertia. To change the natural frequency, change the balance between force and inertia. You have another spring to try (it has a white tab) and you can vary the number of washers. You might have to place the photogate on the table as shown at right for some oscillator setups.

a. Describe and test a way to increase the natural frequency of your oscillator. Increasing the natural frequency makes the oscillator go faster.

b. Describe and test a way to decrease the natural frequency of your oscillator. Decreasing the natural frequency makes the oscillator go slower.

c. Create a data table to summarize your results.

d. Write a brief paragraph to explain *why* the changes you made to the oscillating system affected the natural frequency. Refer to evidence from your data table to support your explanation.

6 Natural frequency and guitar strings

a. Guitar strings are great examples of oscillators. To change the natural frequency of any oscillator, you must change the balance between force and inertia. A guitar player tightens and loosens individual string tensions to make each guitar string sound a certain way. When a string is tightened, its frequency increases, and the pitch of the string is higher. Explain why a guitar string's frequency increases when the tension increases, in terms of how the balance between force and inertia is changed.

b. If you look closely at an ordinary guitar, you will see that some strings are fatter and more massive than others. The more massive strings (at the bottom of the photo at right, showing a right-handed guitar) have a low pitch when you pluck them, and the smaller strings have a higher pitch. Explain why this is so in terms of changing the balance between force and inertia. You can refer to data you collected in Part 5 to support your answer.

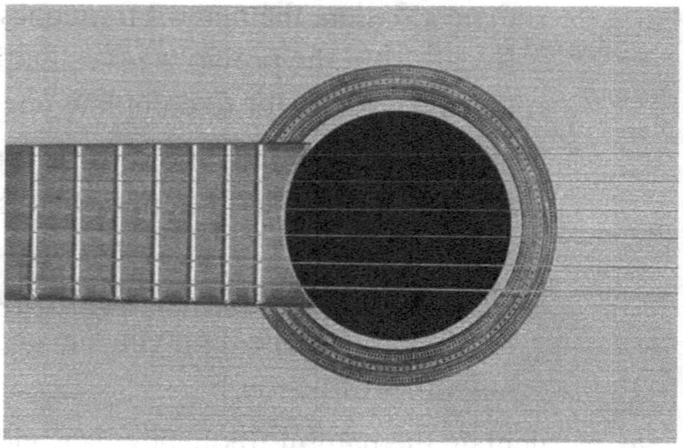

13.3 Resonance and Energy

What is resonance and why is it important?

In ordinary linear motion, Newton's second law allows you to calculate the acceleration if you know the net force and mass. With harmonic motion, a new twist occurs. Newton's laws still apply; however, the frequency at which a force is applied often makes a tremendous difference in the resulting motion. This investigation is about resonance, which can allow very small forces to create very large motions.

In this investigation, you will:

- measure the natural frequency of a pendulum.
- drive a pendulum with a small force at its natural frequency and multiples of the natural frequency.
- relate resonance to natural frequency and energy.

Materials List

- Steel bob with string attached
- Physics Stand
- Photogate
- DataCollector
- Sound and Waves Generator
- Wiggler

1 What is resonance?

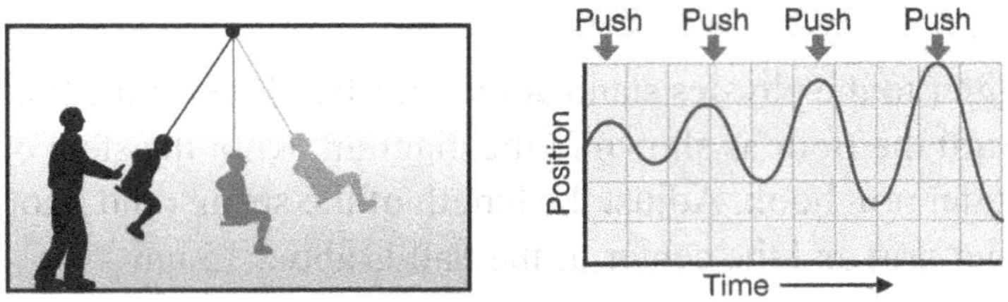

Think about the last time you pushed a child on a swing. If you do it right, each additional push causes the swing to increase its amplitude. Over time, a very large motion builds, although each single push by itself is only a small force. No single push is enough to get the swing to its maximum height. The large motion builds from a succession of pushes applied at the same point in each cycle.

The diagram below shows a useful way to think about the interactions that take place when a person pushes a swing. The person supplies a *driving force* to the swing. The swing is a system that can respond to the force in different ways. If the frequency is just right, the swing's amplitude builds. A large amplitude of motion is a strong response to the force applied. At other frequencies, the swing may not build a large motion. When no large motion occurs, we say the response of the system is weak.

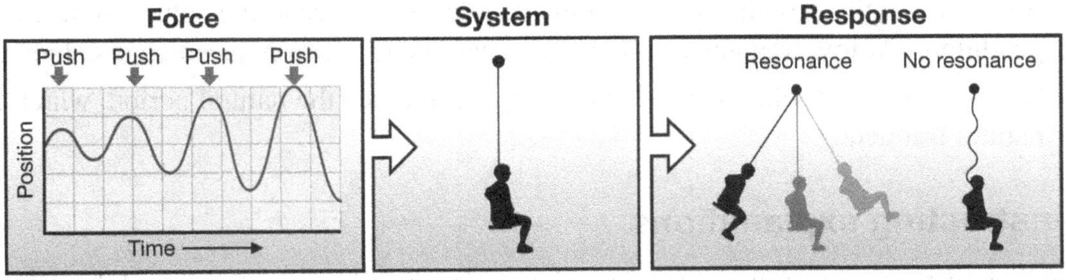

The response of a swing to a periodic push is an example of *resonance*. Resonance happens when the force you apply to a system matches its natural frequency (or a multiple of that natural frequency). When you apply a force matched to the natural frequency of a system, you can get a very strong response. Resonance is what we call the especially strong response we find when an oscillating force is applied to a system at its natural frequency.

2 Creating resonance in a system

A pendulum is a simple oscillator that has a natural frequency. You discovered in a previous investigation that the natural frequency of a pendulum depends on the length of the string. In this part of the investigation, you will observe what happens when you drive a pendulum with a small force at its natural frequency. Because the frequency is very low, we will actually measure the period instead. Remember, the frequency is 1 divided by the period.

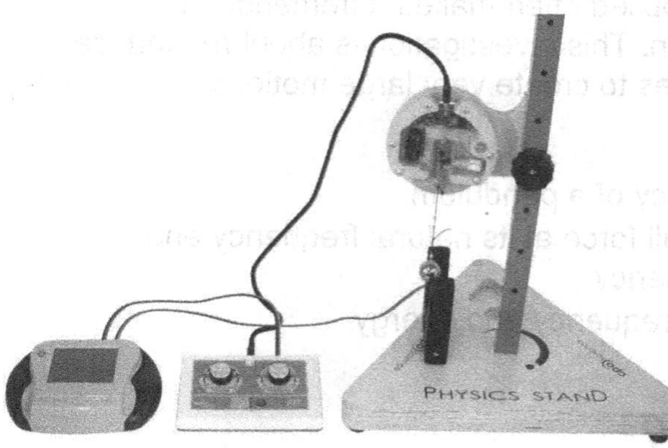

1. Attach the wiggler to the physics stand at the fourth hole from the bottom. The wiggler arm should face down toward the floor as shown in the diagram. Knot the steel bob's string and slip the knot behind the wiggler arm hook. Adjust the length of the string (and knot) so the distance between the tip of the wiggler arm and the center of the ball is about 15 cm.
2. Place a photogate on the base of the physics stand as shown. You will use the photogate and DataCollector to measure the period of the pendulum. This works best if you position the photogate so the pendulum breaks the light beam only once per swing. Gently start the pendulum swinging, record the period of the pendulum, and calculate the natural frequency.
3. Move the photogate to input B of the DataCollector. Connect the wiggler to the wave generator, and plug the wave generator into DataCollector input A. Set the wave generator to Waves, and the wiggler arm should start moving back and forth. The DataCollector should measure a period of about 0.1 seconds, giving a frequency of 10 Hz. The wiggler arm applies a periodic force to the pendulum as it moves back and forth.
4. Vary the frequency control and observe what happens to the motion of the pendulum as the frequency of the driving force from the wiggler gets close to the natural frequency of the pendulum. At low frequencies, it is more accurate to measure period and calculate frequency.
5. See what happens when you drive the wiggler at twice the natural period, which is one-half the natural frequency.

3 Constructing explanations

a. Explain how the force applied by the wiggler causes the response of the pendulum. Your answer should make direct reference to your observations and explain why the natural frequency is important.

b. Make a rough sketch of a graph showing amplitude versus period. Your *x*-axis (period) should range from zero to at least twice the natural period. The graph is *not* a straight line or simple curve.

4 Energy and harmonic motion

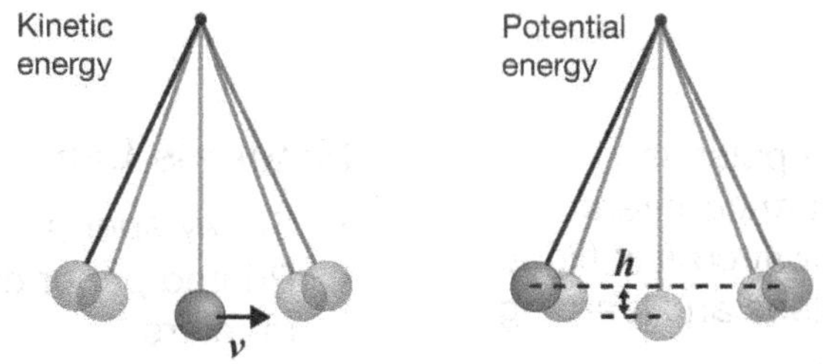

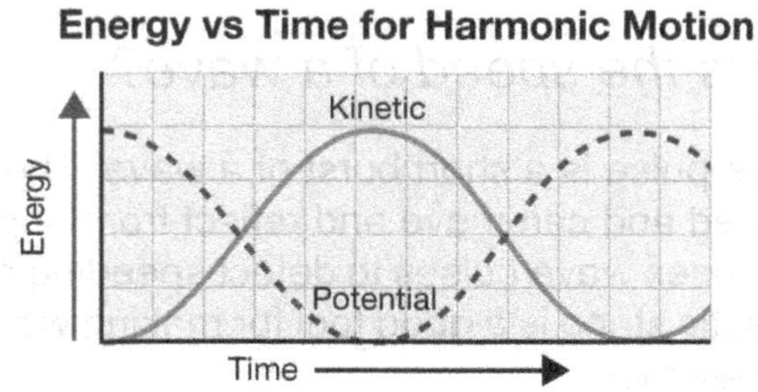

Harmonic motion involves both potential energy and kinetic energy. At the highest point of the cycle, the pendulum is momentarily stopped. It has no kinetic energy because the speed is zero. The pendulum *does* have potential energy, though, because it is raised above its equilibrium position. At the low point of the cycle, all the potential energy has been converted to kinetic energy. The pendulum has its highest speed at the bottom of the cycle. As the pendulum swings through the low point, it climbs up again, converting its kinetic energy back into potential energy.

Oscillators such as a pendulum or a mass on a spring continually exchange energy back and forth between potential and kinetic. The total energy is a combination of potential and kinetic energy. A graph of energy versus time shows the exchange. The potential energy is low when the kinetic energy is high and vice versa.

a. If the *frequency* of an oscillator is increased, what happens to the total energy? Your answer should state whether you think the total energy goes up, goes down, or stays the same. Explain the physical reasoning behind your answer.

b. If the *amplitude* of an oscillator is increased, what happens to the total energy? Your answer should state whether you think the total energy goes up, goes down, or stays the same. Explain the physical reasoning behind your answer.

5 Resonance and energy

Think about resonance in terms of the flow of energy. When the force is matched to the natural frequency, energy is efficiently transferred from the wiggler to the pendulum. As a result, the total energy (potential plus kinetic) increases.

If the force is at a frequency different from the natural frequency, the energy transfer is not very efficient. As a result, the amplitude of the motion stays small.

Why doesn't the amplitude of the driven pendulum keep increasing? There are two explanations. The first one is friction: friction increases as the amplitude and frequency of the motion increase. When the energy lost to friction equals the energy supplied by the wiggler, the amplitude of the motion stops growing. If the wiggler is turned off, you can see the effect of friction quite quickly. The swinging of the pendulum slows down and comes to a stop. The decrease in energy caused by friction in an oscillator is called *damping*.

The second reason the amplitude reaches a limit is that the natural frequency of a pendulum has a small variation with amplitude. As the amplitude increases, the natural frequency shifts so that it no longer matches the frequency of the force from the wiggler.

14.1 Waves and Wave Pulses

What is the speed of a wave?

A wave pulse is a short burst of a wave. The wave pulse is launched and can move and reflect from objects it encounters. Radar uses wave pulses to detect speeding cars or aircraft in flight. An elastic string is a good tool for making wave pulses and learning how they move.

In this investigation, you will:

- create and observe wave pulses.
- measure the speed of a wave pulse with a camera.

Materials List

- Snaky spring
- Printed paper distance markers
- Video camera
- Small C-clamp
- 20-N spring scale
- Masking tape
- Metric ruler

1 Setting up

1. You need about 2–3 meters of table space to make your wave pulses. Print and cut out the markers and tape them down to your table every 0.2 meters.
2. Slip the clamping foot of the c-clamp though one end of the snaky spring and clamp it to the edge of the table.
3. Take the free end of the spring and stretch it until it indicates a 5–6 N force. Wrap a piece of masking tape around the spring to mark the edge of the table. Stretching the spring to the mark will always return it to the same tension.
4. Hold the spring firmly about 20 centimeters from the edge of the table so the tape mark lines up with the table edge and the spring is flat and level with the table surface. To make a wave pulse, pull down on the spring about 10 centimeters and release. You will see the wave pulse race down the table top.
5. Observe the wave pulse as it moves. What happens at the far end when the pulse hits the C-clamp?
6. Try changing the tension in the spring by stretching it more. Observe whether changing the tension has any effect on the speed of the wave pulse.

Make markers and tape them them 0.2 m apart.

Hook one end of the snaky spring through the C-clamp and clamp it to the table.

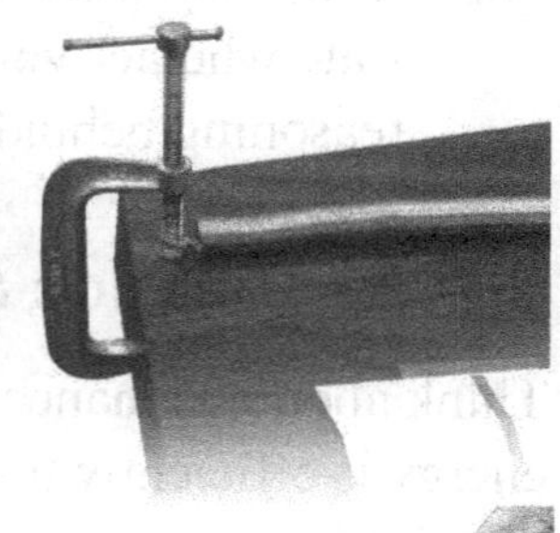

Use a spring scale to set the tension and mark the table edge on the snaky spring with tape.

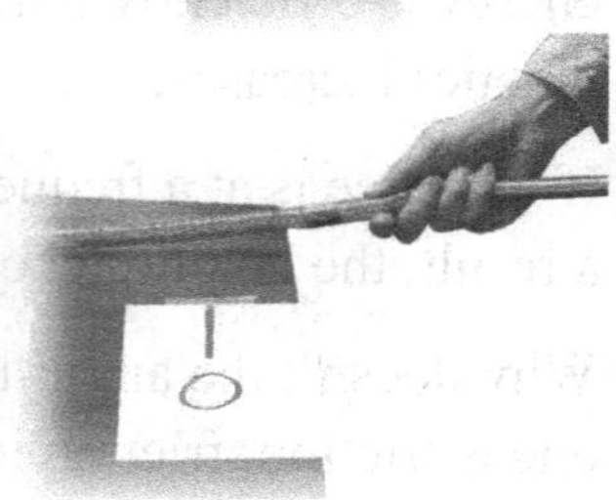

This is how to make a pulse on the snaky spring.

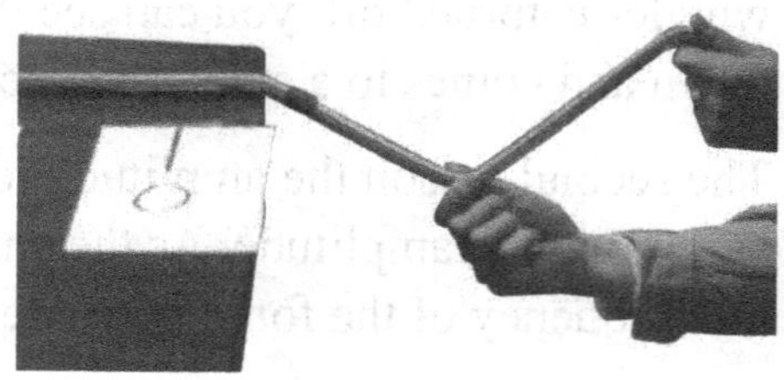

2 Taking the data

Video cameras do not record a smoothly-changing scene; rather, they take a rapid sequence of still pictures. Camcorders record at 30 frames per second and cell-phone video might be 10, 12, 15, 24, or 30 frames per second. This investigation will use successive frames of video to measure how fast the wave pulse moves. To do the experiment, you will need to know the frame rate of your camera. This should be in the owner's manual, or you can use a stopwatch and measure it yourself.

1. One or two people from your group should make the wave pulses and a third should film them against the markers.
2. Use the single frame advance function to count how many frames it takes the wave pulse to advance a certain distance (for example, 1.4 meters). The distance will vary depending on your camera's field of view and how well you can see the wave pulse.
3. Repeat the experiment several times to get an average.
4. Try at least two different spring tensions that are far apart, such as 5 N and 10 N.

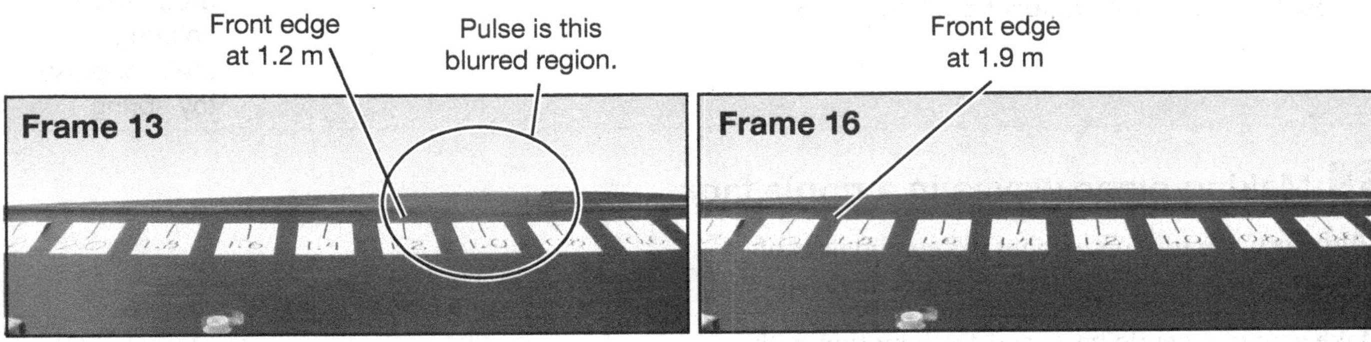

30 frames per second = 1/30 s per frame

$$t = 3 \text{ frames} \times \frac{1 \text{ s}}{30 \text{ frames}} = 0.1 \text{ s}$$

Table 1: Wave pulse data from video analysis

Spring tension (N)	Distance moved (m)	Frames	Time (s)	Speed of pulse (m/s)

3 Reflecting on what you observed

a. Determine the time it takes the pulse to move using the frame rate for your camera and the number of frames you counted.

b. Describe how the speed of a wave differs from the speed of a moving object such as a car. (*Hint*: What is it that moves in the case of a wave?)

c. How did the tension affect the speed of the wave pulse?

d. From what you know about forces, explain why higher tension makes the waves move faster.

14.2 Motion and Interaction of Waves

How do waves move and interact with things?

Waves are oscillations that spread out from where they start. A ball floating on water is a good example of the difference between a wave and ordinary harmonic motion. If you poke the ball, it moves up and down (A). The oscillating ball creates a wave on the surface of the water that spreads outward, carrying the oscillation to other places (B). A second ball floating farther away also starts oscillating as soon as the wave reaches it (C). The wave started by an earthquake can travel all around the world and reach places far away from where it began.

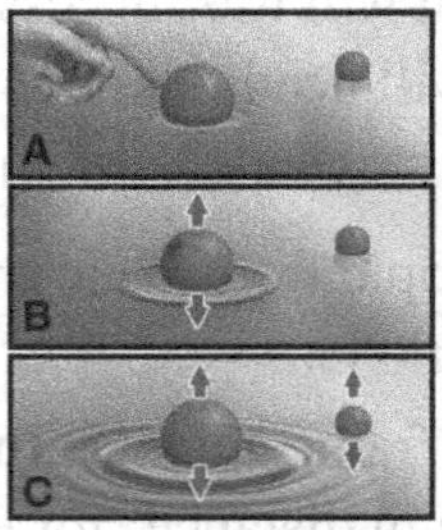

In this investigation, you will:

- create and study waves in water.
- observe how waves can pass through openings and bend around corners.
- observe how waves reflect from boundaries.

Materials List

- Wave tray
- PVC tube that comes with wave tray
- 2 Aluminum blocks that come with wave tray
- Gallon of water mixed with 1 ounce blue or green food coloring
- Metal or plastic toy spring

1 Making plane waves in a ripple tank

Fill your wave tray with about 0.5 centimeters of colored water. The color helps you see the waves.

Make a single, gentle back-and-forth motion with the PVC tube and launch a wave that goes across the tray. The PVC tube makes nearly straight plane waves.

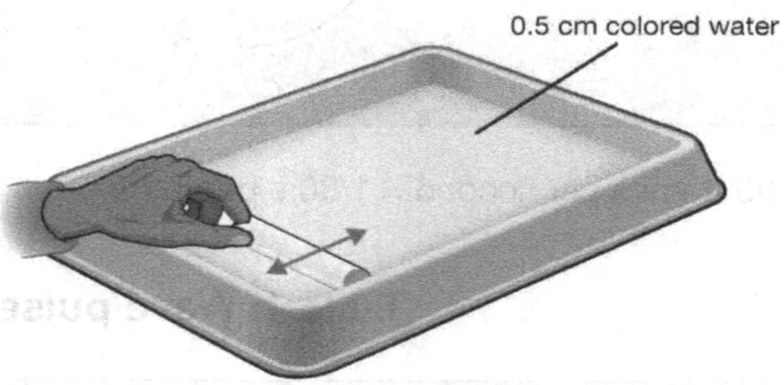

a. Draw a sketch that shows the wave front of your plane wave. Also on your sketch, draw an arrow that shows the direction the wave moves.

b. Is the wave front parallel or perpendicular to the direction the wave moves?

c. Would you consider your water wave to be a transverse wave or a longitudinal wave?

2 Circular waves

Next, poke the surface of the water with your fingertip. Disturbing a single point on the surface of the water makes a circular wave that moves outward from where you touched the water.

a. Draw another sketch that shows the circular wave fronts and include at least four arrows that show the direction in which each part of the wave moves.

b. At every point along the wave, are the wave fronts more parallel or perpendicular to the direction in which the wave moves?

3 Diffraction

1. Put the aluminum blocks in your tray so they block the whole width except for a 1-centimeter opening near the center.
2. Make a plane wave that moves toward the center. What happens to the part of the wave that goes through the opening?

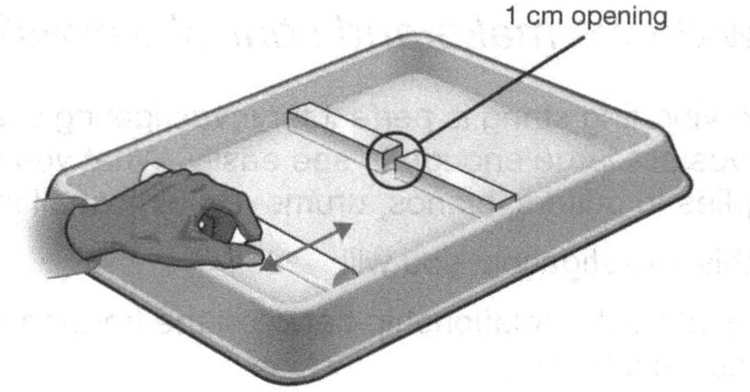

Diffraction is a process that reshapes waves as they move through and around openings or corners. Because of diffraction, waves spread out after passing through openings or around corners.

a. Sketch the shape of the wave fronts before and after the opening.

b. Does the wave change shape when it passes through the opening? If you see any change, your answer should say what kind of shape the wave changes into.

4 Reflection

1. Use the PVC tube to make a plane wave that moves at an angle toward the edge of the tray.
2. Observe what happens to the wave as it hits the edge.

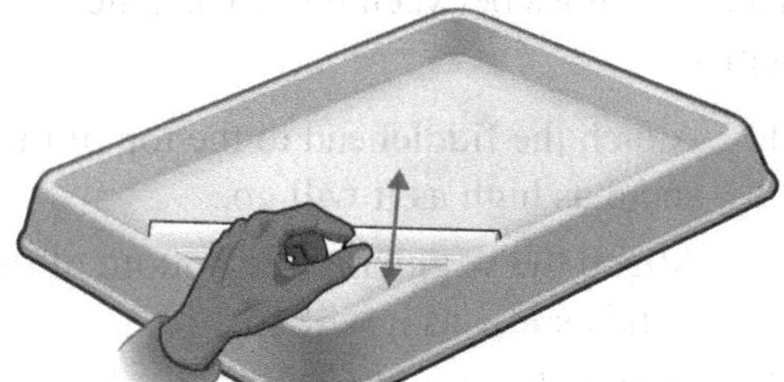

Reflection is the process of waves bouncing off obstacles like the side of the tray. When a wave reflects, it can change direction and may also change in shape.

a. Draw a sketch that shows what happens to the wave front when it hits the side of the tray.

b. Draw an arrow showing the direction of the wave approaching the side.

c. Draw another arrow showing the direction of the wave after it reflects from the side.

d. Do you see any relationship between the incoming and outgoing arrows?

5 Constructing explanations

a. You can easily hear a person talking through a crack in the door although you cannot see them. Do any of your observations provide a clue to how sound can get through tiny cracks?

b. Ocean waves can get many meters high. Big waves on the ocean tend to occur on very windy days. Explain how wind might contribute to making big waves. Use a sketch in your explanation.

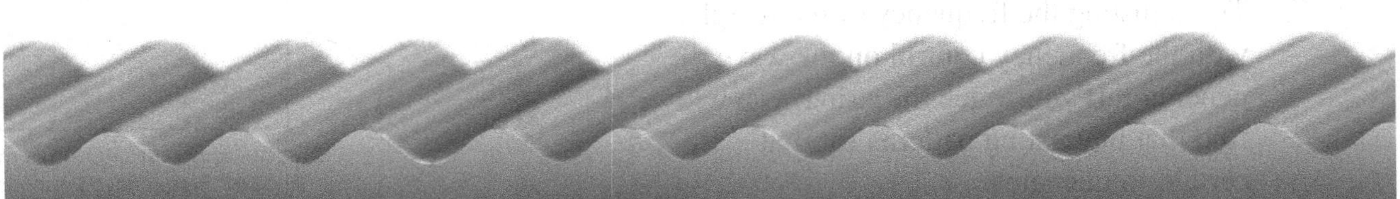

14.3 Natural Frequency and Resonance

How do we make and control waves?

The vibrating string is perfect for investigating waves because the waves are large enough to see easily. What you will see and learn applies to guitars, pianos, drums—to almost all musical instruments.

In this investigation, you will:

- explore the relationship between the frequency of a wave and its wavelength.

Materials List

- Sound and Waves kit
- Physics Stand
- DataCollector

1 Setting up the investigation

Connect the DataCollector to the Sound and Waves generator as shown in the diagram. The telephone cord connects the DataCollector and wave generator. The black wire goes between the wave generator and the wiggler.

1. Attach the fiddlehead to the top of the stand as high as it will go.
2. Attach the wiggler to the bottom of the stand as low as it will go.
3. Stretch the elastic string a little (5 to 10 centimeters) and attach the free end to the fiddlehead. Loosen the knob until you can slide the string between any two of the washers. *Gently* tighten the knob just enough to hold the string.
4. Turn on the DataCollector and be sure to plug in the AC adapter.
5. Set the wave generator to waves using the button. The wiggler should start to wiggle back and forth, shaking the string.
6. Set the DataCollector to measure frequency. You should get a reading of about 10 Hz, which means the wiggler is oscillating back and forth 10 times per second.
7. Try adjusting the frequency of the wiggler with the frequency control on the wave generator. If you watch the string, you will find that interesting patterns form at certain frequencies.

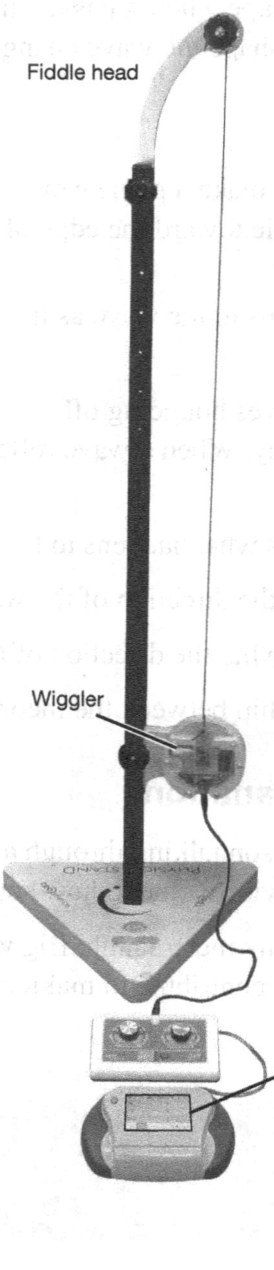

Attaching the elastic string to the fiddlehead

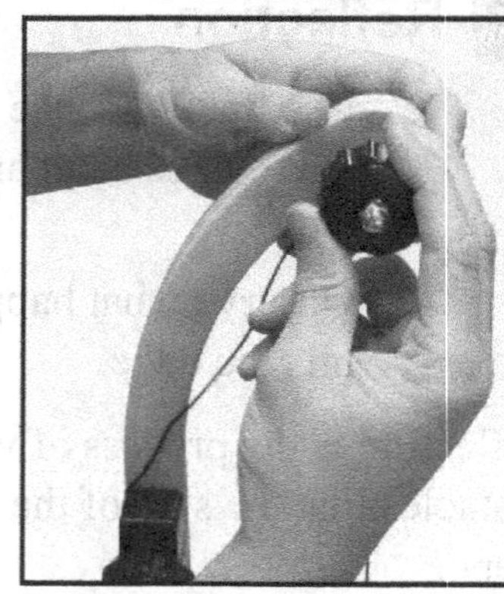

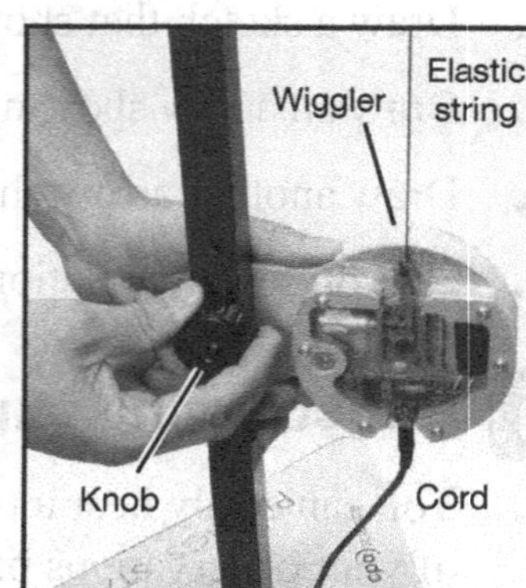

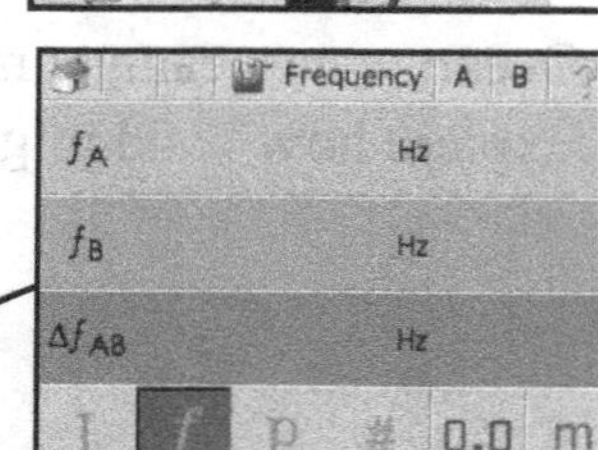

DataCollector – In CPO Timer mode, choose "Frequency(f)" function.

2 Waves on the vibrating string

The first five harmonics of a vibrating string

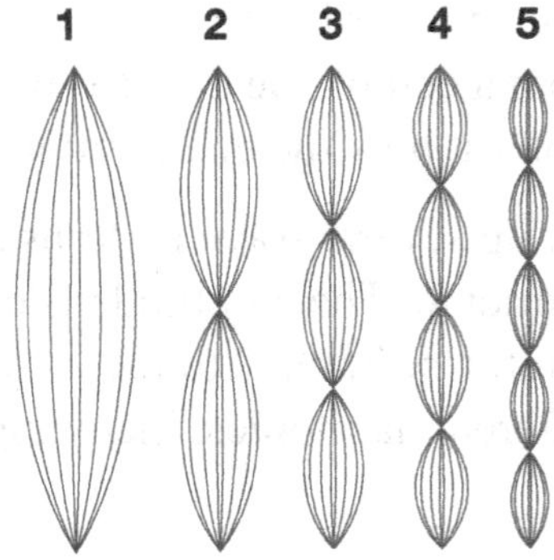

The *frequency, f,* is the rate at which the string shakes back and forth, or oscillates. A frequency of 10 Hz means the string oscillates 10 times each second.

At certain frequencies, the vibrating string forms wave patterns called *harmonics*. The first harmonic has one bump, the second harmonic has two bumps, and so on.

The *wavelength*, λ, is the length of one complete wave. One complete wave is two "bumps." Therefore, wavelength is the length of two bumps. The string is one meter long. If you have a pattern of three bumps, the wavelength is two-thirds of a meter since three bumps equal one meter and a wave is two bumps.

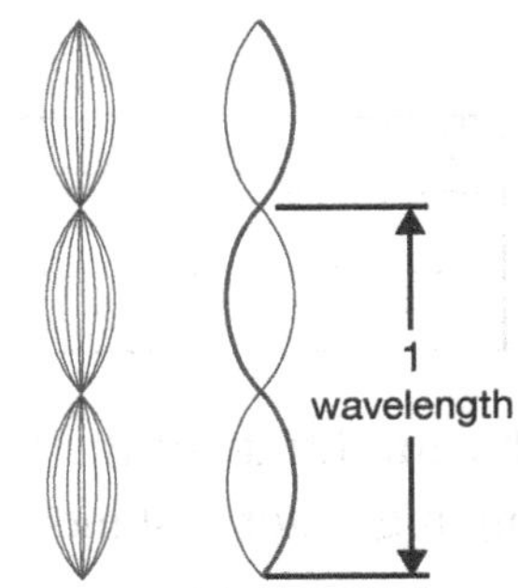

The *amplitude*, A, of the wave is the maximum amount the string moves away from its resting (center) position. A *node* is a point where the string does not move. An *antinode* is a point where the amplitude is greatest. You can measure the wavelength as the distance separating three consecutive nodes.

Adjust the frequency to obtain the first eight to 10 harmonics of the string and record the frequency and wavelength for each one in Table 1. You should fine-tune the frequency to obtain the largest amplitude before recording the data for each harmonic. Look for harmonics two through six before looking for the first one. The first harmonic, also called the *fundamental*, is hard to find with exactness. Once you have the frequencies for the others, they provide a clue for finding the frequency of the first harmonic.

Table 1: Frequency and wavelength of first six harmonics

Harmonic #	Frequency (Hz)	Wavelength (m)	Frequency times wavelength
1			
2			
3			

3 Analyzing the data

a. In one or two sentences, describe how the frequencies of the different harmonic patterns are related to each other.

b. Why is the word *fundamental* chosen as another name for the first harmonic?

c. Give an equation relating frequency (f) and wavelength (λ) that best describes your observations.

d. If the frequency increases by a factor of two, what happens to the wavelength?

e. Propose a meaning for the number you get by multiplying frequency and wavelength.

4 Frequency, amplitude, and energy

Waves are useful because they carry energy from one place to another. The energy of a wave can also carry information such as a voice signal from a cell phone or a TV picture. The next part of the investigation looks at how the energy of a wave is related to frequency and amplitude.

Set up several wave patterns and measure the amplitude for each harmonic. The amplitude is one-half the width of the wave at its widest point. Use Table 2 to record your measurements. Measure at least five different harmonics, including the sixth or higher.

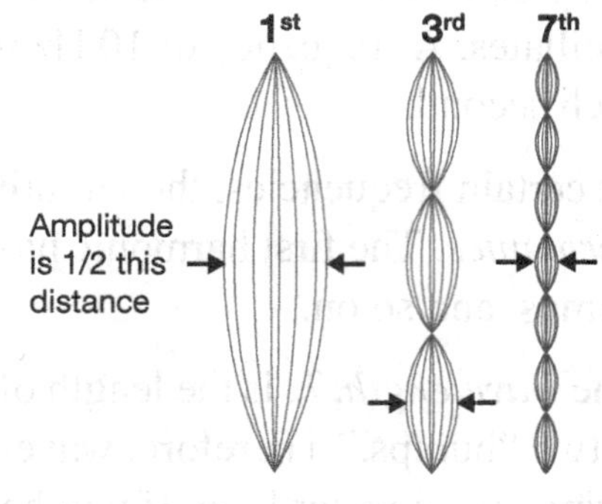

Table 2: Frequency and amplitude

Harmonic	Frequency (Hz)	Amplitude (cm)

If the amplitude of the wave is larger, the wave has more energy because it takes more force to stretch the string a greater distance.

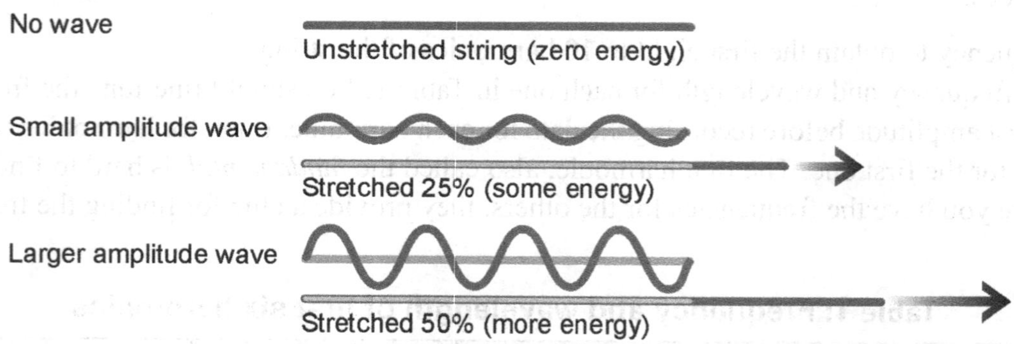

5 Analyzing and interpreting the data

The wiggler applies energy to vibrate the string. Assume that the wiggler supplies about the same amount of energy at all frequencies.

a. Make a graph showing how the amplitude changes with frequency.

b. What happens to the amplitude of the waves as their frequency increases?

c. How does the energy of a wave depend on its frequency? For a given frequency, the amplitude of a wave depends on energy. More energy means larger amplitude. Assume the wiggler supplies the same amount of energy to each wave, independent of frequency. Use your observations of amplitude and frequency to propose a relationship between frequency and energy.

d. Suppose you had two waves of different frequencies but the same amplitude (see diagram at right). Which has more energy, the lower-frequency wave or the higher-frequency one?

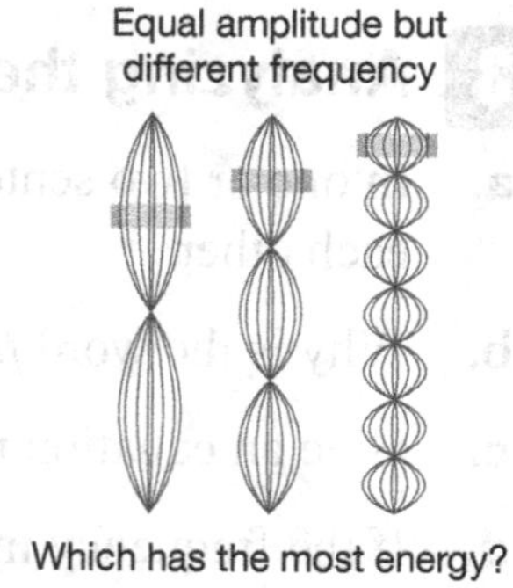

15.1 Properties of Sound

What is sound and how do we hear it?

The ear is a remarkable sensor. We can hear sound waves that change the air pressure by only one part in a million.

In this investigation, you will:

- learn about the range of frequencies the ear can detect and how small a difference in frequency we can perceive.
- use techniques for making experiments on human perception reliable.

Materials List

- Sound Generator
- Speakers
- DataCollector

1 How high can you hear?

The accepted range of frequencies the human ear can hear ranges from a low of 20 Hz to a high of 20,000 Hz. There is tremendous variation within this range as people's hearing changes greatly with age and exposure to loud noises.

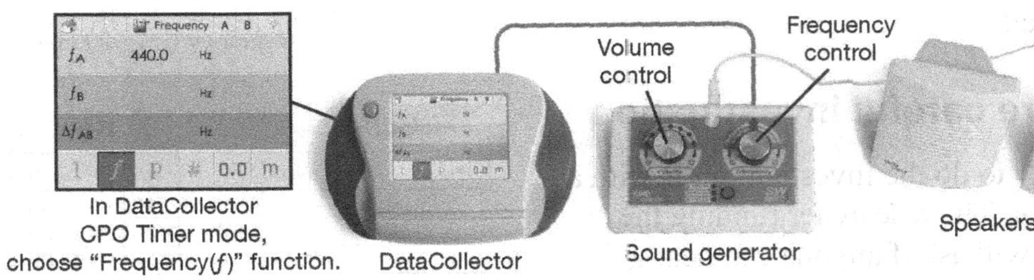

1. Connect the speakers to the sound generator. Connect your sound generator to a DataCollector. Choose Timer mode and select the Frequency function. You should hear a sound and the DataCollector should display a frequency near 440 Hz.
2. The sound generator has dials for frequency and volume control. Try adjusting the frequency and see how high and low it goes.

See if you and your group can agree on frequencies where you hear the sound as low, medium, high, or very high frequency. Use Table 1 to record the frequencies of sound that you perceive as low, medium, high, or very high. Don't try to be too exact because the words *low*, *medium*, and *high* are not well defined. It is difficult to agree exactly on anything that is based completely on individual human perception.

Table 1: Perceived sound frequencies

Description	Frequency (Hz)
Low	
Medium	
High	
Very high	

2 Testing the upper frequency limit of the ear

For this simple experiment, your teacher has a sound generator that can make frequencies up to 20,000 Hz. When the teacher asks, raise your hand if you can hear the sound. Do not raise your hand if you cannot hear it. Someone will be appointed to count hands and survey the class to see what fraction of students can still hear the sound.

a. The objective of the test is to determine how high a frequency people can hear.

b. Make a histogram showing your class response to frequencies between 10,000 and 20,000 Hz. You should have 10 bars, one per 1,000 Hz. Each student who raises a hand is counted as a positive response on the graph.

c. Do you think the method of counting raised hands is likely to give an accurate result? Give at least one reason you believe the method is either good or bad.

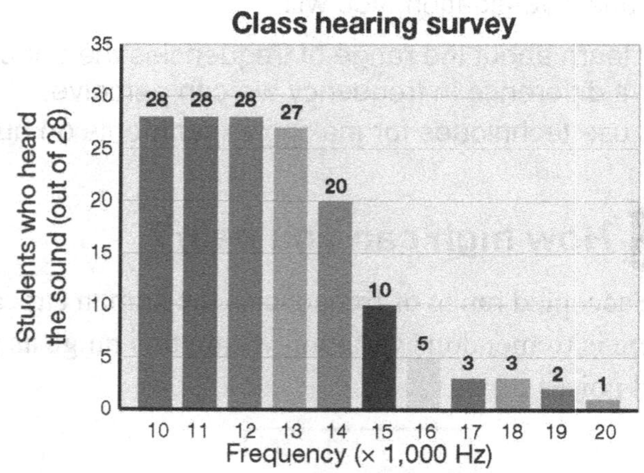

3 Doing a more careful investigation

1. Another way to do the investigation is with a hidden ballot. The researcher running the experiment will ask if anyone can hear a certain frequency of sound. You will check *yes* or *no* on a piece of paper. The researcher may or may *not* play the sound. Each frequency will be played five times and the five repetitions will be mixed up so there is less chance for error. Everyone in the class does one response survey.

Survey

#	Yes	No
1	✔	
2		✔
3		✔
4	✔	
5	✔	
6		✔
7	✔	
8		✔
9	✔	
10	✔	
11	✔	

Key

#	Frequency	Played: Yes	Played: No
1	12,000	✔	
2	16,000		✔
3	10,000		✔
4	12,000	✔	
5	14,000	✔	
6	14,000		✔
7	18,000	✔	
8	20,000		✔
9	10,000		
10			

2. Compile the data from the survey sheets and record it in Table 2. Plot another histogram showing only people whose choices matched the key all five times at each frequency. It is hard to fake a response (or guess) because you have to choose correctly five times for each frequency. This kind of experiment is called a double-blind test, meaning neither you nor the researcher can see anyone else's response. The results from a double-blind experiment are much more reliable than those from other forms of surveys. Researchers use the double-blind method to test new medicines.

Table 2: Class hearing experiment results

Correct responses	10,000 Hz	12,000 Hz	14,000 Hz	16,000 Hz	18,000 Hz	20,000 Hz
5						
4						
3						
2						
1						

4 Probability, chance, and investigations

A very good way to ensure accurate results in a survey test is to make it improbable that anyone could get the correct response purely by chance. A true-false test is not enough to rule out getting a result by guessing because on each test you have a 50 percent chance to guess right. That means one out of every two times you could get the right response just by guessing!

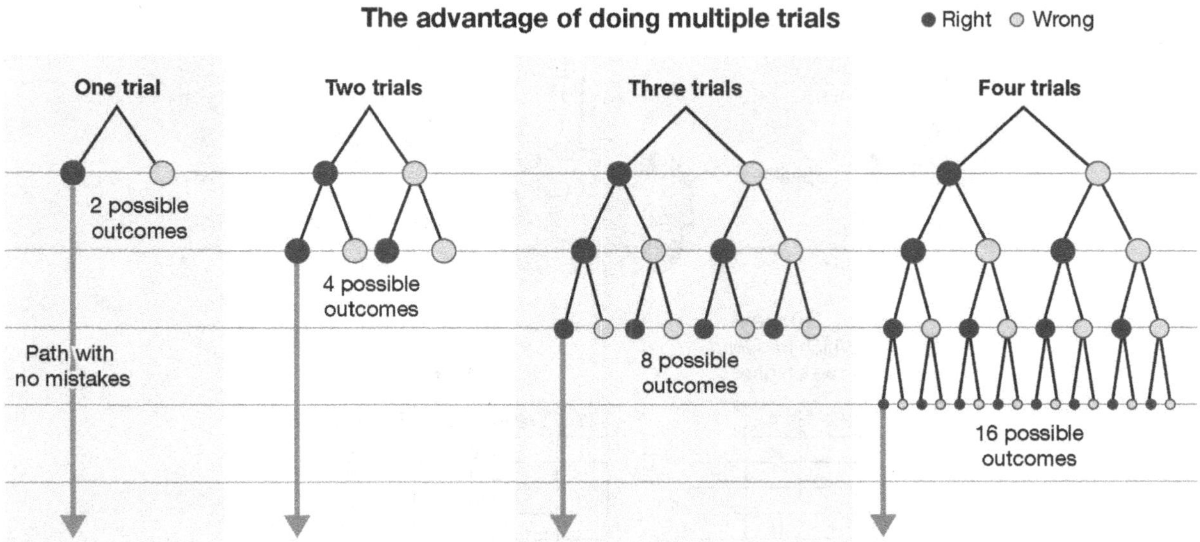

The diagram above shows a decision tree for an experiment with one to four trials. With each additional trial, the total number of possible outcomes increases by two. There is only one path with no mistakes. With two trials, you have one right path out of four choices. That means there is only a 1-in-4 chance someone could guess twice correctly. With three trials, there is only a 1-in-8 chance of guessing correctly three times in a row. With four trials, those chances diminish to 1 in 16.

a. What is the chance of guessing correctly every time with five trials?

b. If 100 people did a true-false test with five trials and everybody guessed, how many people would be likely to make five correct choices in a row?

c. You design an experiment to compare three sets of stereo speakers (A, B, and C). The same song plays sequentially on each set of speakers and people pick whether the first, second, or third sounded best. You set up a screen so your listeners cannot see the speakers. To rule out chance, you scramble the order in which you play each speaker and repeat the test three times. You figure if a person picks the same speaker all three times, it really must sound better than the other two. What is the chance that someone could randomly select the same set of speakers three times in a row?

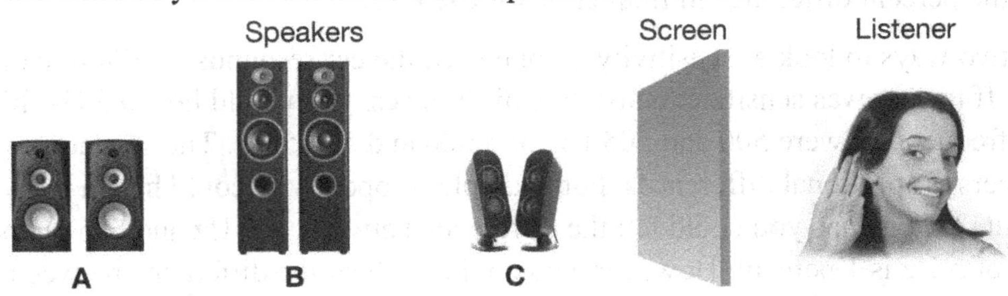

5 Perceiving differences in frequency

Can you tell the difference between a sound with a frequency of 400 Hz and a sound at 401 Hz? The next experiment is to test the ability to distinguish if one sound has higher frequency than another. In this experiment, the researcher will play two frequencies and you mark which one is higher.

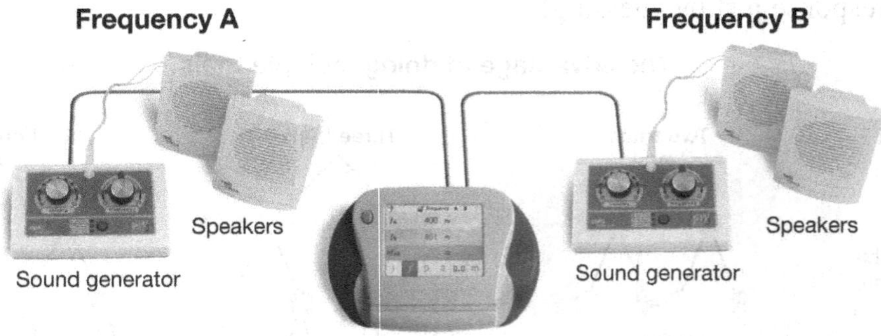

Survey
Which frequency was higher?

#	A	B
1	✔	
2		✔
3		✔
4	✔	
5	✔	
6		✔
7	✔	
8		✔
9	✔	
10	✔	
11	✔	

Key

#	Frequency A	Frequency B
1	400	401
2	404	400
3	390	400
4	1,000	1,001
5	1,050	1,000
6	995	1,000
7	5,005	5,000
8	5,050	5,000
9	4,500	
10		

To analyze the results, you want to know how many people got the right answer for each frequency range. Make a data table like Table 3 that is large enough to hold all your results and fill it in with the data from your class.

Table 3: Perceiving differences in sound frequencies

Frequency A (Hz)	Frequency B (Hz)	Frequency diff. (Hz)	Percent difference	Correct responses
1,000	995	5	0.5%	7
1,000	1,050	50	1%	15
1,000	1,001	1	.1%	8

a. Calculate the percent difference in frequency for each test.

b. There are two ways to look at sensitivity. In one way, the ear responds to *absolute* differences in frequency. If the ear was sensitive to absolute differences, you would hear a 5 Hz difference no matter if the two frequencies were 500 and 505 Hz, or 5,000 and 5,005 Hz. The second possibility is that the ear only hears proportional differences. For example, suppose you could hear a proportional difference of 1 percent. That means you could tell the difference between 500 Hz and 505 Hz because the difference of 5 Hz is 1 percent. However, you could *not* hear the difference between 5,000 and 5,005 Hz because 5 Hz out of 5,000 Hz is a proportional difference of only 0.1 percent. To hear a 1-percent difference from 5,000 Hz, the second frequency would have to be 50 Hz higher or lower. Which model does the data support?

15.2 Sound Waves

Does sound behave like other waves?

What experimental evidence is there that sound has the properties of waves? The fact that sound can produce interference with a characteristic wavelength is strong evidence that sound is a wave.

In this investigation, you will:

- experiment with a sound wave interference phenomenon called "beats."
- use interference to measure the wavelength of sound.
- explore resonance.

Materials List

- Sound Generator
- Speakers
- DataCollector
- Meter stick
- Tuning fork

1 Beats

When you listen to a pair of speakers, two sounds reach your ear at the same time. What you hear is the combination of the two waves.

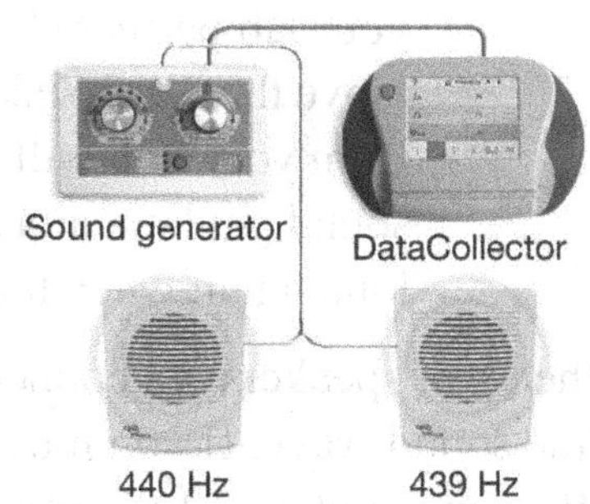

1. Set up the sound generator, speakers, and DataCollector as shown at right. Choose Timer mode and select the Frequency function.
2. When the sound generator is in Sound mode, you will hear a base frequency at 440 Hz coming from both speakers. Switch the unit from Sound mode to Beats mode by pressing the Mode button.
3. In Beats mode, turning the frequency dial changes the frequency of the sound produced by the second speaker from 440 Hz to 439 Hz to 430 Hz. The first speaker continues to play the base frequency. In Beats mode, adjust the frequency to read 439 Hz.
4. Stand back and listen to the sound when both speakers are at equal volume.
5. Cover one speaker with your hand and turn it away so you can listen to the uncovered speaker by itself. Swap speakers and do the same so you can listen to the other speaker by itself. Can you tell the difference between the sound produced by each speaker?
6. Listen to both at equal volume again and listen to the combination.
7. Adjust the frequency between 439 and 430 Hz. Listen to the combinations of sound.

When two waves are close, but not exactly matched in frequency, you hear beats. Beats are an example of the interference of sound waves. The sound gets alternately loud and soft as the waves drift in and out of phase with each other. Sometimes they are in phase and the result is twice as loud. A moment later they are out of phase and they cancel out, leaving periods of quiet. The alternation of loud and soft is what we hear as beats.

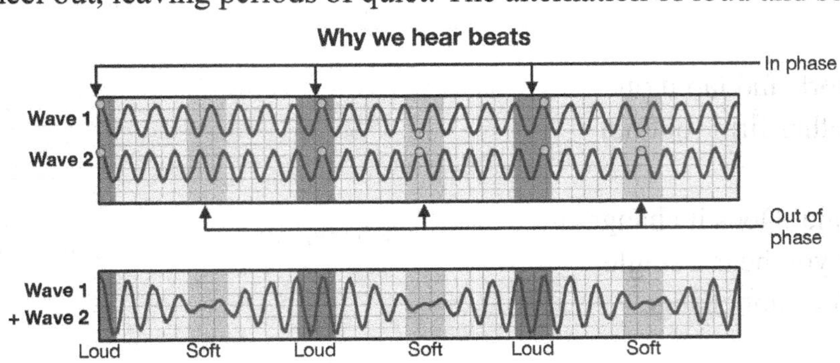

2 Interference

Suppose you have two identical sound waves and you are standing where you can hear them both. For certain positions, one sound wave reaches your ear in the opposite phase from the other wave and the sound gets softer. Move over a little and the two sound waves add up to get louder. These effects are another example of interference and are easy to demonstrate.

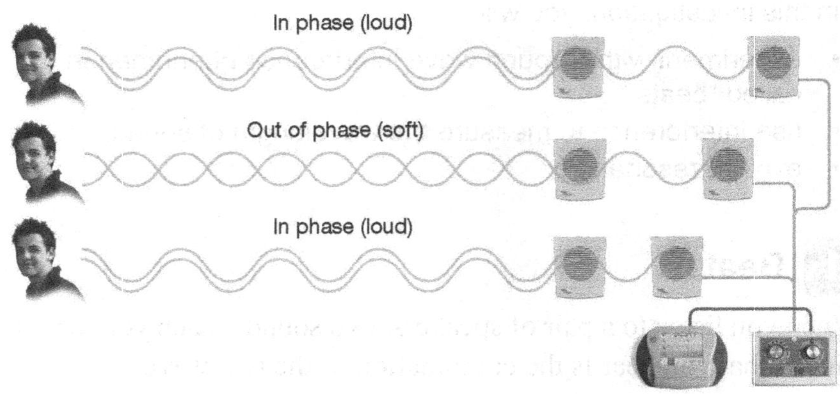

1. Put the sound generator back in Sound mode. Place one speaker about 1/2 meter behind the other.
2. Set the frequency between 400 and 800 Hz.
3. Stand 3 or 4 meters in front of one speaker and have your lab partner slowly move the rear speaker away from you. You will hear the sound get loud and soft and loud again when the distance between speakers has changed by one wavelength.

When two speakers are connected to the same sound generator in Sound mode, they both make the exact same sound wave. If you move around a room, you will hear places of loud and soft volume based on the difference in phase of the sound waves produced by the two speakers.

a. Try to make an approximate measure of the wavelength of sound by changing the separation of the two speakers. The speakers have been moved one wavelength when the sound heard by the observer has gone from loudest to softest and back to loudest again. For this to work, you need to keep the observer and both speakers in the same line.

b. Interference can be bad news for concert halls. People do not want their sound to be canceled out after they have bought tickets to a concert! Why do we rarely hear interference from sound systems, even when they only have two speakers?

3 Resonance

Many objects that can create sound also have natural frequencies and resonance. When they vibrate, these objects can produce a characteristic sound at their natural frequency. A tuning fork is a good example.

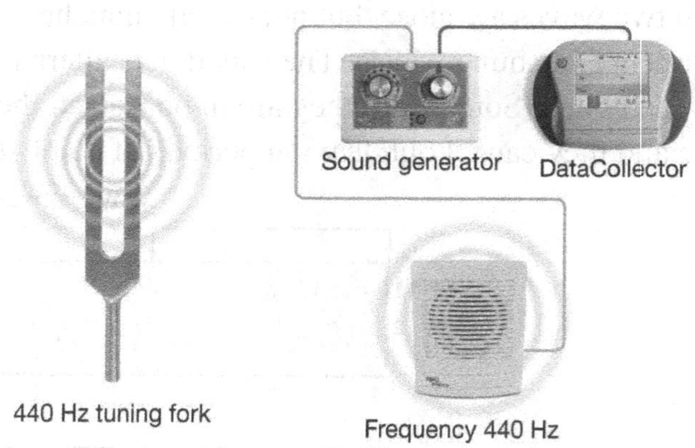

1. Select a tuning fork and tap it on your knee or another firm (but not hard) surface.
2. Listen to the sound. Does it change in frequency, or do you hear a single frequency that does not change?

3. Use the sound generator to measure the frequency of the resonance by matching the frequency of the sound generator with the sound you hear from the tuning fork. When you get to within about 15 Hz of the frequency, you will hear the rapid oscillation of beats as the frequencies created by the tuning fork and the sound generator get closer in value but are not yet exactly the same. As the frequency from the sound generator and the tuning fork get closer together, the beats become slower. The frequency of sound from the speakers is the same as the sound from the tuning fork when you no longer hear any beats.
4. Try several different tuning forks and use the chart below to record the resonant frequencies.

Table 1: Resonant frequencies

Tuning fork description	Measured resonant frequency (Hz)	Labeled resonant frequency (if any, in Hz)

4 Constructing explanations

a. Did you observe any relationship between the size (or shape) of the tuning fork and the frequency at which it was resonant?

b. What range of frequencies did you hear that seemed to match the frequency of the tuning fork? Give your answer in the form of a range written, for example, like 429 Hz–451 Hz.

c. Strike the tuning fork and hold the bottom end against a hard, thin surface such as a window. Does the sound get louder, softer, or remain unchanged? Explain what you hear by describing what might be happening between the tuning fork and the surface it touched.

15.3 Sound, Perception, and Music

How is musical sound different from other types of sound?

We rarely hear only one sound wave at a time. Our brains and ears are constantly processing sound from many sources, at many frequencies and intensity levels. In fact, the meaning in sound comes from patterns in frequency and loudness. Speech is actually a pattern of frequencies and amplitudes of sound that you have learned to interpret as words with meaning. Musical sounds have special relationships based on the frequencies they contain.

Materials List

- Sound Generator
- Speakers
- DataCollector

In this investigation, you will:

- use several sound generators to study musical chords.
- explore the science and math of music theory.

1 Chords

Musical notes are different frequencies of sound. Over thousands of years people have found combinations of frequencies that sound good together.

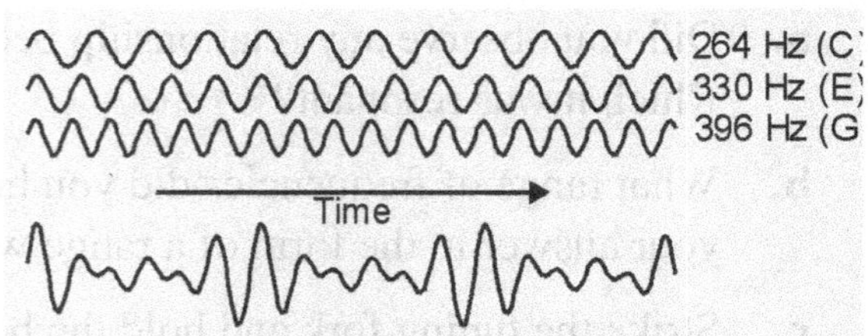

Combination of 264 Hz, 330 Hz,and 396 Hz (C major chord)

1. Set up your sound generator, speakers, and DataCollector.
2. Turn down the volume so that you cannot hear the sound but you can still read the frequency.
3. Each group in the class will be given a different frequency to play. Tune your frequency until you are within +/– 1 Hz.

Your instructor will tell each group to turn different frequencies up and down so they can be heard together. Don't change the frequency, just adjust the volume up or down when you are asked.

a. Describe the sound of the four frequencies 264 Hz, 330 Hz, 396 Hz, and 528 Hz when you hear them together.

b. Describe the sound of the four frequencies 264 Hz, 311 Hz, 396 Hz, and 528 Hz when you hear them together.

c. Contrast the two sounds. Does one sound more happy or sad compared with the other? Does one sound spookier than the other? Which combination reminds you more of spring? Which of fall?

2 Consonance and dissonance

When two frequencies don't combine well, we call it *dissonance*. Beats are the biggest cause of dissonance. When two sounds are within about 15 Hz of each other, the beats are perceptible and create dissonance. The opposite of dissonance is *consonance*. When different frequencies sound good together, we call it consonance. Sounds that *don't* make beats tend to combine more smoothly and are usually consonant. Work with another group to explore how far apart two frequencies have to be to sound consonant together.

a. Give three examples of frequency pairs that are dissonant.

b. Give three examples of frequency pairs that are consonant.

3 The musical scale

Almost every culture has a musical scale. The notes of a scale are a special set of frequencies that are related to each other by ratios. Musicians can play together in a band because all their instruments are tuned to the same musical scale. The table below is an example of the C-major scale that is common to many songs.

Musical instruments are devices designed to play specific frequencies of sound, usually from the musical scale. Instruments work on the principle of resonance. For example, the natural frequency of a chime is proportional to its length, just like a vibrating string. Longer chimes have lower natural frequencies because they are resonant at longer wavelengths. Because frequency and wavelength are inversely related, a chime that must vibrate at a frequency 5/4 higher than another chime must be 4/5 as long. This proportionality rule is the basis for constructing almost all musical instruments.

	C	D	E	F	G	A	B	C
C major scale								
Frequency (Hz)	264	297	330	352	396	440	495	528
Ratio to C-264	1/1 $\frac{264}{264}$	9/8 $\frac{297}{264}$	5/4 $\frac{330}{264}$	4/3 $\frac{352}{264}$	3/2 $\frac{396}{264}$	5/3 $\frac{440}{264}$	15/8 $\frac{495}{264}$	2/1 $\frac{528}{264}$
Length ratio	1	$\frac{8}{9}$	$\frac{4}{5}$	$\frac{3}{4}$	$\frac{2}{3}$	$\frac{3}{5}$	$\frac{8}{15}$	$\frac{1}{2}$

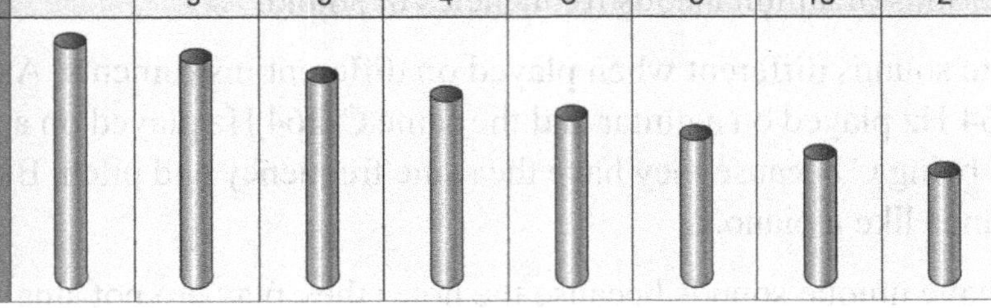

a. Find two notes from the C major scale that sound dissonant when played together.

b. Find two notes that sound consonant when played together.

c. Suppose you wish to design a musical chime to play the notes C, E, and G. If the chime that plays the note C is 1 meter long, how long should you make the chimes that play the notes E and G?

4 Octaves

When the frequency doubles, the new note has the same name. The note with a frequency of 440 Hz is named A. The note with a frequency of 880 Hz (double 440) is also named A. Notes that are an *octave* apart are double in frequency. The note with a frequency of 220 Hz is also an A because two times 220 is 440 Hz (A). Two notes where one is half the frequency of the other are also an octave apart.

a. Calculate the missing frequencies for each of the notes below using the rules for the octave.

The octave below middle C

Note	C	D	E	F	G	A	B	C	D	E	F	G	A	B
Frequency								264	297	330	352	396	440	495

5 The interpretation of sound

Think about reading one single word from a story. You recognize the word, but it does not tell you much about the story. When you read the whole story you put all the words together to get the meaning. The brain does a similar action with different frequencies of sound. A single frequency by itself does not have much meaning. The meaning comes from patterns in many frequencies together. You probably noticed that a single frequency such as 440 Hz sounds “thin” while a combination of frequencies sounds much richer. Speech contains thousands of simultaneous frequencies of sound.

The same note sounds different when played on different instruments. As an example, suppose you listen to the note C-264 Hz played on a guitar and the same C-264 Hz played on a piano. A musician would recognize both notes as being C because they have the same frequency and pitch. But the guitar sounds like a guitar and the piano sounds like a piano.

Instruments have unique sounds because the notes they play are not single pure frequencies. The most important frequency is still the fundamental note (for example, C-264 Hz). The variation comes from the harmonics. Remember, harmonics are frequencies that are multiples of the fundamental note. We have already learned that a string can vibrate at many harmonics. The same is true for all instruments. A single C from a grand piano might include 20 or more different harmonics. A good analogy is that each instrument has its own recipe for the frequency content of its sound.

The rate at which loudness builds and falls off also influences how we hear a sound. The rise time is the time it takes to reach maximum loudness. The fall time is the time over which the sound dies away. Rise time and fall time are related to resonance and damping in an instrument, and are different for each instrument. Rise and fall times are also different for each harmonic, even from the same instrument! Higher harmonics have faster rise times and shorter fall times.

a. Describe the construction of the human ear. What are the three major components? What part converts the vibration of a sound wave into signals that are carried by the auditory nerve?

b. Research how many different frequencies of sound the human brain and ear can respond to. As a comparison, the eye contains about 120 million photoreceptors that each see a dot of brightness. The brain assembles the individual signals (dots) into a picture. The ear works in a similar way. The brain assembles a sonic “picture” from the signals received on the auditory nerve.

c. Describe how an electronic keyboard or synthesizer is able to sound like many different instruments at the touch of a button.

d. Suppose a 1.2-meter string vibrates at a frequency of 440 Hz. What length of string would vibrate one octave higher? What length would vibrate one octave lower? Assume the tension in the string is the constant.

e. *Challenge!* The western musical scale has eight notes—do-re-mi-fa-so-la-ti-do—where the eighth note is an octave above the first. A piano keyboard has 12 keys spanning an octave. Describe why there are five extra keys (the black keys) in each octave. Your answer should be a short paragraph.

f. *Challenge!* What sounds "good" to your ear is largely due to the sounds you heard during your childhood years. Different cultures use different musical scales. Some scales emphasize dissonance, others emphasize consonance. Research the musical scale of a different culture and describe the relationships between the notes.

16.1 Properties and Sources of Light

What are some useful properties of light?

It is hard to imagine living in a world without light. Light has many properties that make it important for all living things.

In this investigation, you will:

- observe interesting examples of reflection and refraction of light.
- compare the speed of light with the speed of sound.
- calculate the approximate intensity of the Sun's light.

Materials List

From Optics kit:

- Laminated sheet
- Red filter cap
- Mirror
- Prism

Other materials:

- Index card
- Small piece of paper
- Marker
- Calculator
- Ruler

1 Fooling the brain with light

Prisms, mirrors, and lenses are devices that can affect the way light travels from an object to your eyes. Your brain perceives an object to be where the light from that object *appears* to come from. The object may not be where the light appears to come from, but that is where your brain tells you the object is.

Place folded paper so it hides the cap from view.

Now arrange the mirror so you can see the cap, even though it is hidden from view behind the paper.

1. Set one of the filter caps on the laminated graph sheet. Fold a large index card in half lengthwise and set it on the graph sheet so that it blocks your view of the cap.
2. A mirror is a shiny surface that reflects light very well. Arrange the mirror so that you can see the cap on the other side of the folded paper.

a. Draw a sketch that uses lines to show how the light from the cap reaches your eyes by reflecting from the mirror.

b. Where does the cap appear to be when you see its reflection in the mirror: at the surface of the mirror, in front of the mirror, or behind it? Give evidence for your answer.

2 Seeing reflection and refraction at the same time

Both reflection and refraction often occur when light hits a boundary between materials, such as the boundary between glass and air. The amount of light reflected or refracted depends on the angle at which you are looking relative to the surface.

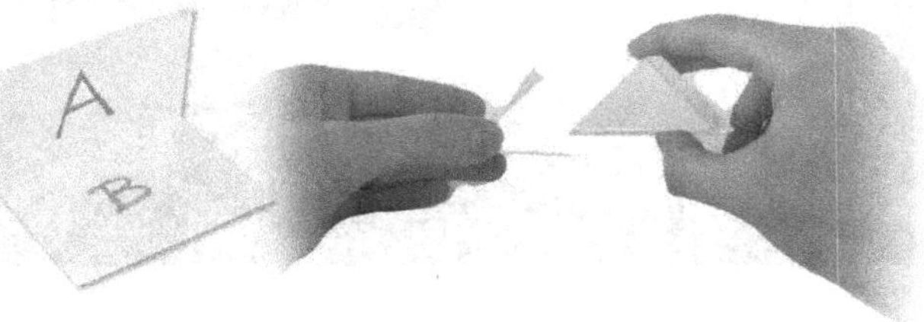

The image in the prism changes as you move your head.

1. Take a piece of paper or index card about the size of a business card and fold it in half lengthwise. Draw the letter A on one side of the line and the letter B on the other side (see photo above).
2. Fold the paper in two and wrap it around one of the corners of the prism that is *not* a right angle.
3. Look into the prism. Move your head up and down to change the angle at which you look.

a. Draw a diagram showing the path of the light when you see the letter A.

b. Draw a diagram showing the path of the light when you see the letter B.

c. Is the image in the prism always reflected or refracted, or can there be both reflection and refraction at the same time? Give evidence from the prism experiment to support your answer.

3 The speed of light

Light travels so fast that it is hard to comprehend the interval of time it takes to get from one place to another. Sound also travels fast, but not as fast as light. You can perceive effects due to the limited speed of sound. For example, you can see a distant lightning bolt before you hear the thunder clap. The time delay between seeing lightning and hearing the thunder is due to the time it takes sound to travel to your ears. The speed of light is so fast (3×10^8 m/s) that there is no apparent delay between the appearance of a distant lightning bolt and having the light reach your eyes so you can see it.

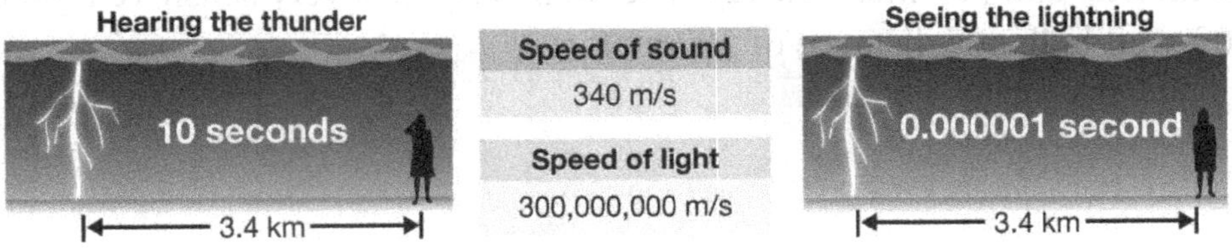

What about light that you see from a distant star? Distances on Earth are very small compared to distances between Earth and stars. Because of this, light on Earth seems to reach your eye instantly, even from what seems like a distant place. But what about the light from very distant stars? How long does it take light to reach your eye from the Sun? As we look out into the universe, we are also seeing backward in time because the light from distant stars and galaxies has traveled many years before reaching human telescopes. Using this technique, we are able to see more than one billion years back in time.

a. Find out how many kilometers Earth is from the Sun, on average.

b. Calculate the time it takes light to travel this distance.

c. Calculate how far light would travel in one second.

4 The inverse square law and light from distant stars and galaxies

Depending on where you live, the intensity of visible sunlight reaching your eyes is 300–600 watts per square meter at noon on a clear day in summer. The stars you see in the night sky are actually as bright as the sun (or brighter) but they are extremely far away.

The Inverse Square Law for Light Intensity

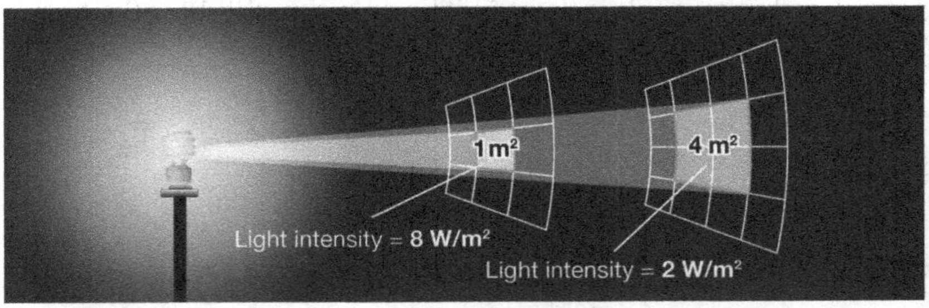

The light from a point source spreads out in a sphere centered on the source. The intensity of light falling on the inner surface of the sphere is equal to the total power of light divided by the surface area of the sphere. Suppose a source of light has a power (P) in watts. The surface area of a sphere is given by the formula $A = 4\pi r^2$. Therefore, the intensity of light a distance (r) away from a point source is given by the intensity formula below. When power is in watts and area is in square meters, the intensity (I) has units of watts per square meter (W/m^2).

INTENSITY

$$I = \frac{P}{4\pi r^2}$$

Intensity (W/m^2) — I; **Power** (W) — P; ***Surface area of a sphere with radius, r*** (m^2) — $4\pi r^2$

a. After the Sun, the next closest star to Earth is the Alpha Centauri star system. This triple-star system is about 4.36 light years away. A light year is the distance light travels in one year, or 9.46×10^{12} km. Use this information and the inverse square law for light intensity to find the approximate intensity of light from this star system that reaches Earth.

b. A typical 100-watt light bulb makes about 20 watts of light and 80 watts of heat. Calculate the intensity of light from the bulb by using the light power (20 W) and the intensity formula for distances of 1 meter, 100 meters, and 500 meters away. The human eye is a very sensitive light detector; you can still see a 100-watt bulb at a distance of 500 meters.

16.2 Color and Vision

How do we see color?

All the colors of visible light can be created artificially using a combination of three primary colors: red, blue, and green. You will use a white light source and color filters to discover what happens when you mix different colors of light. You will also learn how those filters work.

In this investigation, you will:

- show that white light can be made from red, green, and blue light.
- explain the colors we see in terms of subtracting colors from white light.

Materials List

From Optics kit:

- Red, green, and blue flashlights with holders
- Laser flashlight
- Lens with light blue holder
- Diffraction glasses

1 Mixing primary colors of light

1. Slide all three flashlights into their own holders.
2. Turn the flashlights on.
3. Connect the red and green flashlights by sliding their holders together using the rail and slot connectors on the side.

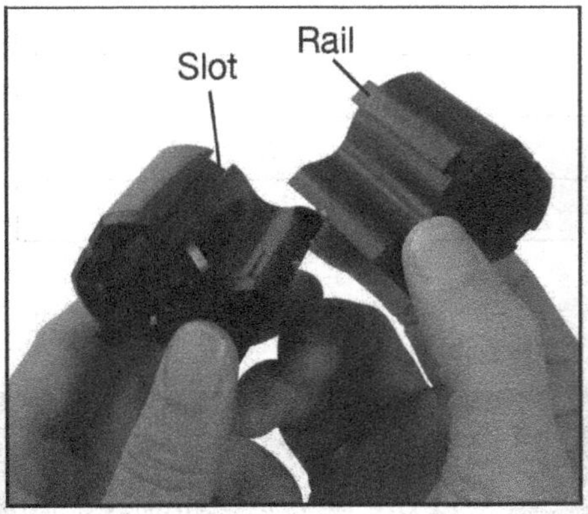

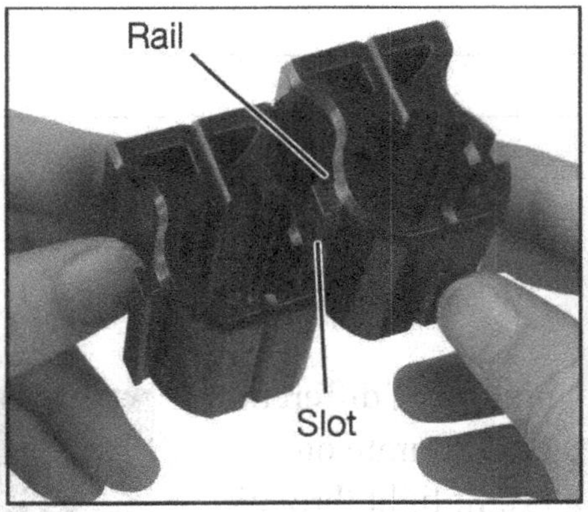

4. Place the blue flashlight on top of the red and green lights, making a small pyramid stack. Set the blue light on top of the other two with the holder on its side, so the rail on the holder fits in the small groove created between the holders of the red and green lights.

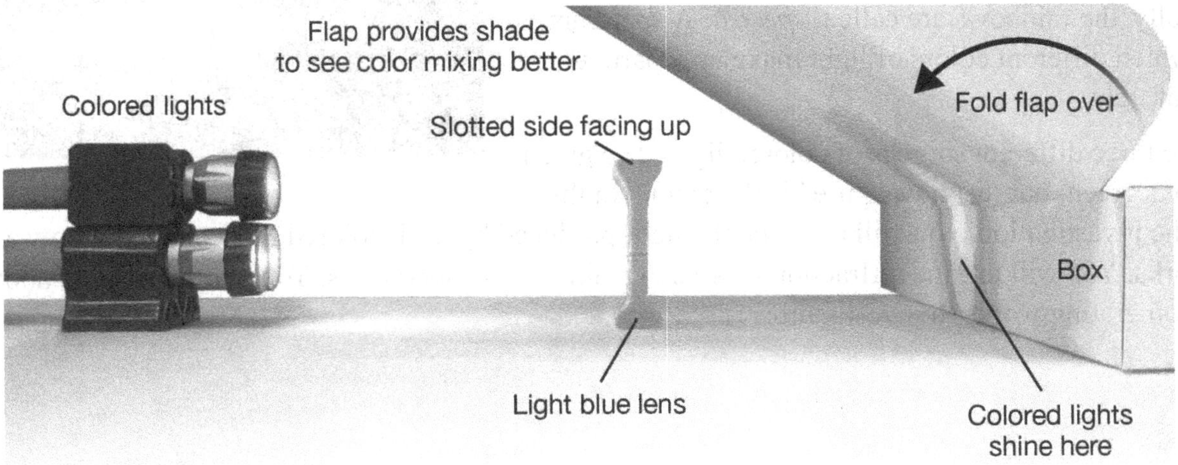

5. Set the light blue lens just in front of the lights so they shine through it. Place the lens so the slotted side is facing up. This will increase the visibility of the color mixing.
6. Set the white box that the Optics with Light & Color kit comes in on the opposite side of the lens from the lights. Fold the top of the box over to shade the area where the three colored lights are shining on the side of the box.
7. Slowly move the lens away from the lights and toward the box until you see the three spots of color (red-green-blue) overlap on the box.

2 Constructing explanations

Use Table 1 to record the answers to the following questions about your observations.

a. What color do you see when you mix red and green light?

b. What color do you see when you mix green and blue light?

c. What color do you see when you mix blue and red light?

d. What color is produced when all three colors of light are equally mixed?

Table 1: Mixing primary colors of light

LED color combination	Color you see
Red + green	
Green + blue	
Blue + red	
Red + green + blue	

3 Breaking apart light

Most of the light we see is made of a mixture of different colors. The diffraction grating glasses can separate out those different colors. If you look at a bright light through the diffraction grating glasses, you see rainbows on all sides. The rainbows spread out all the colors that are present in the light coming through the bright spot in the center. Technically, the rainbows are called *spectra.* A spectrum shows which different colors of light make up a particular sample of light.

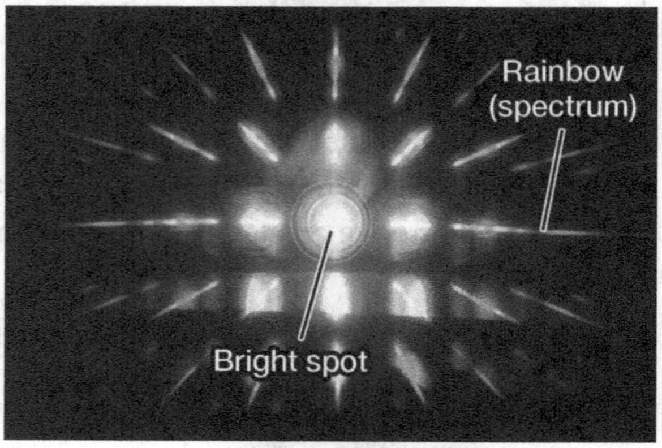

You have three different sources of colored light. The green light looks green, but just how "pure" is the green? In this part of the investigation, you will examine the light produced by each colored LED and learn how a *color filter* works. You will use the diffraction glasses to make your observations. You will learn more about how a diffraction grating works in investigation 18.2.

1. Look at the light produced by each different-colored flashlight (red, green, and blue) through the diffraction grating glasses. Be sure to observe one at a time, and record your observations by answering 4a below.
2. Take the filter cap off of one of the flashlights. Look through the diffraction glasses at the light produced by the white LED and record your observations by answering 4b below.
3. Shine the red laser onto the box. Look at the spot on the box through the diffraction glasses and answer 4c below.

Safety Note: DO NOT LOOK DIRECTLY AT THE LASER BEAM.

4 Constructing explanations

a. Describe the similarities and differences you observed in the spectra from the red, blue, and green LEDs. You may want to use colored pencils to sketch the colors in the spectrum.

b. Describe what you saw looking at the white LED. Compare the spectrum from the white LED with the spectra from red, green, and blue. You may want to use colored pencils to sketch the colors in the spectrum.

c. Describe the spectrum you saw looking through the diffraction grating glasses at the spot made by the red laser on the screen. How is the spectrum of the red laser different from the spectrum of the red LED?

d. Based on your observations, explain how the colored filters transform the white light of the LEDs inside the lamps into red, green, and blue.

16.3 Photons and Atoms

How does light fit into the atomic theory of matter?

Can you name several things that create light? The Sun, burning logs, glowing neon gas in a sign, and light bulbs are common examples. What do all these sources of light have in common? They are all made of atoms. How do atoms create light? Work through this investigation to find out.

In this investigation, you will:

- experiment with a photoluminescent material.
- explore the quantum theory of light.

Materials List

From Optics kit:

- Laminated graph sheet
- 3 Flashlights
- Red, green, and blue filter caps

1 How is light produced?

Almost all the light you see is produced by atoms. When atoms absorb energy, electrons rise to higher energy levels. When the electrons fall back to their lower energy state, they may release energy in the form of light.

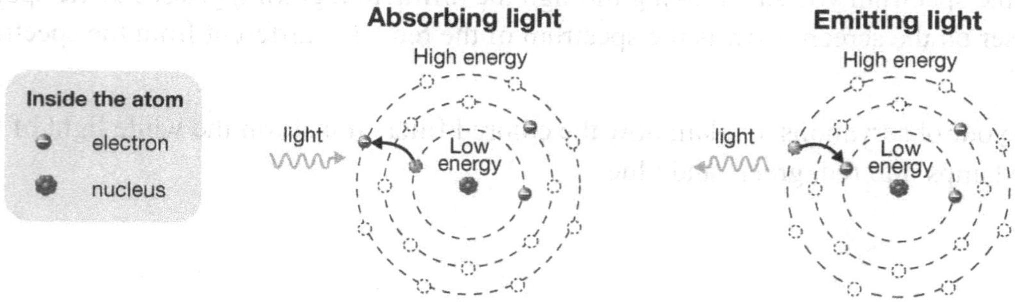

In some elements, it takes time for the energized electrons to fall back and give up their energy. These elements store energy and give off light slowly over a period of time. This is how some kinds of glow-in-the-dark materials work. Embedded in the material are chemical substances called "phosphors." When light energy hits the phosphors' atoms, electrons absorb energy. When the electrons fall back, they slowly release the stored energy and the material glows until all the electrons have returned to the lowest energy level.

2 Examine the effects of light on glow-in-the-dark material

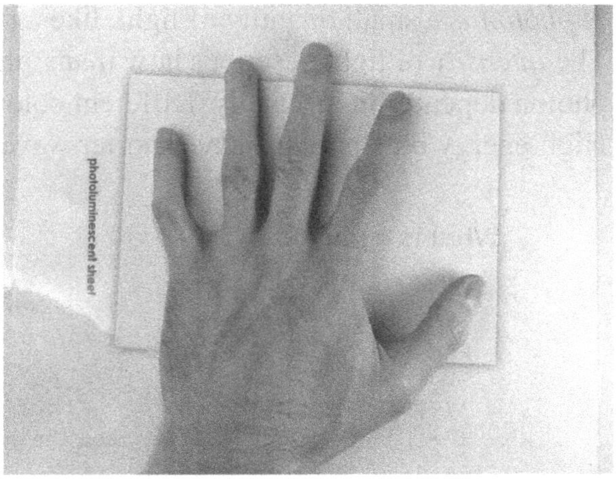

1. Place the laminated graph sheet on your table, graph side facing up. With all the room lights on, turn the sheet to the other side and quickly cover part of the material you find on the back with your hand.
2. Keep your hand in one place on the material for a minute, then turn the lights off (the room needs to be quite dark) and remove your hand. Record your observations.
3. Next, expose the material to light while completely uncovered.
4. Turn off the light and wait a minute, then place your hand over part of the material.
5. Remove your hand and record your observations.

In answering these questions, think in terms of light and energy. Explain what happens to the energy in each of these situations:

a. Why didn't the material glow where your hand covered it (step 2)?

b. What happened when your hand was allowed to rest on the glow-in-the-dark material after it was already glowing? (steps 4, 5)?

3 Analyzing the effect of different colors of light

In Part 2, you used a source of white light (your room lights) to add energy to the phosphorus atoms in the glow-in-the-dark material. White light is a mixture of all colors of the rainbow. In this section, you will determine what happens when more restricted colors of light are used to add energy to the phosphorus atoms. You need to make a table to record your observations.

1. Allow the glow-in-the-dark material to stop glowing by leaving it in the dark for a few minutes.
2. Switch on the flashlight with the red cap and shine it on the glow-in-the-dark material from a distance of about 10 centimeters. Wait 15 seconds, take the flashlight away, and record your observations.
3. Try the same experiment again with the red flashlight five centimeters away and then again with the flashlight placed right on the surface. Decreasing the distance increases the intensity of the light without changing its color.
4. Repeat the procedure with the green flashlight. Record your observations.
5. Repeat the procedure with the blue flashlight. Record your observations. You will answer questions about your observations in Part 5.

4 The quantum theory of light

A *photon* is a small quantity of light, like a particle. You can think of a photon like a short burst of a wave. The *intensity* of light describes how many photons per second are produced (or absorbed). The *color* of each photon depends on its energy. Different colors of light are produced by photons with different energies. High-energy blue photons have shorter wavelengths than low-energy red photons.

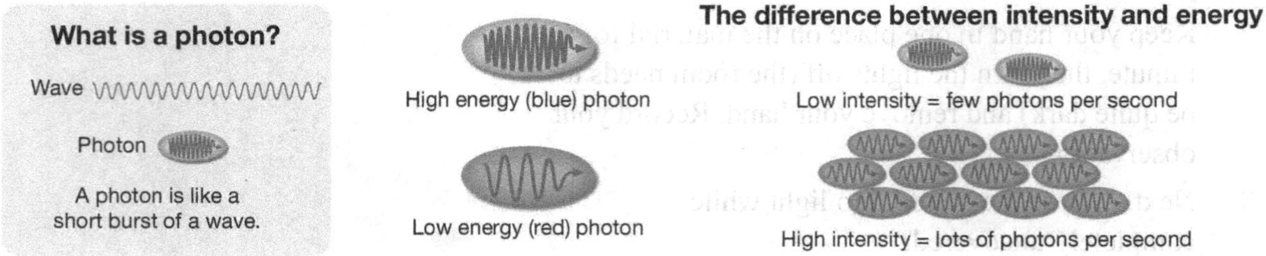

To glow, an electron in a phosphorus atom must first absorb energy from a photon of light. That means a photon must have enough energy to boost the electron up one whole level to be absorbed. If the photon does not have enough energy, the phosphorus atom cannot absorb it.

5 Constructing explanations

a. Based on the observations you made in Part 3, what color light has the highest energy? Your answer should state how your observations support your conclusion.

b. The red flashlight *might* have made the photoluminescent material glow more brightly than the green flashlight. This is a discrepant (disagreeing or conflicting) event. Why is this unexpected? How could this happen? Propose an explanation. If time allows, you can look at the different light sources through the diffraction grating glasses that are in your kit. Observing the colored lights through these glasses will give you clues as to why the red flashlight might have made the material glow brighter than you expected.

c. Intuitively, you might think that the more intense the light you shine, the more brightly the phosphorus should glow. Explain how your observations support or refute this hypothesis.

d. How does what you observed support the quantum theory of light and atoms? (*Hint*: What would have happened if electrons were free to absorb *any* energy rather than just *certain* energies?)

17.1 Reflection and Refraction

How do we describe the reflection and refraction of light?

We observe the law of *reflection* every day. Looking in a mirror, we see ourselves reversed left to right. Our sense of sight depends on light reflected from objects around us. Light rays can also bend when they cross an interface between two different materials. The bending of light rays by a boundary between materials is called *refraction*. Prisms and lenses use refraction to manipulate light in telescopes, binoculars, cameras, and even your eyes.

In this investigation, you will:

- take a closer look at reflection.
- apply geometry to predict exactly where reflected light goes.
- determine the rules for how and to what degree light is refracted by glass and water.

Materials List

From Optics kit:

- Laminated graph sheet
- Laser flashlight and holder
- Mirror
- Prism

Miscellaneous:

- Water soluble marker
- Protractor
- Graph paper and pencil

1 Observing the law of reflection

How to trace the beam of the laser

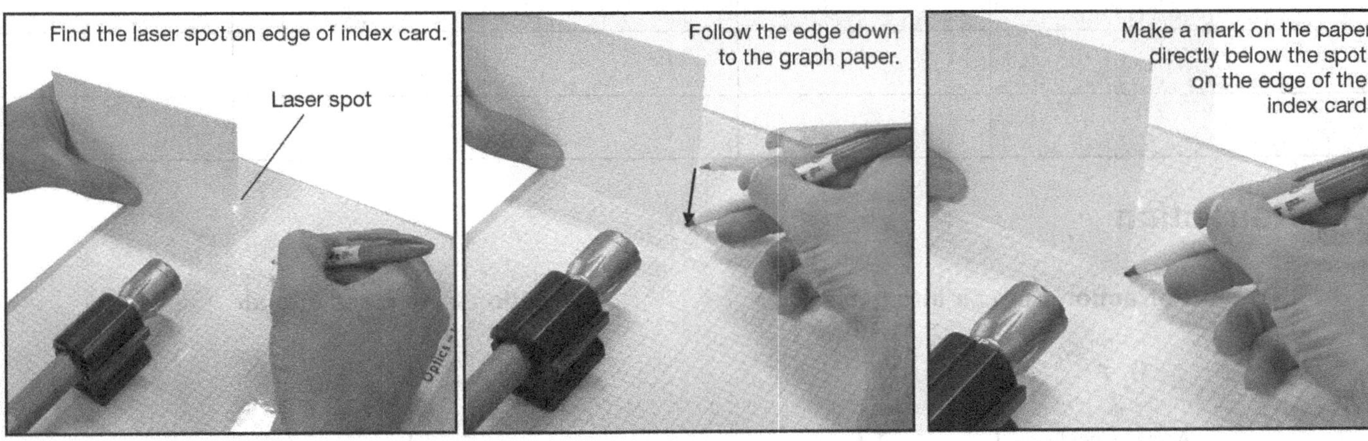

1. Place your laminated graph sheet on a flat surface and align the laser so the beam follows one horizontal line across the paper.
2. Set the mirror on the laminated sheet so the light beam from the laser hits its shiny surface at an angle. Draw a line on the graph paper, marking the position of the front face of the mirror.
3. Use a fine point water soluble marker and an index card to trace the incident and reflected light rays from the laser. See the photos above and at right for clarification.
4. Repeat these steps with the mirror set at a different angle. Do the experiment for at least four different angles. Use different colored markers or label your incident and reflected rays so you don't get them confused.

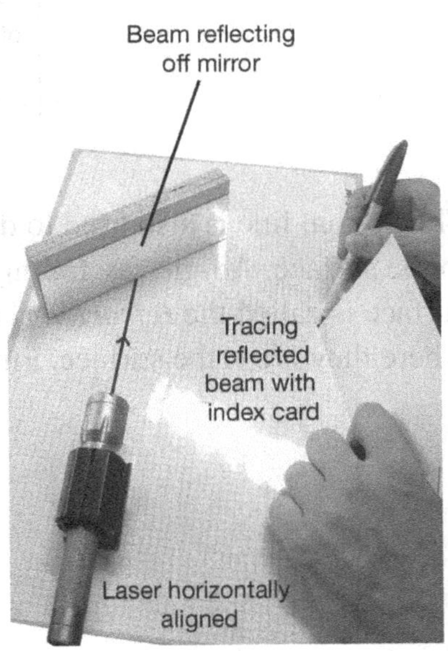

2 Using your model

A diagram showing how light rays travel is called a *ray diagram*. Lines and arrows on a ray diagram represent rays of light. The *incident* ray travels to the mirror: the *reflected* ray travels away from the mirror.

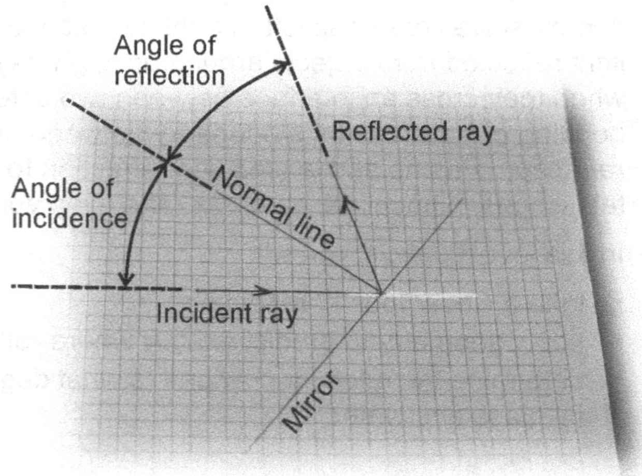

a. For each ray diagram, draw a line perpendicular to the mirror surface at the point where the rays hit. This line is called the *normal line*.

b. Use a protractor to measure the angle between the normal and the incident and reflected rays. Record your measurements in Table 1.

c. Write down your own statement of the law of reflection, describing the relationship between the angles you measured.

d. A laser shines at a mirror at an angle of incidence of 75 degrees. Predict its angle of reflection. After predicting, test your prediction. Were you right?

Table 1: Angles of incidence and reflection

	Diagram #1	Diagram #2	Diagram #3	Diagram #4
Angle of incidence				
Angle of reflection				

3 Refraction

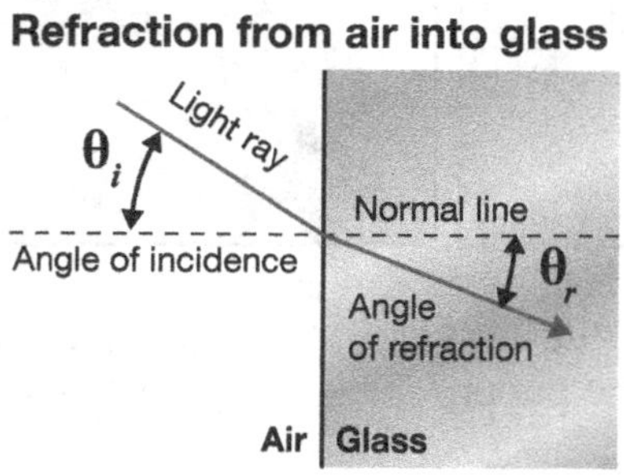

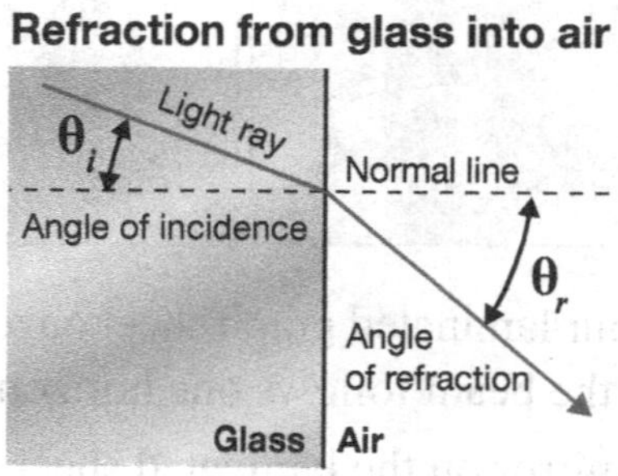

The normal line is also used to describe how refraction works. Remember, the normal is a line perpendicular to the surface. A light ray falling on a surface is called the *incident ray.* The light ray passing through the surface is called the *refracted ray.* You may assume these light rays move in straight lines except at the point where they cross the surface. Find the incident and refracted rays on the diagram above.

The *angle of incidence* is the angle between the incident ray and the normal line. The *angle of refraction* is the angle between the refracted ray and the normal line. Snell's law of refraction states the relationship between the angle of incidence and the angle of refraction.

SNELL'S LAW

$$n_i \sin\theta_i = n_r \sin\theta_r$$

- θ_i: **Angle of incidence** (degrees)
- θ_r: **Angle of refraction** (degrees)
- n_i: **Index of refraction on incident side of boundary**
- n_r: **Index of refraction on refracted side of boundary**

The incident and refracted rays are defined in terms of the direction the light is traveling as it crosses the surface between two materials. Going from air into glass, the incident ray is in air and the refracted ray is in glass. Going from glass into air, the incident ray is in glass and the refracted ray in is air.

4 Tracing rays through the prism

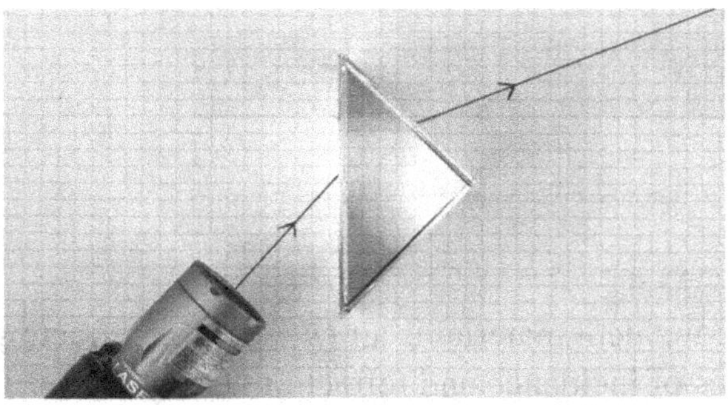

Create an angle of incidence of at least 25°; trace beam going into and out of prism.

A prism is a solid piece of glass with polished surfaces. Prisms are useful for investigating how light bends when it crosses from one material into another.

1. Place the prism on a piece of graph paper as shown at left. Shine the laser so the beam comes out the opposite short side. The angle of incidence should be at least 25 degrees.
2. The beam is entering the prism from the air and passing through the prism into the air again. Using a sharp pencil and an index card, carefully trace the path of the laser beam as it enters and exits the prism.
3. Remove the laser and prism from the paper.
4. Draw the lines connecting the beam through the glass as shown at right. Now, identify the incident/refracted pair of rays involved when the beam passes from the air to the glass. Next, identify the incident/refracted pair of rays involved when the beam passes from the glass to air.

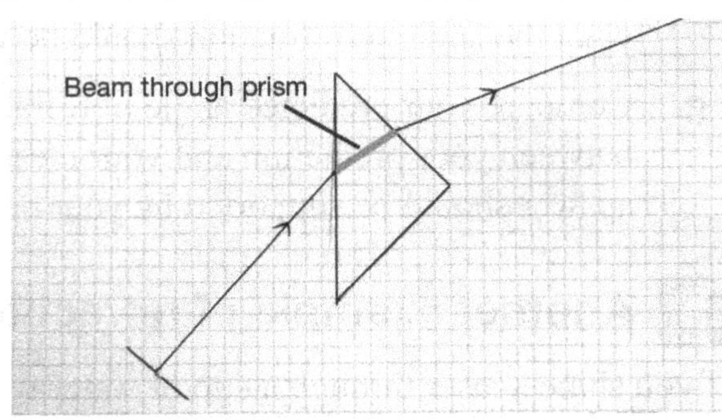

Connect the beam that travels through the prism (shown here in green).

5 Finding the index of refraction

The *index of refraction* is a property of a material that describes its ability to bend light rays. Air has an index of refraction of 1.0. The index of refraction for different types of glass ranges from 1.4 to 1.6. The higher the index of refraction, the more the material bends light.

a. Draw the normals to the two faces of the prism the beam passed through as shown at right.

b. When light goes from a low-index (air) to a higher-index (glass) material, does it bend toward the normal or away from the normal?

c. When light goes from a high-index (glass) to a low-index (air) material, does it bend toward the normal or away from the normal?

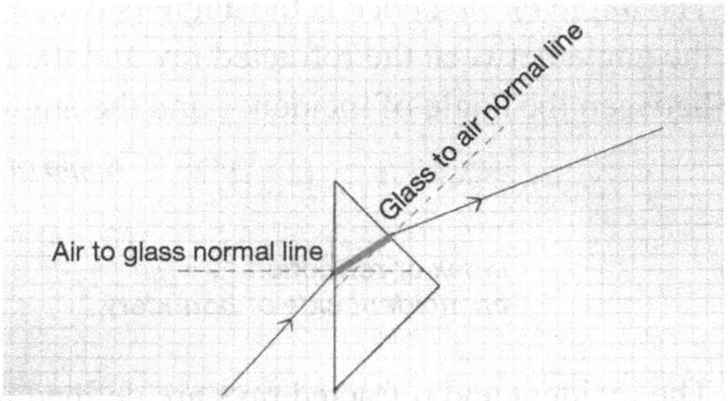

Draw the normal lines perpendicular to each surface where the beam enters and exits.

d. Use the two normals and a protractor to determine the angles of incidence and refraction for both surfaces crossed by the light beam. Use Table 2 to record the angles.

Table 2: Angles of incidence and refraction

	Angle/incidence	Angle/refraction
Going from air to glass		
Going from glass to air		

e. Apply Snell's law to the light ray entering the prism. The incident material is air ($n = 1$); the refracting material is glass ($n = n_g$). Calculate the sines of the angles of incidence and refraction. Use your calculation to determine the index of refraction of glass (n_g).

f. Apply Snell's law to the light ray leaving the prism. Using the index of refraction for glass, predict what the angle of refraction should be when the laser beam goes from glass to air.

g. Compare your predicted angle of refraction to the angle you measured. Comment on any differences between your prediction and your measurement. Do your observations support Snell's law? Your answer should be supported by your observations of the laser beam.

6 Another example of refraction

If you shine a laser through a cup of water with a drop of milk in it, the beam will be visible. In this mini-experiment, you will see what happens when you shine the laser from air to water and back to air.

1. Fill a clear plastic cup about halfway with water. Add a drop of milk to the water.
2. Set the cup on the laminated graph sheet and trace around the base of the cup.
3. Shine the laser through the cup so it passes off-center, as shown in the photo. Use an index card and water soluble marker to find and mark the beam going into and out of the cup.
4. Remove the cup and laser. Connect the sections of the beam that passes through the cup.

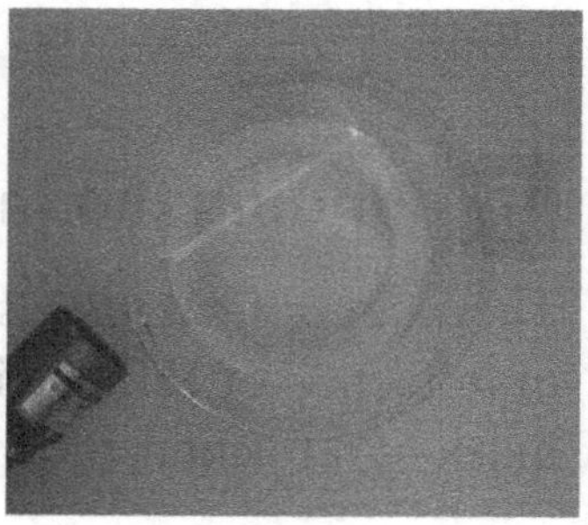

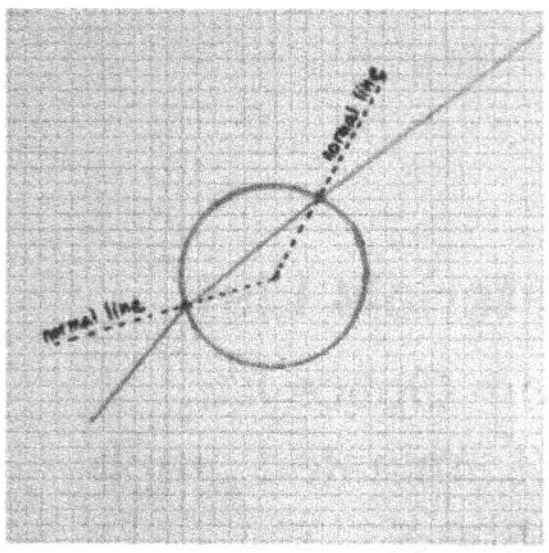

a. Draw the normal lines to the surface of the cup at the points where the light ray enters and exits the cup.

b. When the light is going from air into water, does the ray bend away from the normal or toward the normal?

c. When the light is going from water back into air, does the ray bend away from the normal or toward the normal?

d. Based on your answers to the previous questions, which has a higher index of refraction, air or water?

7 The critical angle of refraction

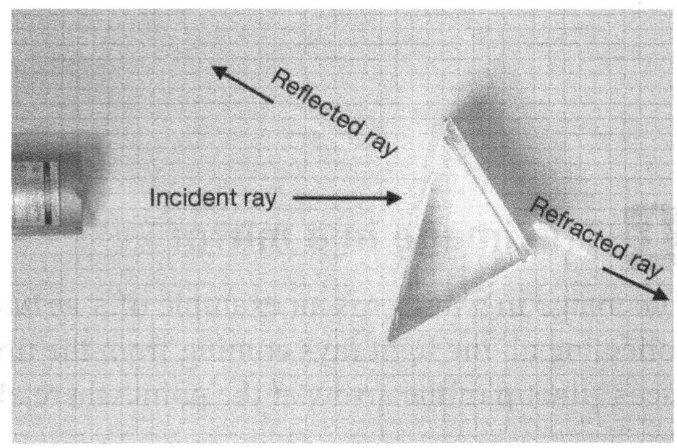

When light is going from a high-index material into a lower-index material, the ratio n_i/n_r is greater than 1. That means the angle of refraction (θ_r) must be greater than the angle of incidence (θ_i). When the angle of refraction becomes greater than 90 degrees, it becomes—reflection! The *critical angle* is the angle of incidence that makes the angle of refraction exactly 90 degrees. When the angle of incidence exceeds the critical angle, light is reflected because the angle of refraction becomes greater than 90 degrees. Scientists and engineers use the term *total internal reflection* to describe light reflecting back into a high-index material from a boundary with a low-index material.

1. Shine the laser into the long side of the prism. Observe what happens as you change the angle of incidence by rotating the prism. For some angles, the laser is reflected and exits the prism to the left. For other angles, the laser comes out on the right.
2. Try to identify the angle at which the laser beam makes the transition between refracting and reflecting when it reaches the boundary between glass and air.

a. You already experimented with the prism's critical angle of refraction, in Investigation 16.1, Part 2. You created a small card with the letters "A" and "B" on each half. You then looked at the card through the prism and noticed that the letter you could see changed as you changed your viewing angle. Explain what was happening in that experiment, using what you now know about the critical angle of refraction.

b. Fiber optics is an important tool in optical technology. A fiber optic is like a wire for light. You can bend a fiber optic and the light will still come through! The basis for the conduction of light by fiber optics is total internal reflection. Do some research on how fiber optics work. How is critical angle important to this technology? What are fiber optics used for?

17.2 Mirrors, Lenses, and Images

How do mirrors and lenses form images?

A lens uses the refraction of light to bend light rays to form images. A mirror also forms an image, but with reflected light instead of refracted light.

In this investigation, you will:

- use the laser flashlight to trace light rays from a lens to determine its focal length.
- show how ray diagrams are used to predict where images form with lenses and mirrors.

Materials List

From Optics kit:

- Laminated graph sheet
- Light blue lens
- Dark blue lens
- Mirror
- Laser flashlight and holder

Other:

- Water-soluble marker

1 The image in a mirror

The image in a mirror is an example of a *virtual image*. When you see the tip of an arrow, your eye is collecting all the light rays coming from the tip of the arrow. Since light travels in straight lines, your brain "sees" the tip of the arrow at the point where all the light rays seem to come from.

When you observe the tip of the arrow in a mirror, the light rays are reflected. The reflected rays *appear* to come from somewhere behind the mirror. You see a virtual image of the arrow reflected in the mirror because your brain "sees" the arrow where the light rays *appear* to come from instead of where they actually come from.

1. Draw a line with a water-soluble marker on the laminated graph sheet where you will place the mirror. Place the reflecting surface of the mirror along this line. Draw a 1-centimeter arrow on the graph paper about 3 centimeters away from your line. The arrow should be parallel to the line.

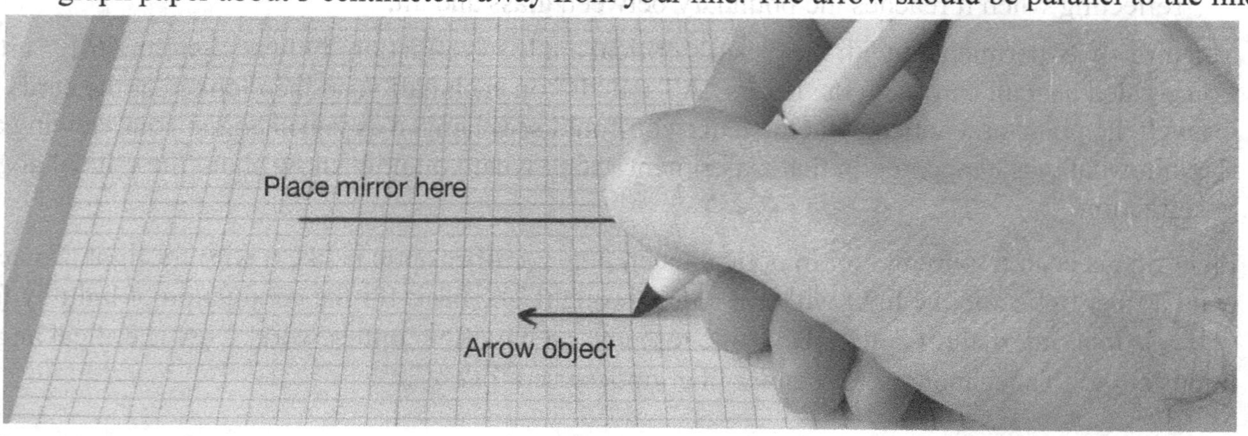

2. Move your head until you can see the reflection of the arrow in the mirror. The image of the arrow appears to be behind the mirror.
3. Hold your marker straight up with the point on the tip of your arrow. Use the marker to set the laser beam so it passes right over the tip of your arrow and hits the mirror. Trace the laser beam using the index card technique you learned in the previous investigation. Trace the incident and reflected beams.

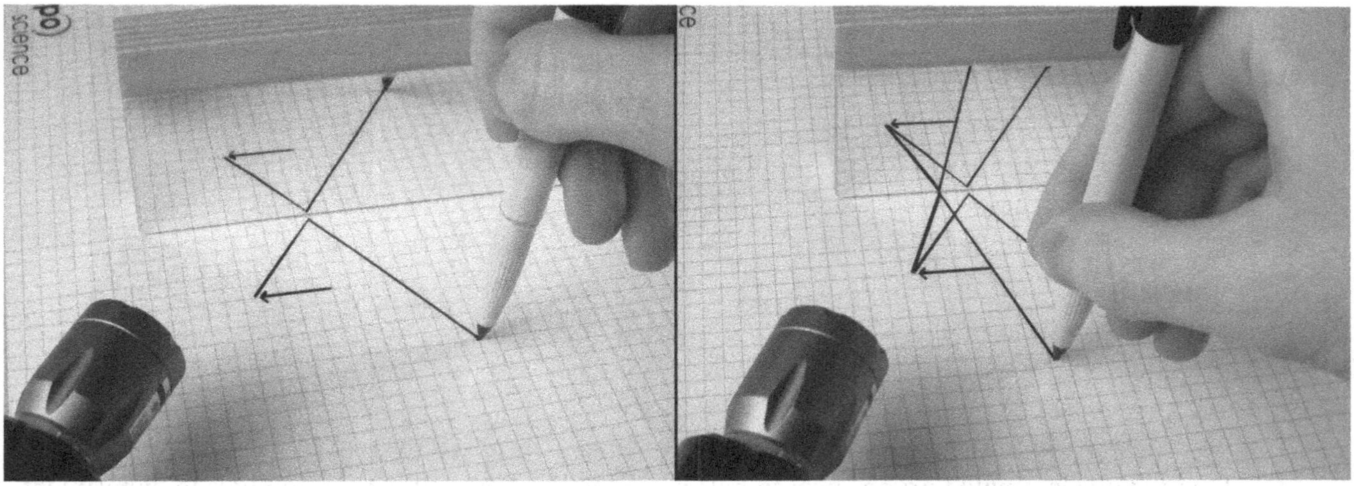

4. Move the laser so the beam passes over the tip of your arrow from a different angle, but still hits the mirror. Trace this second beam as you did in step 3.

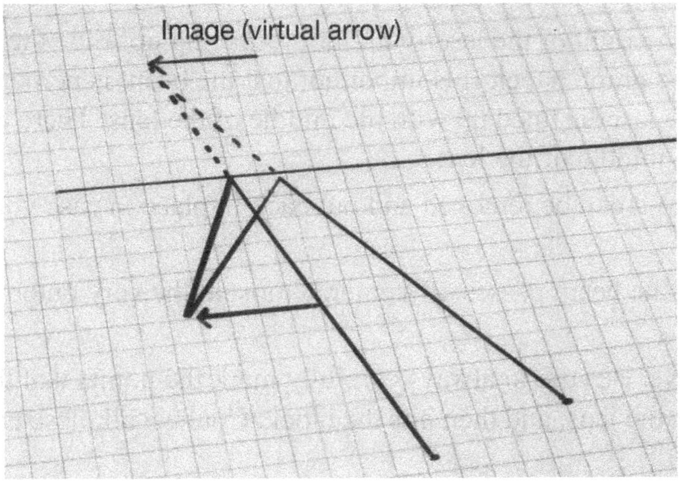

5. Remove the mirror and use a ruler to extend the two reflected rays. They should meet in a point on the other side of the line from where the real arrow is. The meeting point for the reflected rays is where you see the image of the tip of the arrow. The image forms where all rays that leave the same point on an object come together again.

Safety Tip: *Never look directly into a laser beam. Some lasers can cause permanent damage to your eyes.*

2 Refracting light through a lens

Like a prism, a converging lens bends light. Because the shape of a lens is curved, rays striking different places along the lens bend different amounts. The laser allows us to follow the path of the incident and refracted rays. Rays that approach a lens *parallel to the axis* meet at a point called the *focal point*. The distance between the center of the lens and the focal point is called the *focal length*.

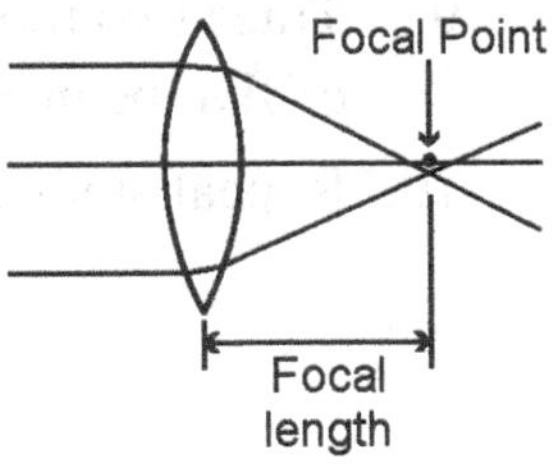

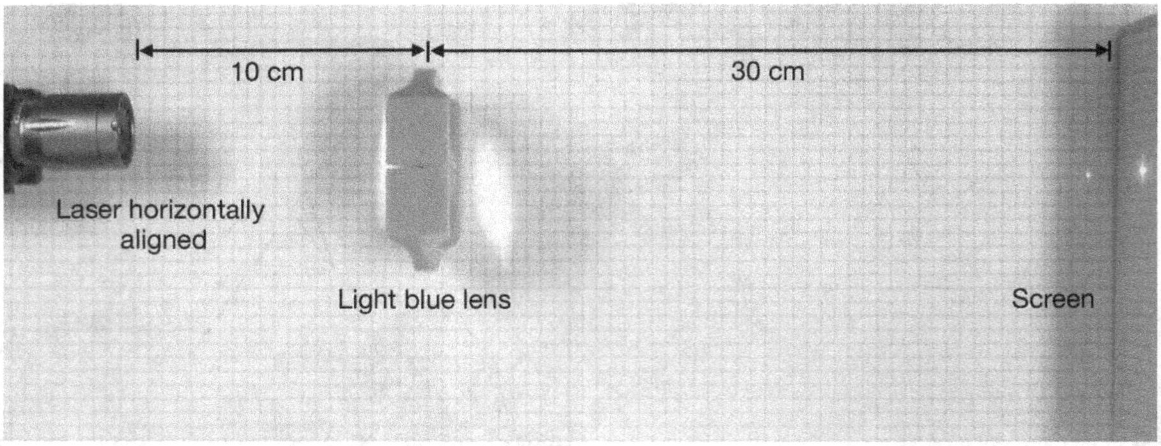

1. Divide your laminated graph sheet in half horizontally. You will use the top half to experiment with the light blue lens, and the bottom half will be reserved for the dark blue lens.
2. Place the laser on the edge of the laminated graph sheet and shine the laser so it follows a horizontal grid line across the paper.
3. Place the light blue lens 10 cm to the right of the laser with the slot facing up. Line the lens up vertically using the grid lines on the graph paper. It is important that the beam is perpendicular to the lens. Make sure the beam of the laser is lined up with the middle of the lens. There are lines on the side of each lens indicating the middle of the lens.
4. Trace around the base of the lens so it can be removed and put back in place in case you need to move it to complete ray tracing.
5. Shine the laser through the lens so the beam passes off-center, almost at the very outer edge of the lens.
6. Trace the incident and refracted rays. Be sure to always carefully mark the points that the beam exits the laser, enters the lens, exits the lens, and then hits the block. Connect all these points to see the path of the beam.
7. Realign the laser with a different horizontal grid line parallel to the original beam and closer to the center of the lens. Again, trace the path of the beam before and after it passes through the lens.
8. Realign the beam so it passes directly through the center and trace the beam again.
9. Trace two more beams passing through the lens on the other side of the center of the lens for a total of five beams. Label the beams 1 through 5 on both sides of the lens.
10. Repeat steps 2 through 8 with the dark blue lens on the bottom half of your graph paper.

3 Constructing explanations

a. Feel the glass surface with your fingers and note the shapes of the lenses. How are they different?

b. Draw a quick sketch of the shape of each lens itself with no stand from a side view. Label each lens.

c. Describe the paths of the rays before and after they traveled through the light blue lens. Include the words *refract*, *converge*, and *diverge* in your description.

d. What is the focal point of a lens? Mark the focal point on the light blue lens ray diagram.

e. What is the focal length of the lens? Measure the focal length of the light blue lens.

f. Describe the paths of the rays before and after they traveled through the dark blue lens. Include the words *refract*, *converge*, and *diverge* in your description.

g. One lens is referred to as a diverging lens, and the other a converging lens. They are also sometimes referred to as convex or concave. Research these terms and explain which is which.

4 The image from a single lens

The image from a distant light source forms at a place that is one focal length away from a single lens. This provides a convenient way to measure the focal length.

1. Find a wall at least 5 meters away from a lamp or sunlit window. Tape a piece of white paper to the wall to create a screen for seeing the image.
2. Hold the light blue lens at different distances from your screen. Try distances between 15 and 30 centimeters.
3. You will see a sharp image of the lamp or window on the screen when your lens is exactly one focal length away from the wall. Use this technique to determine the focal length for the lens.
4. Images can be smaller or larger than the object that created them. Images can also be right side up or inverted.

a. Was the image created by a single lens smaller or larger than the object?

b. Was the image right side up or was it inverted?

c. What is the focal length of the light blue lens, using this method?

d. How does the focal length you found in this part of the investigation compare to the focal length you found from your ray diagram for the light blue lens in Part 3?

17.3 Optical Systems

How are the properties of images determined?

Geometric optics describes a way to use scale drawings and geometry to analyze optical systems. Rays of light are represented by lines. Virtual rays are represented by dotted lines. Images form where the rays leaving a point on the object come together again. The images are *real* when actual light rays meet and *virtual* when virtual rays meet.

In this investigation, you will:

- predict where images form with lenses.
- use the thin-lens formula to predict how and where images are formed by a single convex lens.

Materials List

From Optics kit:

- Flashlight and holder
- Clear filter with letter *F*
- Light blue lens
- Dark blue lens
- Laminated graph sheet

Miscellaneous:

- Water-soluble marker
- Ruler or straight-edge

1 Projecting an image with a lens

You can think about a lens as collecting a cone of light from each point on an object. For a perfect lens, all the light in the cone is bent so it comes together at a point again to make the image. This is how movie projectors take an image on film and project it onto a screen.

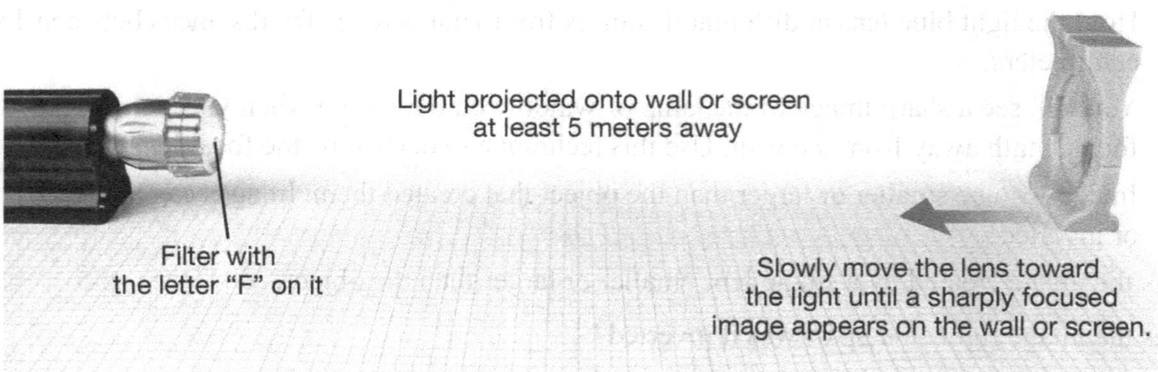

1. Shine the flashlight (with the letter *F* filter attached) horizontally on the laminated sheet.
2. Set the light blue lens about 35 cm away from the light.
3. Shine the light at a distant wall at least 50 centimeters away. If one is not available, affix a piece of white paper to the wall as your projection screen. Slowly move the lens toward the light until you see a sharp image of the "F" on the wall or screen. Have one group member check the projected image closely while the lens is slowly moved to find the exact place the lens needs to be to make the "F" come into focus.
4. When you have the image in sharp focus, measure the object distance and image distance and record them in Table 1. The object distance is measured from the front of the light to the middle of the lens. The image distance is measured from the middle of the lens to the front of the screen.
5. Fill in the rest of Table 1. For image orientation, record whether the image is inverted. For image height, measure the height of the image with a ruler to the nearest millimeter. The magnification is the image height divided by the object height (the height of the letter *F* on the filter cap is the object height).

6. Next, try projecting the image at a wall or screen that is farther away. Go even farther away for a third trial. (*Hint*: You will only be able to get a clear image if the separation between the lens and the screen is more than twice the focal length of the lens.) Record your data in Table 1.

Table 1: Projecting an image with a lens

Object distance (cm)	Image distance (cm)	Image orientation	Image height (mm)	Magnification

2 Analyzing and interpreting data

The image forms in a very specific place that depends on the lens and also on the location of the object relative to the lens. Tracing rays using geometric optics will show you what is happening. You will make a scale ray diagram out of the first row of data in Table 1. Once you complete the ray diagram, you can compare your theoretical (diagram) values with your measured values from Table 1.

1. Using the laminated graph sheet and a fine-point, water-soluble marker, make a scale drawing showing the positions of the object and lens from the first trial in Table 1. Measure and mark the near and far focal point of the light blue lens. (You found this value in the previous investigation.)
2. Draw an arrow for an object (you pick the height; 6 centimeters for example) at an object distance that corresponds to your first trial in Table 1. Your drawing should look like the diagram below. You must decide how many centimeters each box on your grid will equal, then keep that value throughout the exercise.

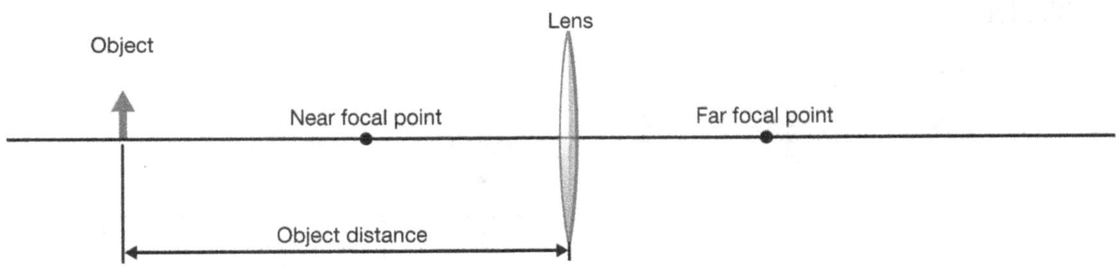

3. Draw three rays from the tip of the arrow using these rules:

- A ray parallel to the axis is bent to pass through the far focal point.
- A ray passing through the near focal point emerges parallel to the axis.
- A ray passing through the center of the lens is not deflected at all.

Drawing the three principle rays

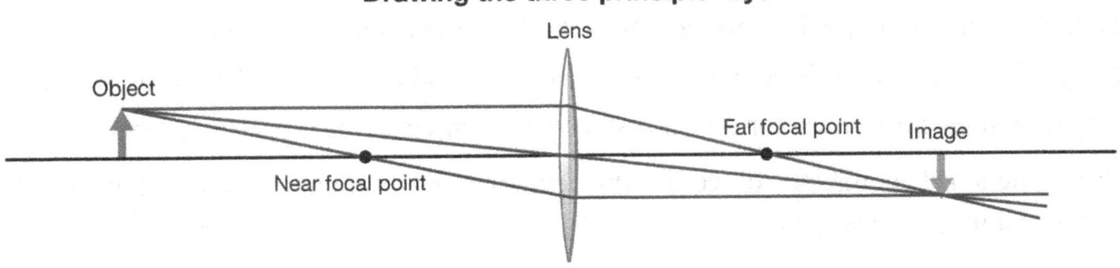

4. Where the rays meet is where the image forms.

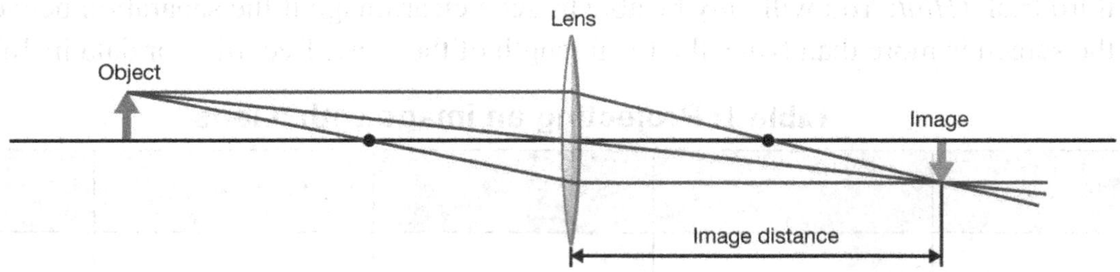

5. Measure the image distance from your drawing, and the height of the image (length of the image arrow from the optical axis to the tip). Record your measurements in Table 2.

Table 2: Result from ray diagram

Object height (cm)	Object distance (cm)	Image distance (cm)	Image orientation	Image height (cm)	Magnification

a. How does your ray-tracing prediction compare with your actual measured image? Write one or two sentences comparing measured and calculated data.

3 The thin-lens formula

Geometric optics are useful for understanding how lenses work, but there are much faster mathematical ways to predict the locations of objects and images. The thin-lens equation provides a way to calculate where images form given the positions and focal lengths of all objects and/or lenses in the system. The thin-lens equation is a good approximation as long as the object and image distances are much greater than the thickness of the lens.

THIN LENS FORMULA

$$\frac{1}{d_o} + \frac{1}{d_i} = \frac{1}{f}$$

d_o Object distance (cm), d_i Image distance (cm), f Focal length (cm)

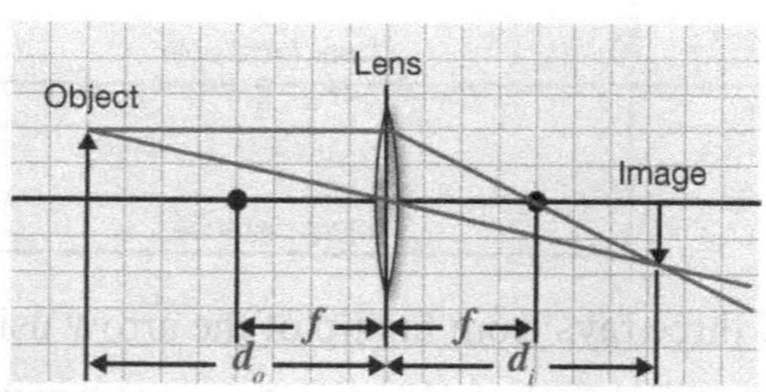

When using the thin-lens equation, distances are either positive or negative depending on a *sign convention*. The equation is written assuming that light goes from left to right. When the object and image appear like the diagram above, all distances are positive.

1. Object distances are positive to the left of the lens and negative to the right of the lens.
2. Image distances are positive to the right of the lens and negative to the left of the lens.
3. Negative image distances (or object distances) mean virtual images (or objects).

The image from one lens becomes the object for the next lens. In this manner, the thin-lens equation can also be used to analyze multiple-lens systems.

4 The image from a single lens

Use the thin-lens formula and the light blue lens to predict the image distance for four different known object distances.

1. For each of the object distances listed in Table 3, use the known focal length for the light blue lens, and plug the values into the thin-lens equation to predict the distance at which an image should form.
2. Place the screen at the predicted image distances and locate the image. How close were your predictions? Record the measured image distance for each trial in Table 3.

Table 3: Using the thin-lens equation to predict image distances

Object distance (cm)	Focal length (cm)	Predicted image distance (cm)	Measured image distance (cm)
30			
40			
50			
60			

a. How close did your prediction of the image come to the actual image? Answer with a percentage.

b. *Challenge*: Use the dark blue lens and the thin-lens equation to predict some image distances. You will need to find the focal length of the dark blue lens. One way to find this is explained in Investigation 17.2, Part 5. (*Hint*: The dark blue lens is a diverging lens. Instead of looking for the projected image in front of the object, you will have to look for the projected image *behind* the object!)

18.1 The Electromagnetic Spectrum

What is the electromagnetic spectrum?

The *electromagnetic spectrum* is a group of waves that all travel at the same speed (3×10^8 m/s) but have different wavelengths and frequencies. The waves in this spectrum are called *electromagnetic waves.*

In this investigation, you will:

- research one type of wave that is part of the electromagnetic spectrum.
- prepare a poster containing the information you discover.
- make an oral presentation to share the information with your class.

Materials List

- Poster board
- Access to the Internet and other research materials
- Poster-making supplies

1 Researching your electromagnetic wave

The electromagnetic spectrum

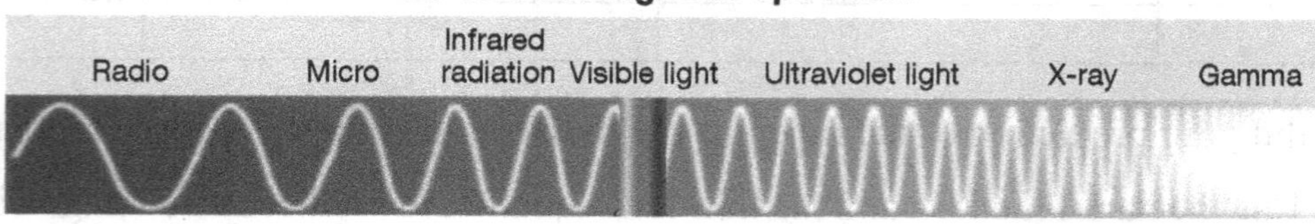

Your teacher will assign your group one type of electromagnetic wave that you will research using the Internet or reference materials. You will construct a poster displaying the information you gather about your type of wave. You will also present what you have learned to your classmates. The posters made by the class will be displayed in the order of the waves that make up the electromagnetic spectrum.

Your research should include the answers to the following questions. Do not limit yourself to these questions; include other interesting information as well. Record the sources you use in a bibliography.

1. What is the range of wavelengths for your type of wave?
2. What is the range of frequencies?
3. What is the source of the wave?
4. Who discovered this type of wave and when?
5. Are these waves easily blocked or can they pass through objects?
6. Do the waves have an effect on people? Are they harmful?
7. Discuss the uses for the waves. These may include inventions that we use in our everyday lives, medical uses, or ways scientists use the waves for research.

2 Communicating your results

a. Organize your information on a poster. You should include drawings where they are appropriate. Make a bibliography of your sources and put it on the back of your poster.

b. Present what you learned to your class. You should be able to discuss the uses for your type of wave.

c. Listen to the presentations made by other groups. Which type of wave do you think is the most useful to us? Why?

18.2 Interference, Diffraction, and Polarization

What are some ways light behaves like a wave?

A diffraction grating is a precise array of tiny engraved lines with small spaces in between. The spaces block light from passing through while the lines allow light through. When you look at a small bright white light source through a diffraction grating, you see two things:

(1) You see the light straight ahead as if the grating were clear glass.

(2) You see several rainbow-colored strips of light emanating from the light source in a starburst pattern.

In this investigation, you will:

- observe and explain how a diffraction grating creates a spectrum.
- observe and explain what happens when a laser shines through a diffraction grating.
- explain the interaction of light and polarizers, using the wave theory of light.

Materials List

From Optics kit:
- Laser flashlight
- Diffraction glasses
- 2 Polarizers

Other materials:
- Meter stick
- Snaky spring
- Permanent marker
- Paper or foam cup

The light scattered by a diffraction grating is an example of a *diffraction pattern*. The precise array of lines and spaces separates the light according to wavelength and produces a spectrum. Most of the light we see is a mixture of many different wavelengths. A spectrum shows how much there is of each wavelength in a sample of light. Because we see different wavelengths as different colors, your eye sees the spectrum as a rainbow.

1 How a diffraction grating works

Each scratched line in the diffraction grating allows light waves to pass through it. The light waves that have passed through all of the scratches on the grating interfere with each other. If the diffracted light waves all have a single wavelength, like from the laser, you see the interference as a pattern of bright and dark bands on the screen. The bright bands correspond to *constructive interference* of the light waves, and the dark bands correspond to *destructive interference*.

The diagram shows a magnified view of the diffraction grating. Distance a is a bit longer than distance b. The difference is $d\sin\theta$ where d is the spacing of the scratches on the grating. We see constructive interference (bright spots) when the difference between a and b is equal to one wavelength of light.

If the light is mixed wavelengths, each wavelength makes its bright spot in a different place. You see a rainbow pattern because each color is bright where its wavelength satisfies the condition $d\sin\theta = \lambda$. This is the basic principle behind a *spectrometer*.

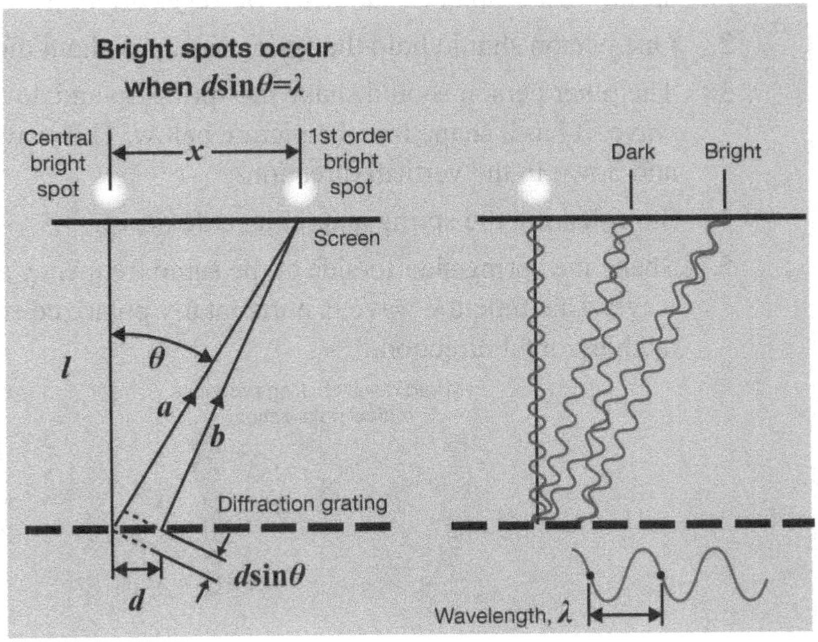

2 Measurement with light waves

Because light behaves like a wave, each type of light has a certain wavelength. You can use the wavelength of light to measure the spacing of the scratches on the diffraction grating. The wavelength of the red laser in your kit is approximately 652 nanometers (652×10^{-9} m).

1. Shine the laser through the diffraction grating glasses against a screen at least a meter away.
2. Measure the distances x and l as shown in the diagram on the previous page. From these distances calculate the angle, θ.
3. Change the distance (l) between the grating and the screen and make several measurements of x and θ. Record your results in Table 1.

Table 1: Diffraction grating data

Distance (l) (m)	Distance (x) (m)	Angle (θ) (degrees)

a. Use the grating equation ($d\sin\theta = \lambda$) to calculate the spacing between the scratches of the grating.

b. A CD has tiny mirrored spots that represent digital music. Because the spots are close together and regularly spaced, the CD acts like a combination of a diffraction grating and a mirror. Reflect the laser from a CD and use the grating equation to determine the spacing between the spots on the CD.

3 Polarization of a transverse spring wave

With a transverse wave, the oscillation can be in different directions relative to the direction the wave moves. Waves on a spring are a good example of *polarization.*

1. Find a partner. Each of you should take one end of the spring and stretch it. Don't let go or the spring will snap back suddenly. *Be very careful with the springs.*
2. One person should hold the spring firmly without moving.
3. The other person should shake the spring up and down at a frequency to get the second harmonic wave. It has a shape like the picture below. This wave is vertically polarized since it oscillates up and down in the vertical direction.
4. Stop shaking the spring and let it settle down.
5. Shake the spring side-to-side at the same frequency and you can also get the second harmonic wave. This time the wave is horizontally polarized since the oscillations travel back and forth in the horizontal direction.

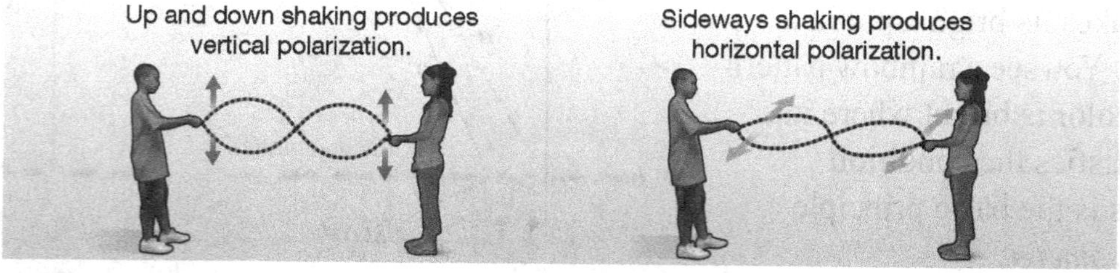

4 Arguing from evidence

a. Describe the motion of the spring using the terms *horizontal polarization* and *vertical polarization.* Your description can be in words, diagrams, or both.

b. Suppose you try to sandwich your spring wave between two boards. What happens to the waves if you make them pass through the narrow space between the boards? If the boards were oriented like the picture below, discuss how the two different polarizations of waves would behave. Which would get through the slot and which would be blocked?

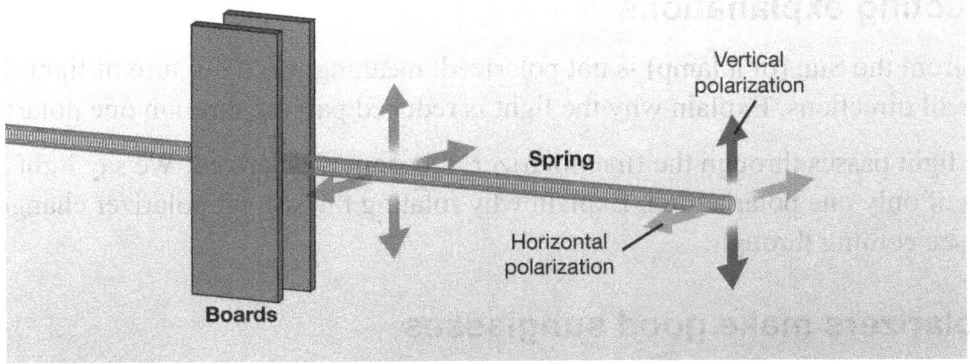

c. Describe the polarization of water waves. Are there two polarizations (as with the spring) or only one? What is it about a water surface that makes it different from a spring?

5 The polarization of light

If you were to observe polarization, it would be strong evidence that light is a transverse wave. A *polarizer* is a material that allows one polarization of light to pass but blocks the other polarization. A polarizer works like the two boards, except the "boards" are long, thin molecules. The molecules let light waves oscillate one way but absorb light that oscillates the other way.

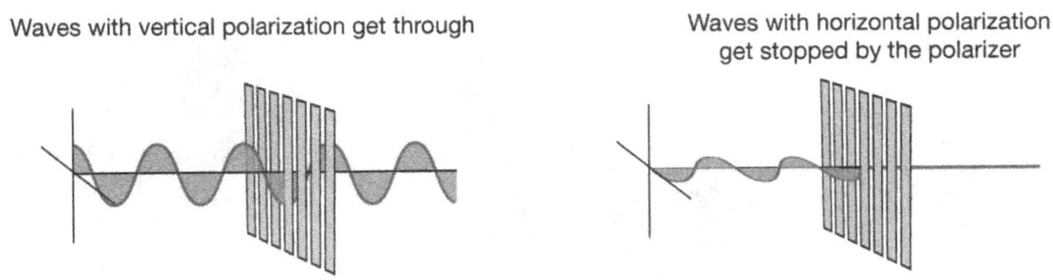

1. Hold up one sheet of polarizer and look through it. Observe the effect of looking through the polarizer. Try rotating the polarizer and see if it makes a difference.
2. Take a second sheet of polarizer and look through it. Observe the effects, just as with the first sheet.
3. Now look through both sheets of polarizer together. Leave one fixed and rotate the other one as shown in the photo on the next page. Observe how much light you see through both polarizers as you rotate the second one.

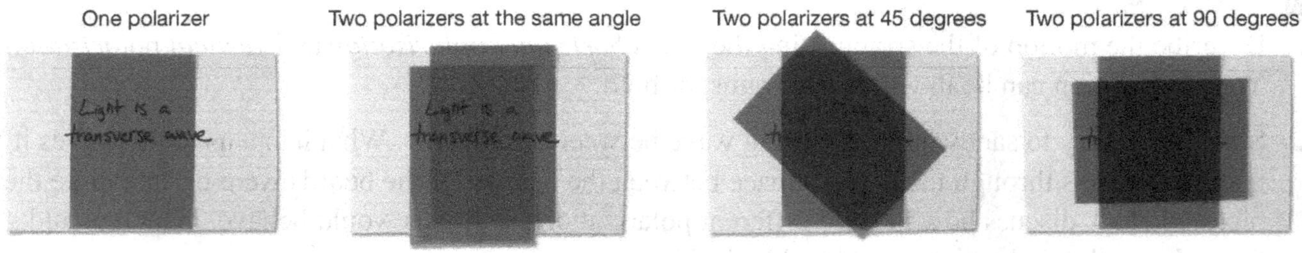

6 Constructing explanations

a. The light from the Sun (or a lamp) is not polarized, meaning it is a mixture of light that is polarized equally in all directions. Explain why the light is reduced passing through one polarizer.

b. When the light passes through the first polarizer, it becomes polarized. We say light is polarized when it consists of only one polarization. Explain why rotating the second polarizer changes the amount of light you see coming through.

7 Why polarizers make good sunglasses

1. Draw some letters on the inside of a paper or foam cup about two centimeters below the rim with a waterproof permanent marker. Fill the cup with water, covering the letters.
2. Place place the cup on a table (or the floor) two meters from a window. When you stand over the cup you should see the letters through the surface of the water.
3. Position yourself so you can see light from the window reflected of the surface of the water. The glare on the surface should prevent you from seeing the letters.
4. Look at the cup (and the glare) through a polarizer as you slowly rotate the polarizer. Notice you can see the letters sometimes, but not all the time.

a. Is there an orientation of the polarizer that allows you to see the letters through the glare? If so, what is it?

b. What does this experiment tell you about the light that is reflected from the surface of the water compared to other light you see?

c. Explain how polarizing sunglasses can stop most of the glare but still allow half of the regular (unpolarized) light to come through.

18.3 Special Relativity

What are some of the implications of special relativity?

Time moves more slowly for an object in motion than it does for an object that is not in motion. In practical terms, clocks run slower on moving spaceships compared to clocks on the ground. By moving very fast, it is possible for one year to pass on a spaceship while 100 years passes on the ground. This effect is known as *time dilation*.

Materials List

- Access to the Internet and other research materials
- Calculator

In this investigation, you will:

- explore some consequences of time dilation.
- calculate the equivalence of mass and energy using Einstein's formula, $E = mc^2$.

Table 1 gives some examples of how time dilation affects the perception of one second of time for observers moving relative to each other at different speeds.

Table 1: Speed and Time Dilation

Speed			Time for the moving observer	Time for the observer at rest
m/s	mph	% of c	(s)	(s)
100	224	3.3×10^{-5}	1.0000	1.0000
1,000	2,240	3.3×10^{-4}	1.0000	1.0000
10,000	22,400	0.0033	1.0000	1.0000
1,000,000	2.24×10^6	0.33	1.0000	1.0000
1.00×10^7	2.24×10^8	3.3	1.0000	1.0006
1.00×10^8	2.24×10^8	33	1.0000	1.0607
2.00×10^8	2.24×10^8	67	1.0000	1.3424
2.80×10^8	6.26×10^8	93	1.0000	2.7981
2.90×10^8	6.49×10^8	97	1.0000	3.9434
2.99×10^8	6.69×10^8	99.7	1.0000	13.6976

1 When does special relativity become important?

a. A high-performance aircraft flies at a speed of 1,340 m/s, or roughly four times faster than the speed of sound (340 m/s). At this high speed, can the effects of time dilation be perceived by a person with an ordinary watch?

b. A rocket traveling to Mars must have a speed greater than the minimum speed required to break Earth's gravitational attraction. This minimum speed is called the escape velocity. Research to find Earth's escape velocity. If the rocket is traveling at Earth's escape velocity, do you think special relativity should be considered when synchronizing clocks between the ship and Earth?

c. The numbers in Table 1 were calculated using a formula proposed by Einstein. Research the formula to identify what the variables mean.

2 Relativity and frames of reference

The word *relative* in the special theory of relativity means that velocities can only be determined relative to a *reference frame*. It makes no sense in physics to say that an object is moving at 1 m/s without providing a reference. For example, you might measure a speed of 1 m/s relative to your lab or classroom. However, Earth rotates and also circles the Sun. A velocity of 1 m/s relative to the lab (on Earth) is *not* the same as a velocity of 1 m/s relative to the Sun.

Consider the following experiment. A person on a train throws a dart at a dart board. Relative to the train, the dart travels a distance of 5 meters to the dart board in 0.5 seconds. The train is moving forward at a speed of 30 m/s.

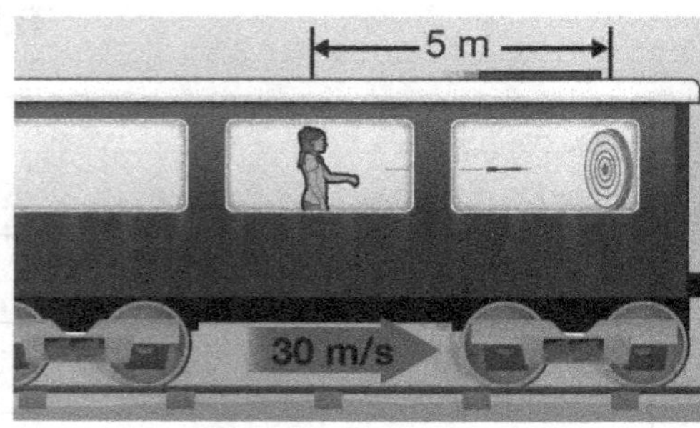

a. There are two reference frames important to understand the motion of the dart. What are they?

b. What is the speed of the dart relative to the train?

c. What is the speed of the dart relative to the ground?

3 The equivalence of mass and energy

EINSTEIN'S MASS-ENERGY FORMULA

$$E = mc^2$$

Energy (J) — E; Mass (kg) — m; Speed of light (3×10^8 m/s) — c

According to Einstein's theory of special relativity, mass and energy are two forms of the same thing. One way to think about mass is as extremely concentrated energy. In fact, according to the mass-energy formula (above), 1 kilogram of mass is equivalent to 9×10^{16} joules of energy.

a. Calculate the amount of energy used by a 100 W light bulb that is left on continuously for one year.

b. Suppose you could extract all of the energy in 1 kilogram of mass with 100 percent efficiency. How long could this amount of energy keep the 100 W light bulb lit? Give your answer in years.

c. Nuclear reactors convert a tiny fraction of the mass of uranium atoms into energy. Assume that 0.07 percent of the mass of uranium is converted to energy. How much energy do you get from 1 kg of uranium in a nuclear reactor?

d. To appreciate the energy obtained from a kilogram of uranium, estimate the electric power used by a city of one million people. Assume that each person in the city uses an amount of electricity equal to 10 light bulbs that use 100 watts each. Calculate the total energy used by multiplying the total power in watts by the number of seconds in one year (you will get a very large number).

e. How many kilograms of uranium must be used in a nuclear reactor to produce this amount of energy?

f. One barrel of ordinary gasoline yields about 5×10^9 joules of energy. Calculate how many barrels of gasoline are required to produce the energy you estimated in d above.

19.1 Electric Circuits

What is an electric circuit?

Electricity is an integral part of our lives. Our homes, stores, and workplaces use many electrical devices such as electric ovens, TVs, stereos, toasters, fans, air conditioners, heaters, and light bulbs. In fact, the use of electricity has become so routine that many of us never stop to think about what happens when we switch on a light or turn on a motor. If we stop to look, we find that most of what is "happening" is not visible. What exactly is electricity? How does it work?

In this investigation, you will:

- construct simple electric circuits using a battery, bulb, switch, and wires.
- draw circuit diagrams using electrical symbols.
- explain the concepts of open, closed, and short circuits.

Materials List

- Electric Circuits components (bulb and holder, battery holder, wires, switch)
- C-cell battery

1 Building circuits with a battery, a bulb, and a wire

a. Using *only* one bulb, one battery, and one wire, find four different ways you can arrange these three parts to make the bulb light. As you work, determine the kinds of connections that are needed to make the circuit work.

b. Record all your circuit attempts. Draw both successful and unsuccessful attempts.

c. Make sure your drawings show the difference between the two ends of the battery, and where the bulb is touching the wire and the battery.

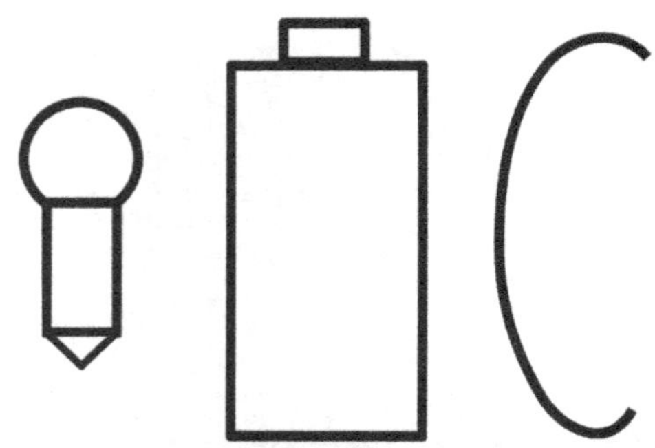

Using simple geometric shapes (circle, rectangle, triangle) and a line, you can draw representations of the light bulb, battery and wire.

d. Explain why you think some configurations work and others do not. Record your first thoughts and impressions—and don't worry if they are right or wrong.

2 Observing how a switch works

Using the electric circuits set, build a circuit that turns a bulb on and off using a switch. When designing your circuit, consider the configurations that worked in Part 1. Your completed circuit should include one battery holder with battery, one bulb holder with bulb, and a switch. The bulb should light in the completed circuit. Check that the switch turns the light bulb on and off.

a. Examine the switch as it turns the bulb on and off. Explain how the switch works. You may use both words and drawings.

c. Nuclear reactors convert a tiny fraction of the mass of uranium atoms into energy. Assume that 0.07 percent of the mass of uranium is converted to energy. How much energy do you get from 1 kg of uranium in a nuclear reactor?

d. To appreciate the energy obtained from a kilogram of uranium, estimate the electric power used by a city of one million people. Assume that each person in the city uses an amount of electricity equal to 10 light bulbs that use 100 watts each. Calculate the total energy used by multiplying the total power in watts by the number of seconds in one year (you will get a very large number).

e. How many kilograms of uranium must be used in a nuclear reactor to produce this amount of energy?

f. One barrel of ordinary gasoline yields about 5×10^9 joules of energy. Calculate how many barrels of gasoline are required to produce the energy you estimated in d above.

19.1 Electric Circuits

What is an electric circuit?

Electricity is an integral part of our lives. Our homes, stores, and workplaces use many electrical devices such as electric ovens, TVs, stereos, toasters, fans, air conditioners, heaters, and light bulbs. In fact, the use of electricity has become so routine that many of us never stop to think about what happens when we switch on a light or turn on a motor. If we stop to look, we find that most of what is "happening" is not visible. What exactly is electricity? How does it work?

In this investigation, you will:

- construct simple electric circuits using a battery, bulb, switch, and wires.
- draw circuit diagrams using electrical symbols.
- explain the concepts of open, closed, and short circuits.

Materials List

- Electric Circuits components (bulb and holder, battery holder, wires, switch)
- C-cell battery

1 Building circuits with a battery, a bulb, and a wire

a. Using *only* one bulb, one battery, and one wire, find four different ways you can arrange these three parts to make the bulb light. As you work, determine the kinds of connections that are needed to make the circuit work.

b. Record all your circuit attempts. Draw both successful and unsuccessful attempts.

c. Make sure your drawings show the difference between the two ends of the battery, and where the bulb is touching the wire and the battery.

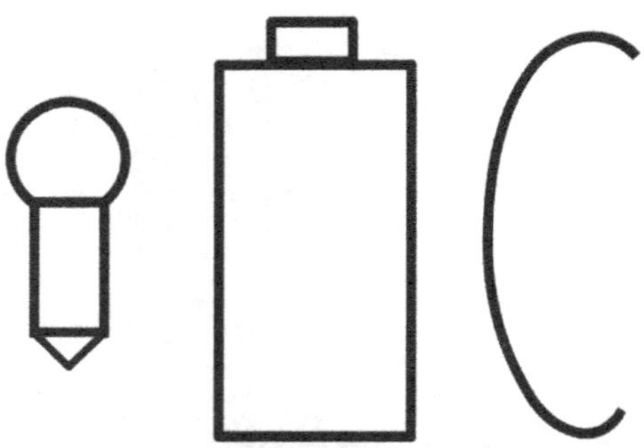

Using simple geometric shapes (circle, rectangle, triangle) and a line, you can draw representations of the light bulb, battery and wire.

d. Explain why you think some configurations work and others do not. Record your first thoughts and impressions—and don't worry if they are right or wrong.

2 Observing how a switch works

Using the electric circuits set, build a circuit that turns a bulb on and off using a switch. When designing your circuit, consider the configurations that worked in Part 1. Your completed circuit should include one battery holder with battery, one bulb holder with bulb, and a switch. The bulb should light in the completed circuit. Check that the switch turns the light bulb on and off.

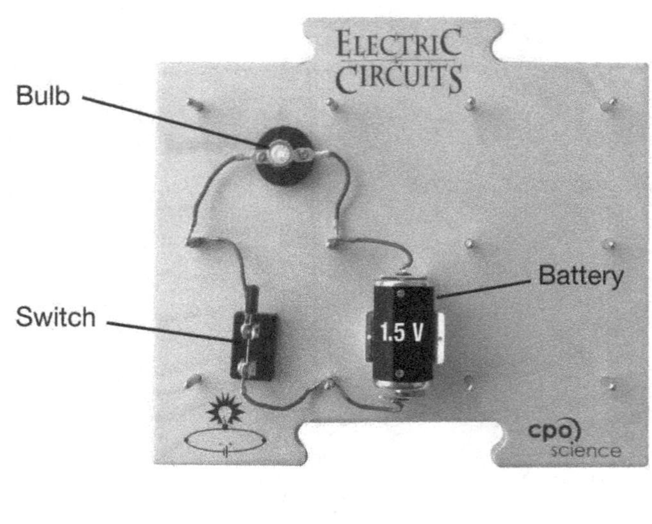

a. Examine the switch as it turns the bulb on and off. Explain how the switch works. You may use both words and drawings.

3 Drawing circuit diagrams

It can be time-consuming to draw pictures of batteries and bulbs every time you make a circuit diagram. People who work with electricity have shortcut methods for drawing electric circuits. All the electrical parts in a circuit are represented by standard pictures called *electrical symbols*. Some electrical parts and their corresponding symbols are shown in the picture at right. Study the symbols and practice drawing them.

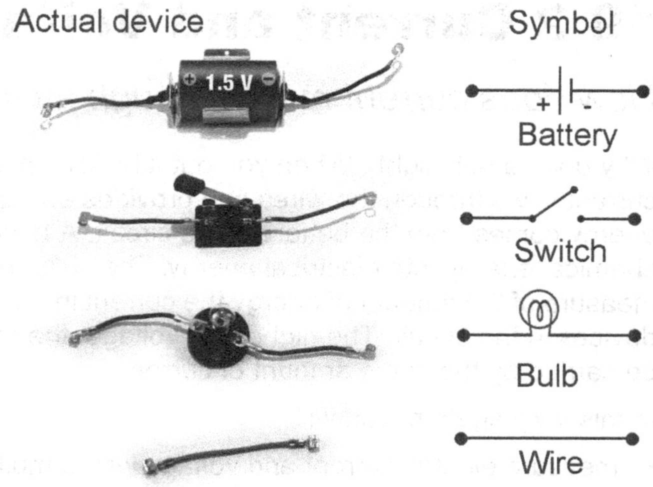

Using these symbols, draw a picture of the circuit you built using the electric circuits set. This type of drawing is called a *circuit diagram*.

4 Constructing explanations

a. Using what you learned in this investigation, describe some materials that electricity can and cannot travel through.

b. The word *circuit* comes from the same root as the word *circle*. Describe the similarities between a circle and the circuits that you built.

c. A circuit that is on and working is sometimes called a *closed circuit*. Based on your observations of the switch, explain what *closed* means in a circuit.

d. A circuit that is off or that is not working is sometimes called an *open circuit*. Based on your observations of the switch, explain what *open* means in a circuit.

19.2 Current and Voltage

How does current move through a circuit?

Why does a bulb light? When you put a battery in a circuit, electric current flows through the wires and provides energy to the bulb. The energy comes from the battery in the circuit. A battery transforms chemical energy into electrical energy. The *voltage* of a battery is a measure of the amount of energy the current in the circuit will carry to the devices in the circuit. The higher the voltage, the more energy that can be carried by the same amount of current.

In this investigation, you will:

- measure electric current and voltage with a multimeter.
- compare current in circuits with one or two bulbs.
- combine two batteries to increase the voltage in a circuit.

Materials List

- Multimeter with red and black leads
- 2 D-cell batteries
- Electric Circuits components (circuit board, 2 bulbs and sockets, wires, switch, battery holder)

1 Building test circuit 1

Build test circuit 1 with a battery, battery holder, bulb, bulb holder, and two connecting wires as shown in the diagram at right. There are two points marked on the diagram, A and B, which are where the connectors are attached to the bulb holder. Make sure that you can easily identify these points in your circuit.

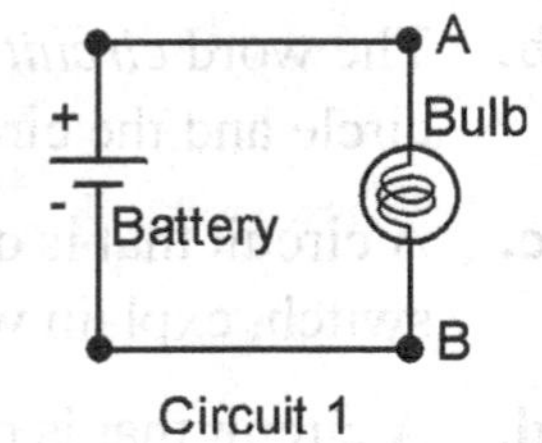

Circuit 1

2 Measuring current through test circuit 1

To measure current, the meter must be placed *in* the circuit so that the current has to flow through it. Follow the instructions below carefully.

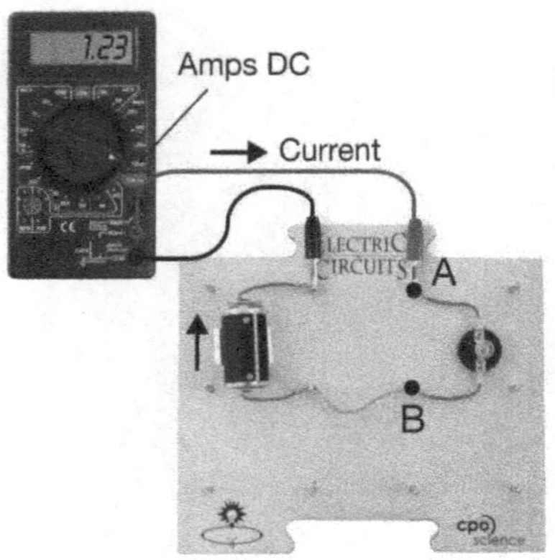

1. Set the multimeter to measure DC amps (current).
2. Remove the connector between the positive terminal of the battery and the bulb holder.
3. Place the red positive lead of the multimeter on the post attached to the positive terminal of the battery.
4. Place the black negative lead on the free post attached to the bulb holder. You should now have a complete circuit. The bulb should light and you should get a reading on the multimeter.
5. Record the current at point A in Table 1.
6. Remove the meter and reconnect the circuit.

7. Break the circuit between the battery and point B.
8. Predict what the current at point B will be. Record your prediction.
9. Following the same procedure you used for point A, use your meter to measure current at point B and record your result in Table 1. The positive lead of the multimeter should be closest to the positive battery terminal and the negative lead should be closest to the negative battery terminal.
10. Record the current at point B.
11. Remove the meter and reconnect the circuit so the bulb lights.

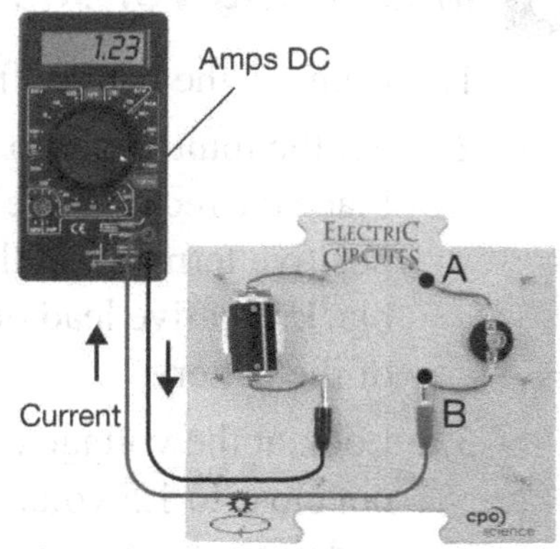

3 Test circuit 2

1. Gather the following additional materials: one battery and battery holder, two bulbs with bulb holders, and three connectors.
2. Build the circuit shown in the diagram at right. Make sure that you can easily identify points A, B, and C where the connectors are attached.
3. Using the same procedure you followed in Part 2, measure and record the current at point A in Table 1.
4. Predict what the current will be at points B and C.
5. Using the same procedure you followed in Part 2, measure and record the current at points B and C and record your results in Table 1.

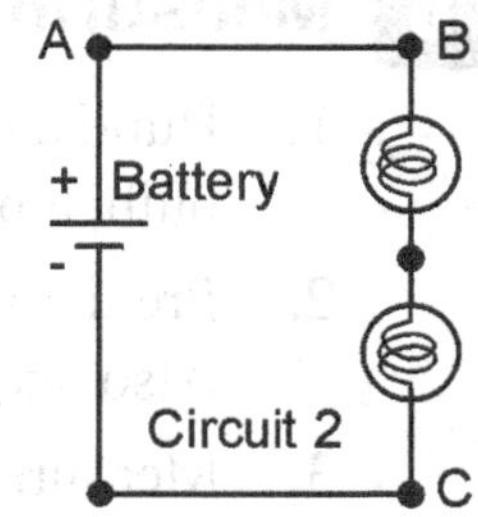

4 Analyzing and interpreting data

a. Review the two current readings for circuit 1. What conclusions can you draw from these results?

b. Review the three current readings for circuit 2. What conclusions can you draw from these results?

c. Compare the current readings in the two circuits. What happened to current when you added a bulb to the circuit?

Table 1: Current measurement data

	Current in circuit 1 (amps)		Current in circuit 2 (amps)
Point A		Point A	
Point B		Point B	
		Point C	

d. In which circuit were the light bulbs brighter? Offer a possible explanation for this.

5 Measuring voltage across a battery

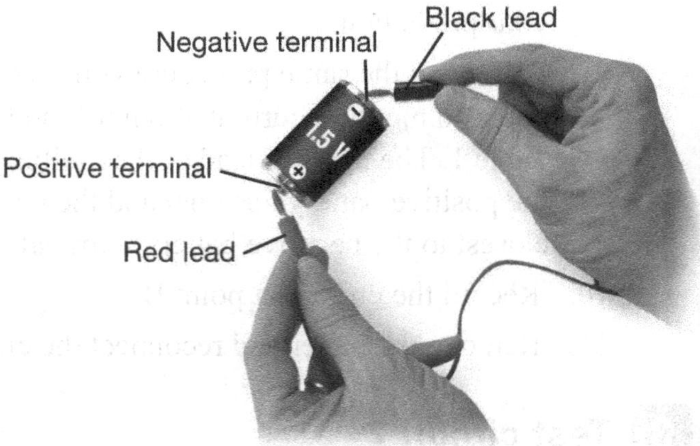

1. Remove the battery from your circuit.
2. Set the multimeter to measure DC voltage. Place the red positive lead of the meter on the positive terminal of the battery. Place the black negative lead on the negative terminal of the battery.
3. Look at the voltage reading. If the number is not close to 1.5 volts, the battery may be dead and you need to get a fresh one.
4. Record the exact voltage of the battery.

6 Measuring voltage across a battery and bulb in a circuit

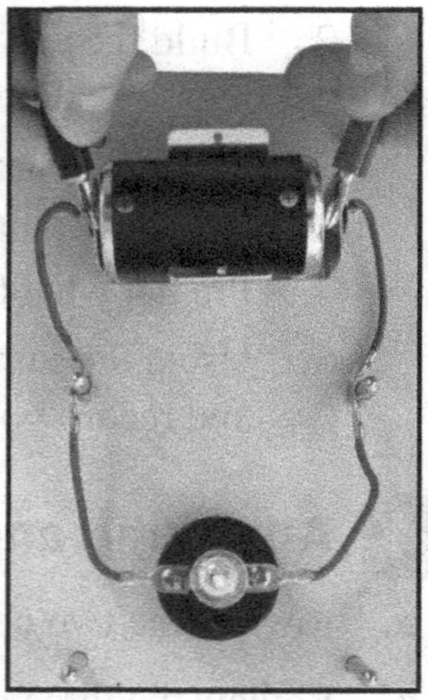

1. Build a circuit with the battery you just tested, a battery holder, one bulb, a bulb holder, and two connectors.
2. Predict what the voltage of the battery will be while it lights the bulb. Also, explain the reasoning behind your prediction.
3. Measure the voltage across the battery as shown at right. *Do not disconnect the circuit.*
4. Record the voltage of the battery while it lights the bulb.
5. Predict what the voltage across the bulb will be while it is lit by the battery.
6. Connect the positive lead to the side of the bulb that is connected to the battery's positive terminal. Connect the negative lead to the side of the bulb that is connected to the battery's negative terminal. Record the voltage across the bulb.

7 Analyzing and interpreting data

a. Were your predictions correct?

b. Was there much difference in the battery voltage when it was not lighting the bulb and when it was?

c. How did the battery's voltage compare with the bulb's voltage? Why do you think this is?

8 Building circuits with two batteries

How many batteries does a small flashlight use? How about a large radio? What happens when you wire batteries together?

Build a circuit with two batteries and two holders, one bulb and holder, and three wires. Connect the two batteries in four different configurations as shown. For each configuration, do the following:

Trial 1

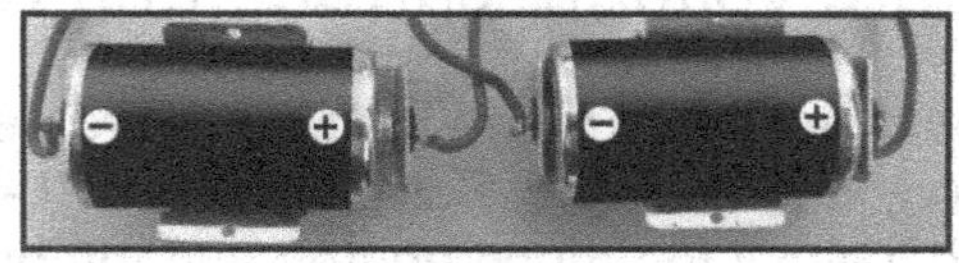

Trial 2

Trial 3

Trial 4

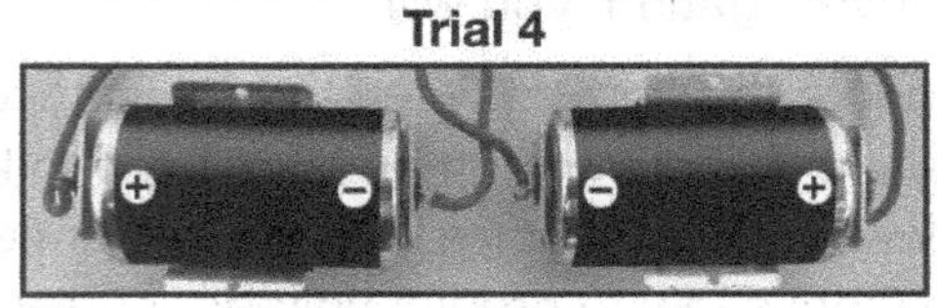

1. Measure voltage across both batteries before adding the bulb.
2. Predict whether the bulb will light, and then connect the bulb and check.
3. Measure voltage across the bulb. Do this by placing one meter lead on one side of the bulb holder and the other meter lead on the other side of the bulb holder.

Examine your data carefully. Then discuss the following questions with your group and record your answers.

a. Describe how you should wire batteries together to light the bulb.

b. Each battery is 1.5 volts. Explain how to calculate voltage when two batteries are connected.

c. Compare the brightness of a bulb in a one-battery circuit to that of a bulb in a two-battery circuit. How is the bulb's brightness related to energy transfer in each circuit?

d. Explain the meaning of the voltage reading across the bulb.

19.3 Electrical Resistance and Ohm's Law

How are voltage, current, and resistance related?

Electrical devices such as light bulbs are designed to use a certain amount of energy each second. For example, a 10-watt bulb uses 10 joules each second. Give it too much energy and the bulb burns out; too little and the bulb gives off no light. Energy is carried by electric current, and devices like light bulbs are designed to use the right amount of current when connected to the proper voltage. To design electrical devices (and use them properly), it is important to understand the relationship between current, voltage, and resistance. German physicist Georg S. Ohm (1789–1854) experimented with circuits to find out how voltage, current, and resistance are related. The relationship he discovered is called *Ohm's law*, and is the basis for understanding how electrical devices regulate how much energy they use.

In this investigation, you will:

- measure the current when the resistance is changed.
- analyze a current versus voltage graph to determine a relationship between voltage, current, and resistance.
- use Ohm's law to predict voltage, current, or resistance when two of three variables are known.

Materials List

- Multimeter with red and black leads
- 1 D-cell battery
- Graph paper
- Electric Circuits components (circuit board, potentiometer, switch, battery holder, wires, 20 Ω resistor, bulb and socket)

1 Resistance and current

A *resistor* is used in a circuit to provide resistance. A *variable resistor* allows you to change its resistance by turning a dial. A variable resistor is also called a *potentiometer*. In this part of the investigation, you will use a potentiometer to control the current in a circuit. The higher the resistance, the lower the current.

Inside the potentiometer is a circular resistor. The dial selects how much of the resistor is connected between terminals A and B. When the dial is turned so the wiper is near terminal A, the resistance is very low, since the current only passes through a small part of the resistor. When the dial is turned so the wiper is near terminal C, the resistance between A and B is high because the current must go through nearly the whole resistor. Many dials you use every day, like dimmer switches, are actually potentiometers.

The inside of a potentiometer

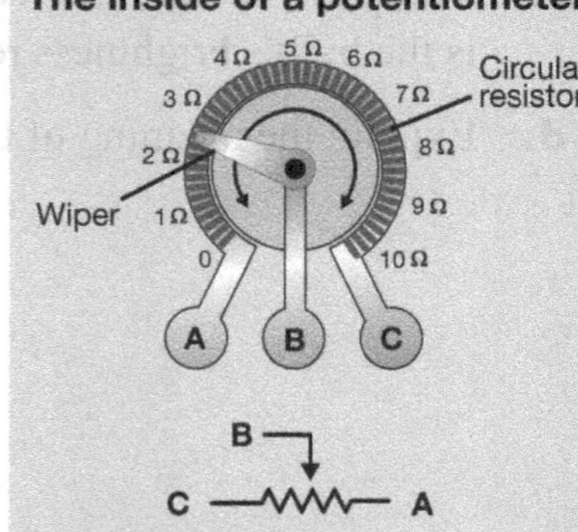

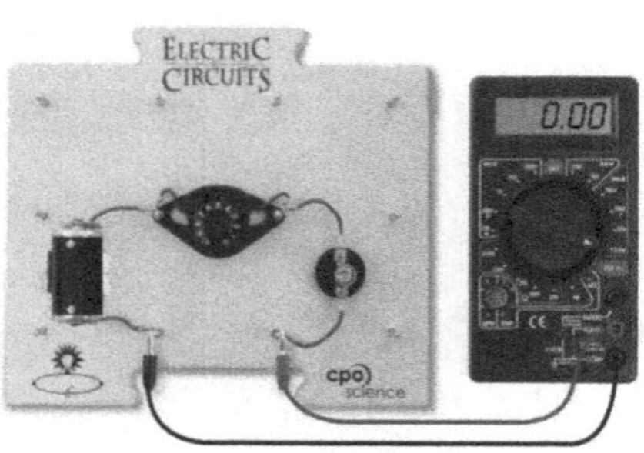

1. Build a circuit with a battery, a bulb, and the potentiometer as shown in the diagram at right. Set the meter to DC amps and measure current in the circuit as shown on the left. (*Note*: The wires in your potentiometer are connected to terminals A and B.)

Build this circuit

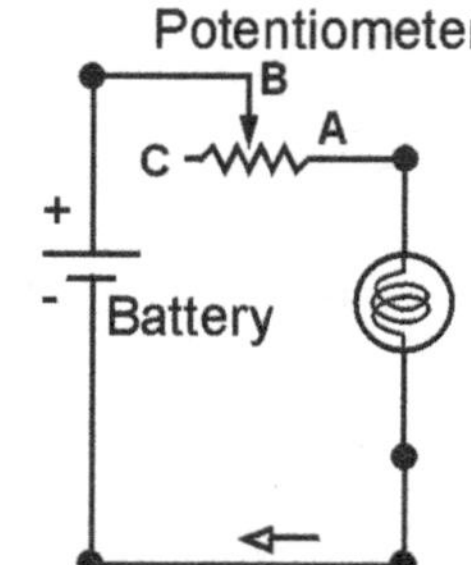

2. Turn the dial of the potentiometer up and down and observe what happens to the current and the amount of light produced by the bulb.

a. Explain why the current changes when the dial of the potentiometer is turned. The dial controls the resistance of the potentiometer, which varies from 0Ω (ohms) to 105Ω.

b. Explain why the bulb is dim or bright depending on the position of the potentiometer. Your answer should include the concepts of resistance, energy, and current.

2 The current and voltage relationship for a fixed resistor

In this part of the investigation, you will use two types of resistors. A *fixed resistor* has a constant resistance. You will use the potentiometer (a variable resistor) to control the current in the fixed resistor. This experiment requires both current and voltage measurement. The meter must be connected in different ways for each type of measurement. To avoid connecting and disconnecting wires for each measurement, use the switch to break the circuit.

1. Build a circuit with one battery, a green resistor, a switch, and the potentiometer as shown in the diagram at right.
2. Start with the dial of the potentiometer turned halfway. For each trial, turn the dial further clockwise.
3. To measure voltage across the resistor, set the meter to DC volts. Close the switch and connect the meter across the fixed resistor. Record the voltage in Table 1.
4. To measure current through the resistor, set the meter to DC amps. Open the switch and touch the meter leads to the *terminals of the switch.* With the switch open, all the current flows through the meter where it can be measured. Record the current in Table 1.
5. Change the setting of the potentiometer and measure current and voltage again. Repeat for at least six settings covering the whole range of the potentiometer.

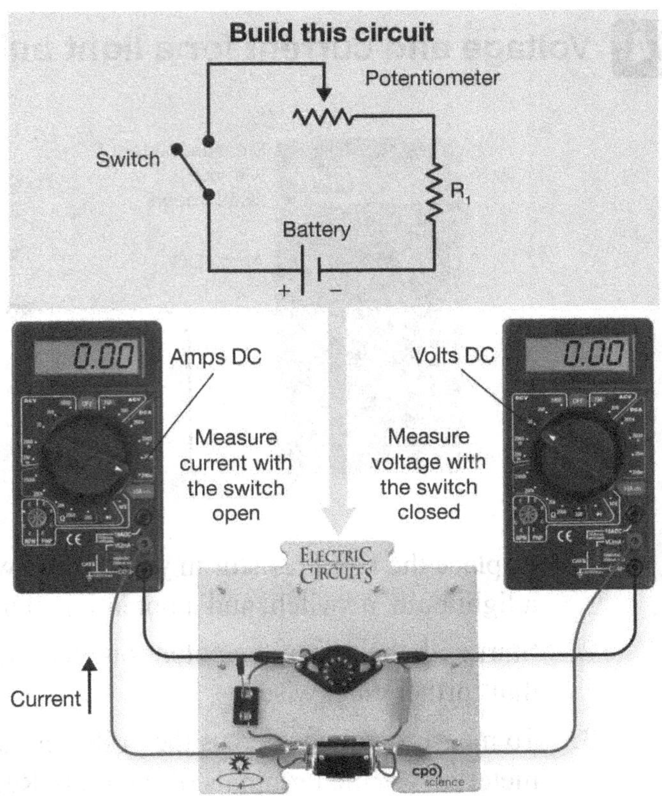

Table 1: Voltage vs. current

Potentiometer setting	Voltage (V)	Current (A)

3 Finding the relationship between voltage, current, and resistance

a. Examine the data in Table 1. Describe in words what happens to the voltage across the fixed resistor as the current decreases.

b. Graph the data from Table 1. Put voltage on the y-axis and current on the x-axis. Label your x- and y-axes and title your graph.

c. Draw a best-fit line that matches the trend of the points on your graph. Calculate the slope of the line.

d. The resistance of the green fixed resistor is five ohms. How does this value compare with the slope of your graph?

e. The equation that relates current, voltage, and resistance is called Ohm's law. Write the mathematical equation for Ohm's law using V for voltage, I for current, and R for resistance.

4 Voltage and current for a light bulb

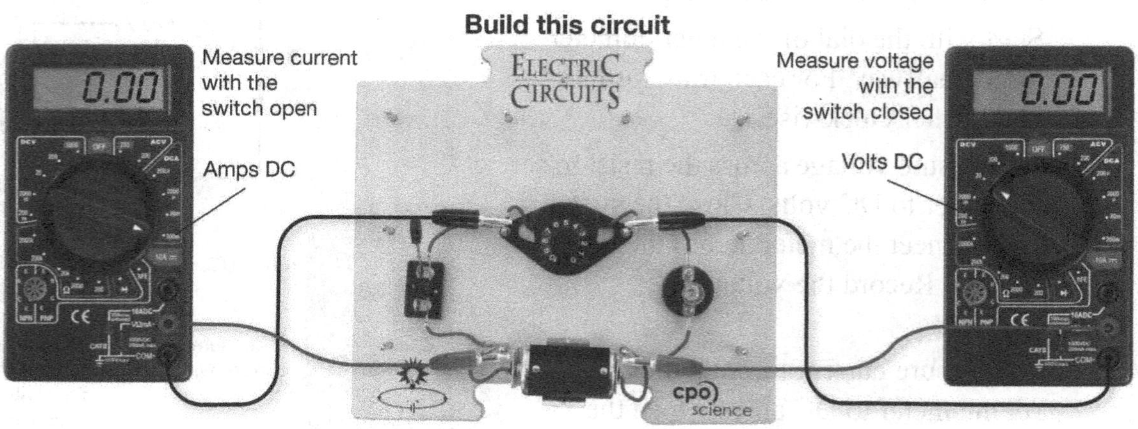

1. Replace the fixed resistor in your circuit with a light bulb.The circuit should contain two batteries, a light bulb, a switch, and a potentiometer.
2. Start with the dial of the potentiometer turned all the way counterclockwise. For each trial, turn the dial further clockwise.
3. To measure voltage across the bulb, set the meter to DC volts. Close the switch and connect the meter across the terminals of the bulb. Record the voltage in Table 2.
4. To measure current through the bulb, set the meter to DC amps. Open the switch and touch the meter leads to the *terminals of the switch.* With the switch open, all the current flows through the meter where it can be measured. Record the current in Table 2.
5. Change the setting of the potentiometer and measure current and voltage again. Repeat for at least six settings covering the whole range of the potentiometer.

Table 2: Voltage vs. current for a bulb

Potentiometer setting	Voltage (V)	Current (A)

5 Analyzing the bulb's current, voltage, and resistance

a. What happened to the brightness of the bulb as you decreased the current?

b. Graph the data from Table 2. Put voltage on the *y*-axis and current on the *x*-axis. Label your *x*- and *y*-axes and title your graph.

c. Compare the shape of the graph for the light bulb to the shape of the graph for the fixed resistor.

d. Use Ohm's law to calculate the resistance of the light bulb for each pair of voltage and current values.

e. How is the resistance of the light bulb related to the amount of current flowing through it?

f. Discuss a possible reason for the change in the bulb's resistance.

20.1 Series and Parallel Circuits

How do series and parallel circuits work?

There are two fundamental ways to connect electrical components in circuits. In a series circuit, the current flows in a single loop and goes through all components. Parallel circuits are complex and not intuitive, so you will need to examine your data carefully to understand how these circuits work.

In this investigation, you will:

- explain the difference between a series circuit and a parallel circuit.
- calculate the total resistance in series and parallel circuits.
- describe what happens to the voltage across each component in a series circuit.
- evaluate the advantages of parallel circuits and series circuits.

Materials List

- Multimeter with red and black leads
- 2 D-cell batteries
- Electric Circuits components (circuit board, potentiometer, switch, 2 battery holders, 3 bulbs and sockets, wires)

1 Voltage and current in series circuits

1. Using two batteries, build the simple circuit with one light bulb and a switch as shown above (left).
2. Set the meter to DC volts. Close the switch and measure the voltage across the bulb by touching the meter's leads to the bulb's terminals. The bulb should be lit. Record the value in Table 1.
3. Set the meter to DC amps. Measure the current by opening the switch and inserting the leads of the meter between the terminals of the switch. Opening the switch forces the current to flow through the meter where it can be measured.
4. Add a second bulb to the circuit as shown above (center). Repeat the voltage and current measurements and record the results in Table 1.
5. Add a third bulb to the circuit (above right). Repeat the voltage and current measurements and record the results in Table 1.

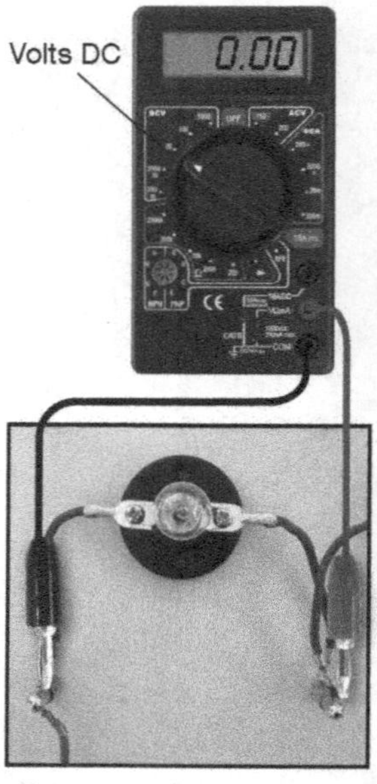

Measure voltage across the terminals of each bulb with the switch closed.

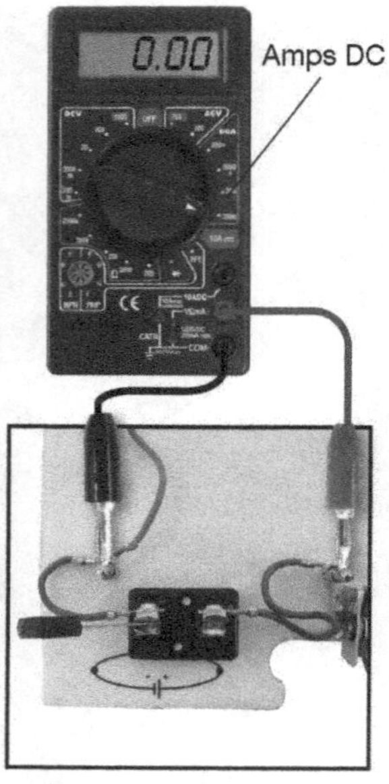

Measure current across the switch terminals with the switch open.

Table 1: Comparing series circuits

	1-bulb circuit	2-bulb circuit	3-bulb circuit
Battery voltage (V)			
Current (A)			
Bulb 1 voltage (V)			
Bulb 2 voltage (V)			
Bulb 3 voltage (V)			

2 Using your model

a. Use the data in Table 1 to explain the relationship between the battery voltage in a circuit and the voltage across the bulb or bulbs connected in series.

b. What did you notice about the brightness of the bulbs as you increased the number of bulbs in the circuit? Why do you think this is? How do the measurements you made support your hypothesis about why the bulbs changed in brightness?

c. What happens to the other two bulbs when one bulb is removed from the three-bulb circuit? Explain why the circuit behaves as it does.

3 Determining the total resistance of resistors in series

a. You will use two 5-ohm resistors and one 10-ohm resistor during this investigation. Predict what the total resistance of the circuit will be when all three resistors are connected in series.

b. Predict the circuit's total current when three volts are applied using two batteries.

Build this circuit.

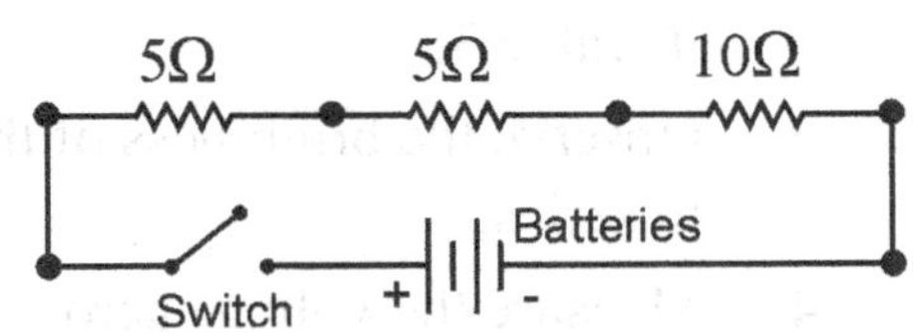

4 Measuring the current and resistance in series circuits

1. Build the circuit in the diagram in Part 3. Use two 5-ohm resistors and one 10-ohm resistor.
2. Measure the battery voltage applied to the circuit. Record the value in Table 2.
3. Measure the current through the circuit. Record the value in Table 2.

Table 2: Current and resistance data for series circuits

Current (amps)	Voltage (V)	Resistance (ohms)

a. Use Ohm's law (in the form $R = V/I$) to calculate the total resistance in the circuit from the combined battery voltage (Table 1) and the current for the circuit (Table 2). Record your results in the bottom row of Table 2.

b. Compare your measurements with your predictions from Part 3. Explain the effect that adding resistors has on the current flowing in a series circuit.

5 Parallel circuits

A parallel circuit has more than one branch so that current can divide and flow "in parallel" among the branches of the circuit.

1. Build a circuit with two batteries, a switch, and three bulbs as shown in the diagram below.

Build this circuit

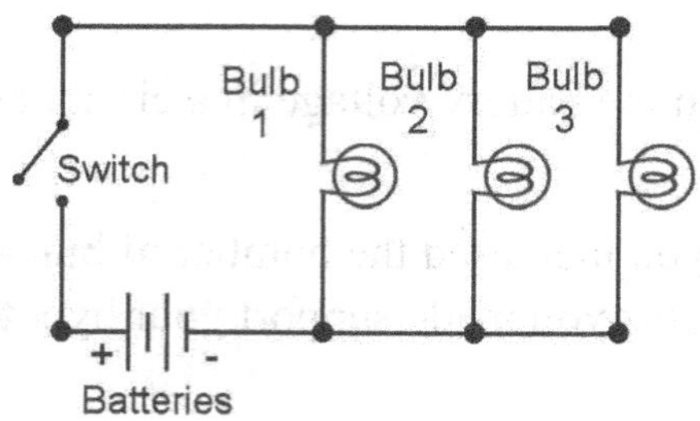

2. Set the meter to DC volts. Close the switch and measure the voltage across the battery by touching the leads of the meter to the terminals of the battery. All three bulbs should be lit. Record the value in Table 3.
3. Observe the brightness of the bulbs compared with their brightness when all three were connected in series.
4. Measure the voltage across each bulb by touching the leads of the meter to the terminals of each bulb separately. Record the values in Table 3.
5. Set the meter to DC amps. Measure the total current in the circuit by opening the switch and inserting the leads of the meter between the terminals of the switch. This forces the current to flow through the meter where it can be measured.

Table 3: Comparing parallel circuits

	Total circuit	Bulb 1	Bulb 2	Bulb 3
Voltage (*V*)				
Current (*A*)				
Resistance (Ω)				

6 Comparing series and parallel circuits

a. Use Ohm's law and the total circuit voltage and current from Table 3 to calculate the total resistance of the circuit with three bulbs in parallel. Record the result in the last row of Table 3.

b. Use Table 4 to summarize the total circuit current and voltage measurements from the single-bulb circuit and the two different three-bulb circuits (series and parallel).

Table 4: Comparing series and parallel circuits

	Single-bulb circuit	3-bulb series circuit	3-bulb parallel circuit
Current (A)			
Voltage (V)			
Resistance (Ω)			

c. Compare the total resistance of the single-bulb circuit with the total resistance of the three-bulb series circuit. Which circuit resistance is greater? Why do you think this is?

d. Compare the total resistance of the single-bulb circuit to the total resistance of the three-bulb parallel circuit. Which circuit resistance is greater? Why do you think this is?

e. How are the branch currents in the parallel circuit related to the total current supplied by the batteries? Explain how you arrived at your answer.

f. Compare the total current in the single-bulb circuit, the three-bulb series circuit, and the three-bulb parallel circuit. Is there an approximate relationship between the currents that also agrees with the observed brightness of the bulbs in each circuit?

g. If you wish to brightly light three bulbs with one battery, should you connect them in series or parallel?

h. Calculate the total current that should flow when three resistors (of 5, 5, and 10 ohms) are connected in parallel as shown. HINT: Does your meter have an internal resistance when it is measuring DC amps? How could you find out? How would this affect the current in the circuit while it is being measured? How should any extra resistance affect your prediction? Build and test the circuit to evaluate your prediction. How close was your prediction to the actual measurements? If your meter does have an internal resistance, redraw the diagram to include your meter and its resistance.

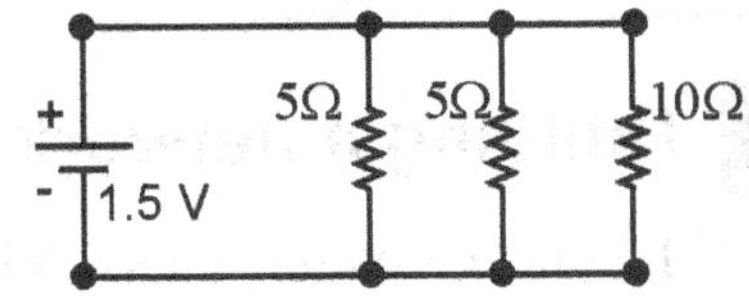

20.2 Analysis of Circuits

How do we analyze network circuits?

In the previous investigation, you worked with series and parallel circuits. When devices are combined in both series and parallel in a single circuit, a network circuit is created. Network circuits require a more careful examination of the exact placement of all circuit elements in the circuit as a whole.

In this investigation, you will:

- build a network circuit.
- determine the total resistance of a three element resistor network circuit.

Materials List

- Multimeter with red and black leads
- 2 D-cell batteries
- Electric Circuits components (2 5-ohm resistors, 1 10-ohm resistor, 1 20-ohm resistor, 2 battery holders, wires, a switch, 3 bulbs and sockets)

1 Determining the resistance of each resistor

You will use four resistors in this investigation: two 5-ohm, one 10-ohm, and one 20-ohm. Label the 5-ohm resistors R1 and R2, the 10-ohm resistor R3, and the 20-ohm resistor R4.

Use the multimeter to measure the resistance of each resistor and enter the results in Table 1.

Measuring resistance

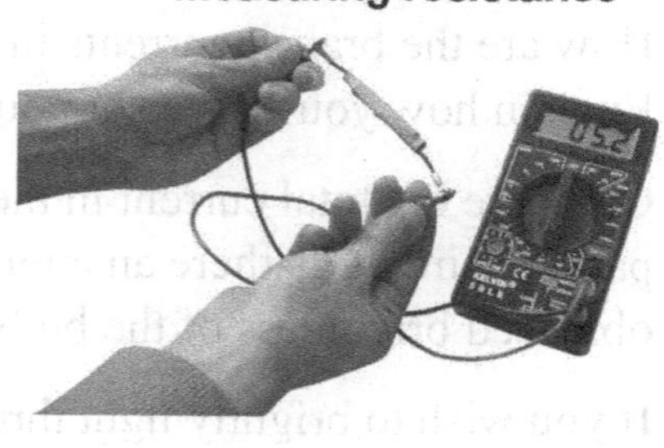

Table 1: Measured resistance values

	R1	R2	R3	R4
Resistance (Ω)				

2 Building a network circuit

1. Use a battery; a switch; resistors 1, 2, and 3; and as many wires as necessary to construct the network circuit shown in the diagram at right.
2. Measure the voltage across the battery. This is the total circuit voltage. Then measure the voltage across each of the resistors. Record your measurements in Table 2.
3. Measure the current through each resistor. Record your measurements in Table 2. You have to break the circuit temporarily to insert the meter and make each current measurement. Be sure the meter is set to DC amps.

Build this circuit.

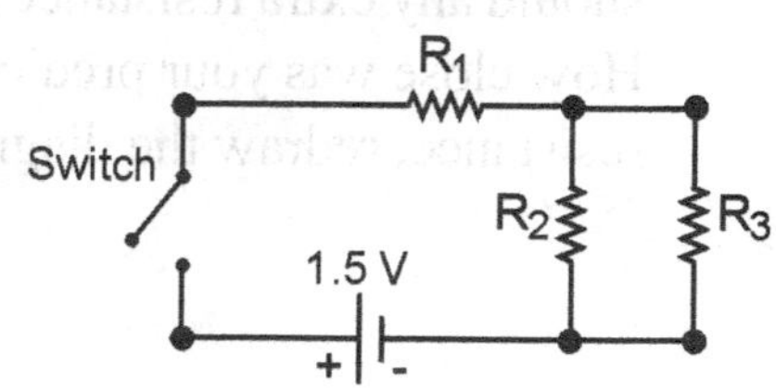

3 Analyzing the circuit

Table 2: Voltage and current data

	Battery (total circuit)	R1	R2	R3
Voltage (volts)				
Current (amps)				

a. Which two resistors are connected in parallel?

b. Which resistor is in series with the other two?

c. How does the voltage across the parallel resistors (R_2 and R_3) compare?

d. How does the voltage across the parallel resistors relate to what you learned about voltages in a parallel circuit in the previous investigation?

e. Voltage is the measure of the amount of energy carried by the current in a circuit. Trace the path of the current from the positive end of the battery, through the circuit, to the negative end of the battery. There are two possible paths because of the parallel part of the circuit, but each path will only pass through two of the resistors. Does the amount of energy lost through the resistors equal the amount of energy supplied by the battery? Is this the case for each possible path? Explain.

f. How does the current flowing through R_1 compare with the current through R_2?

g. How does the current coming out of the battery compare with the sum of the currents flowing through R_2 and R_3? Explain this relationship.

h. Use what you know about series and parallel circuits to calculate the theoretical total resistance of the circuit. The formulas for finding the total resistance are:

Series circuit	Parallel circuit
$R_{tot} = R_1 + R_2$	$\frac{1}{R_{tot}} = \frac{1}{R_1} + \frac{1}{R_2}$

i. Now calculate the total resistance of the circuit using Ohm's law, the battery voltage, and the total circuit current you measured.

j. How does the total resistance calculated using Ohm's law compare with the theoretical total resistance found in h above?

4 Predicting the effect of changing a resistor

a. Replace the10-ohm resistor (R_3) with the 20-ohm resistor (R_4). Use what you have learned about network circuits to predict the total circuit resistance and total circuit current. Show the process you used to make your predictions.

b. Measure the voltage across the battery and each resistor and the total current in the circuit. Use Ohm's law to find the total circuit resistance. You will need to make a data table similar to Table 2.

c. How did the predicted values compare with the measured ones?

5 A circuit puzzle

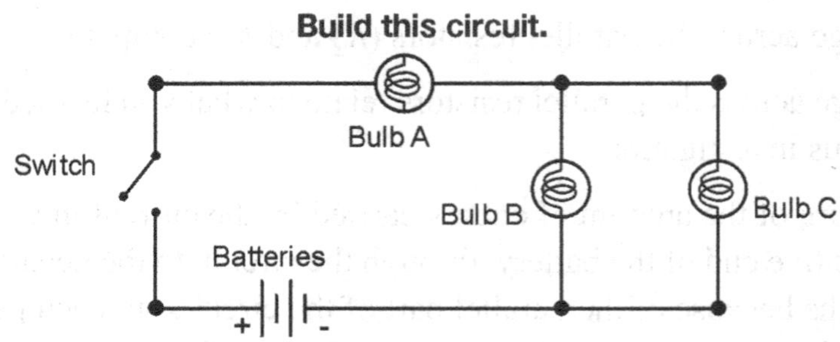

1. Build the circuit shown in the diagram above with 2 batteries, a switch, and three bulbs.
2. Turn the switch on and observe the brightness of each bulb.
3. Temporarily interrupt the current to bulb B by disconnecting one wire from the terminal post.
4. Observe how the brightness of the other two bulbs changes.

a. When bulb B is disconnected, does bulb A get dimmer, brighter, or stay the same?

b. When bulb B is disconnected, does bulb C get dimmer, brighter, or stay the same?

c. Use what you know about series and parallel circuits to propose an explanation for what you observed.

6 *Challenge*: Analyze a four-resistor network circuit

a. Build the circuit shown in the diagram at right.

b. Use what you have learned about network circuits to predict the total circuit resistance and current. Show the process you used to make your predictions.

c. Use the meter to measure the voltage across the battery and each resistor and the total circuit current. Use Ohm's law to find the total circuit resistance.

d. How did the predicted values compare with the measured ones?

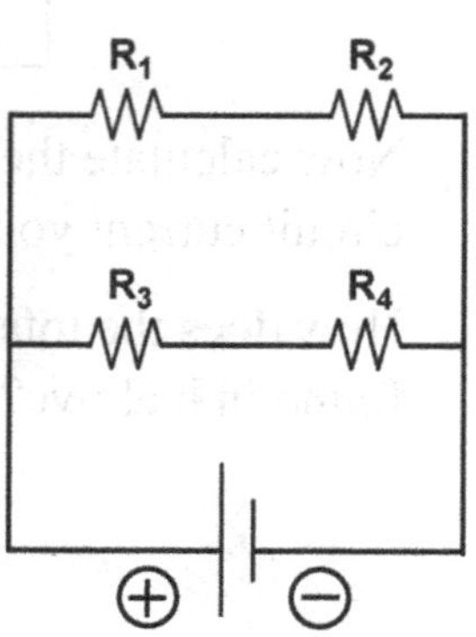

20.3 Electric Power, AC, and DC Electricity

How much does electricity cost?

You have learned how to measure three electrical quantities: voltage, current, and resistance. In this investigation, you will learn about a fourth quantity, power, which you have already studied in the context of mechanical systems (Chapter 11). You will find the power ratings of electrical appliances and use this information to estimate electricity costs.

In this investigation, you will:

- explain power in terms of energy and time.
- calculate power in a DC circuit when given the current and voltage.
- rank various household appliances by the amount of power they use.
- estimate the cost per month of using a common household appliance.
- use dimensional analysis to show that a kilowatt hour is a unit of energy—not a unit of power.

Materials List

- Two or three small appliances labeled with power ratings, such as an iron, toaster oven, electric drill, desk lamp, or hair dryer
- Calculator

1 Find the power rating of home appliances

You will need to complete the first part of this investigation at home. Your assignment is to find five electrical appliances that have a label with the device's power rating in watts (W) or kilowatts (kW). Some appliances you might look at are a blender, coffee maker, toaster oven, microwave, television, hair dryer, space heater, room air conditioner, or computer. The rating is often stamped on the back or the bottom of the appliance. In the example below, the power rating is 1.05 kW.

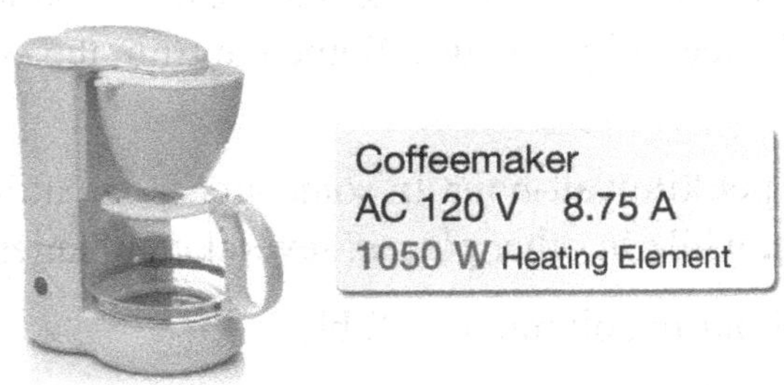

a. Fill out the first two columns of Table 1 with the power rating of each appliance. The value should be in kilowatts. To convert watts to kilowatts, divide the number of watts by 1,000. For example, 1,500 watts is equal to 1,500 ÷ 1,000, or 1.5 kilowatts.

b. Estimate the number of hours the device is used each month. Assume that one month equals 30 days. If your coffee maker is used for a half hour each morning, you would calculate a monthly usage of 15 hours (0.5 hr/day × 30 days).

Table 1: Power rating, usage, and cost of household appliances

Appliance	Power rating in kW	Est. usage hrs. per mo.	kWh per mo.	Price per kWh	Total cost per mo.

2 Estimate the number of kilowatt-hours each appliance uses in a month

a. Multiply the power rating in kilowatts (from column 2) by the number of hours the appliance is used each month (column 3). For example, if you use a 1.4-kilowatt microwave oven for 22 hours a month, multiply 1.4 times 22. Write your answers in column 4 of Table 1 as shown in the sample below:

Appliance	Power rating in kW	Est. usage hrs. per mo.	kWh per mo.	Price per kWh	Total cost per mo.
Microwave	1.4 kW	22 hours	30.8		

3 Determine the monthly cost of using your appliances

Utility companies charge consumers for the number of kilowatt-hours of electricity they use each month. Many houses and apartments have a meter attached to the outside of the building. The meter uses a system of spinning disks to record how much electricity you use. Someone from the electric company reads the meter each month.

a. Find out how much you pay per kilowatt-hour. In some areas, one utility company provides all the electricity to an entire region, while in other places, several companies compete for customers.

b. Write the price per kilowatt-hour in column 5 of Table 1.

c. Calculate how much your household spends to operate each appliance during one month. Multiply the kilowatt-hours per month by the price per kilowatt-hour to determine your cost and record in column 6.

4 Analyze your data

a. Compare your results with those of the other members of your group. List the three appliances that had your group's highest power ratings in Table 2.

b. Think about the function of each appliance listed in Table 2. What kind of work is being done? In other words, electrical energy is being converted into what other type(s) of energy?

c. Do you see any similarities in the kinds of work being done by the three appliances in Table 2? If so, what are these similarities?

d. Suggest one practical way you or another group member could reduce your electricity bills.

e. Discuss the effect of climate on electricity use. What climate factors might influence the month that has the peak electricity use in your area?

f. Name one other factor not related to climate that may influence which month has the highest electricity use in your area.

Table 2: Appliances with the highest power ratings

Appliance	Power rating in kilowatts

5 What do you buy from the electric utility company?

People often use the phrase "power plant" to refer to their local electric (or public utility) company. You may have heard people say that electric companies "sell power" to their customers or that there was a "power shortage" in a particular area. Let's take a look at these phrases from a scientific perspective. What exactly do electric companies sell?

1. Electricity bills are calculated based on the number of kilowatt-hours used per month.
2. Use dimensional analysis to convert one kilowatt-hour into units of watt-seconds:

$$1 \text{ kilowatt•hour} \times \frac{1000 \text{ watts}}{\text{kilowatt}} = 1000 \text{ watt•hours}$$

$$1000 \text{ watt•hour} \times \frac{3600 \text{ seconds}}{\text{hour}} = 3{,}600{,}000 \text{ watt•seconds}$$

3. One watt of power is equal to one joule of energy flowing per second. Substitute units of joules per second for watts in the measurement for watt-seconds:

$$\text{watt - seconds} = 3{,}600{,}000 \text{ joules/second} \times \text{seconds}$$

a. Which units in the last equation cancel?

b. After simplifying the units in the last equation, what fundamental unit remains?

c. Is the remaining unit a measure of energy, work, or power?

d. Do electric companies sell energy, work, or power?

21.1 Electric Charge

How do electric charges interact?

If you have ever felt a shock when touching a doorknob or removing clothes from the dryer, you have experienced the effect of electric charge. Electric charge is a fundamental property of matter. All ordinary matter contains both positive and negative charge. You do not usually notice the charge because most matter has equal amounts of both types of charge. Interesting things can happen when objects lose or gain electric charges, as you will see in this investigation.

In this investigation, you will:

- triboelectrically charge different materials.
- use a triboelectric series to make predictions about charged objects.
- make an electrophorus and explain how it works.

Materials List

From Static Electricity kit:

- Static electricity box
- Piece of wool
- Modeling clay

Other materials:

- Puffed rice
- Aluminum foil
- PVC tube (from wave tray)
- Aluminum block (from wave tray)

1 Triboelectric Series

A tiny imbalance in either positive or negative charge on an object is the cause of static electricity. If two neutral objects are rubbed together, charges may be rubbed off one object and onto the other. This is known as *triboelectric charging*, and it happens to clothes in the dryer and to your socks when you walk on a carpet. A triboelectric series lists materials by how likely they are to hold on to their electrons or give up their electrons. The first items in the list at right tend to give up their electrons and become positive when triboelectrically charged. The last items in the list tend to hold on to their electrons and become negative. If you allow interaction between two materials that are several items away from each other on the list, you could end up with triboelectrically-charged objects. Have you ever rubbed a balloon on your hair? Some types of hair tend to give up electrons to the balloon. When this happens, your hair is positively charged and each strand repels the next, so you have wild hair that sticks up in all directions! Also, the negatively-charged balloon can attract neutrally-charged objects like bits of paper or dust. This is especially dramatic on a dry day, because humidity makes triboelectric charging more difficult.

Triboelectric series

Most Positive
air
human hands, skin
glass
human hair
wool
silk
aluminum
paper
steel
wood
balloon
copper
acrylic
cellophane tape
polyvinyl chloride (PVC)
Most Negative

1. Triboelectrically charge the PVC tube by rubbing one end of the PVC tube vigorously with the piece of wool, at least 20 times.
2. Dip the wool-contacted end into a small dish of puffed rice.

a. Describe what happened when you dipped the PVC tube into the dish of puffed rice.

b. Study the triboelectric series list. Which material, the PVC or the wool, tends to hold onto electrons? Which material tends to give up electrons? Based on this information, what charge builds up on the PVC tube?

c. Using what you know about the charge buildup on the PVC tube, explain why the puffed rice acts the way it does when you dip the PVC tube in the dish.

d. If you wait patiently after dipping the PVC tube into the dish of puffed rice, you will notice something surprising happen to some of the puffed rice that collects on the tube. What happens? What must happen, in terms of charge, for this effect to occur?

2 More triboelectric charging

1. Scatter a dozen or so pieces of puffed rice on your table or desk.
2. Turn the acrylic static electricity box upside-down over the puffed rice.

a. Predict what will happen, if anything, to the puffed rice if you rub the static electricity box with a piece of wool.

3. Create a negative charge buildup on the box by vigorously rubbing it with a piece of wool. Rub the wool over the box quickly at least 100 times. The effect is worth it!
4. Remove the wool and observe the puffed rice for one to two minutes.

b. Describe what happens to the puffed rice.

c. Explain the effect you observe in terms of positive and negative charge.

5. Run your finger over the upside-down tray.

d. Describe what happens, and explain it in terms of charge (refer to the triboelectric series).

6. Wrap and cover a few pieces of puffed rice with a tiny amount of aluminum foil. Add these to the scattered puffed rice already under the tray.
7. Repeat steps 3 and 4.

e. Did the aluminum-covered puffed rice behave differently than the plain puffed rice? Describe what you saw and explain what happened in terms of charge.

f. Can you replace the puffed rice with anything else and see the same effect? Try several different things and describe the results.

3 Investigating with an electrophorus

An electrophorus is a device invented over 200 years ago. A simple electrophorus has a charging surface and a metal plate with an insulating handle. The metal plate becomes charged and you can carry the charge to various places to perform static-electricity experiments. Follow the steps below to make an electrophorus, and then use the triboelectric series and your knowledge of how charges interact to explain how the electrophorus works.

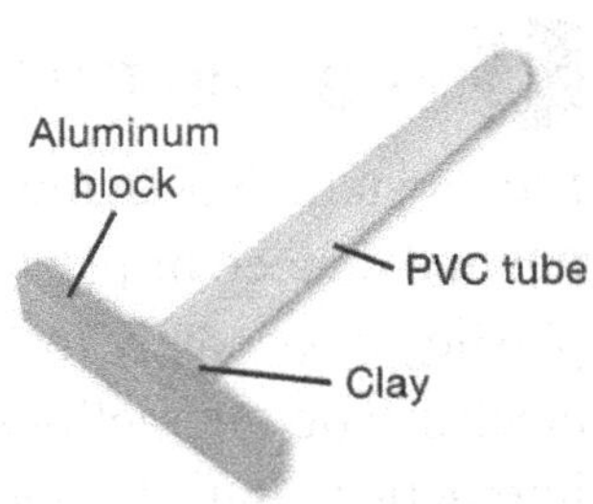

1. Attach the PVC tube to the aluminum block with some modeling clay (see photo at right). The PVC tube is an insulating handle so you can move the aluminum block from place to place without touching it.

2. Vigorously rub a piece of wool over the upside-down static electricity box at least 100 times. This is the charging surface for the electrophorus.
3. Hold the PVC tube handle and place the aluminum block down on the area of the box that you rubbed with the piece of wool. Touch a finger to the aluminum block next to the handle (see photo at right).

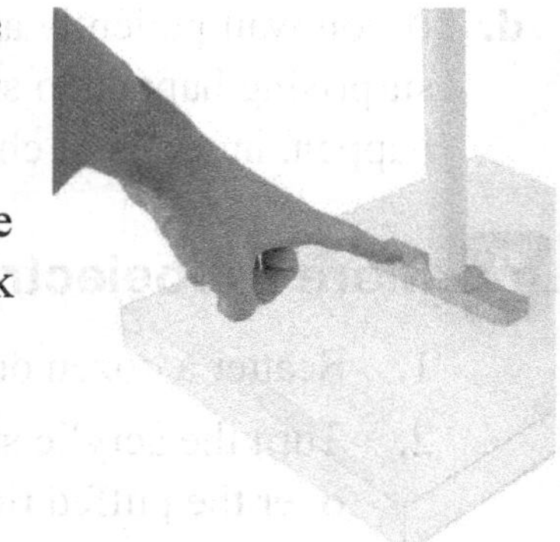

4. Pick up the aluminum block by the handle. Hold your knuckle near the aluminum block and observe what happens. If you can do this in a dark area of the room, it will be more dramatic.
5. Repeat step 3. Use the aluminum block to conduct static electricity experiments of your own. Each time you discharge the aluminum block, you can re-charge it by repeating step 3 (you won't have to repeat step 2 every time).

a. What charge does the box have when you are finished rubbing it with the wool? Explain.

b. Aluminum is a good conductor of electrons. When the aluminum block is first placed on the charged acrylic tray, the electrons tend to move to one area of the aluminum block, leaving one area positively charged and one area negatively charged. When you touch the aluminum block with your finger, the electrons will follow the conductive path to your finger and *leave the block*. With this information, complete the following sequential explanation of how the electrophorus works:

- When the aluminum block is placed on the charged acrylic tray, the electrons move to the _____(top or bottom) of the block, leaving the _____ (top or bottom) of the block positively charged.
- At this point, the aluminum block is still neutral, since no charge has entered or left the block.
- When you place your finger on the aluminum block, the _____ (electrons or protons) are conducted away from the block to your finger, leaving the aluminum block with a _____ (positive or negative) charge.
- When you bring the charged aluminum block near a conductor like a piece of metal or one of your fingers, electrons move from ______ to ______, causing the aluminum block to become neutral again.

c. Why can you recharge the aluminum block on the charged acrylic tray without having to rub wool over the acrylic tray again?

d. Why does the electrophorus work better in a dry environment than it does in a humid environment?

e. *Illustrate* how an electrophorus works, step by step. Show charged areas of objects.

4 Arguing from evidence

Electrostatic interactions are used quite a bit in industry, technology, products, and research. When you unwind a piece of plastic wrap from a roll, tear it off, and place it over a glass bowl, the negatively-charged plastic sticks nicely to the glass (check out the triboelectric series to verify). There are cases when electrostatic interactions are beneficial, like with the plastic wrap, and there are instances when electrostatic interactions must be avoided or reduced.

a. Do your own research to briefly describe one example of a beneficial electrostatic interaction, and a case when the interaction is avoided or reduced. Be sure to cite your sources.

21.2 Coulomb's Law

How strong are electrical forces?

All objects contain charges that exert forces on each other. Like charges repel and unlike charges attract. Objects normally contain equal numbers of positive charges (protons) and negative charges (electrons). You transferred charge from one object to another in the previous investigation. Coulomb's law describes the strength of the force between two charges.

In this investigation, you will:

- investigate the relationship between charge, distance, and force using charged pith balls.
- use Coulomb's law to calculate the force between two charged objects.

Materials List

From Static Electricity kit:

- Static electricity box
- Piece of wool
- Pith balls attached to string
- Thread

Other materials:

- Physics Stand
- Spinner
- Flashlight
- Rubber bands
- Tape (clear)
- White paper
- Miscellaneous objects

1 Investigating forces between charged objects

Pith is the cork-like meaty part of a plant stem. Pith balls are useful in electrostatic experiments because they are very lightweight and can easily be coated with a conductive material, such as aluminum paint. In this investigation, you will use aluminum-coated pith balls that are already attached to a short piece of string. You will set up an apparatus that allows you to view the pith balls' shadows. You will transfer charges to the pith balls and use the shadows to measure changes that occur.

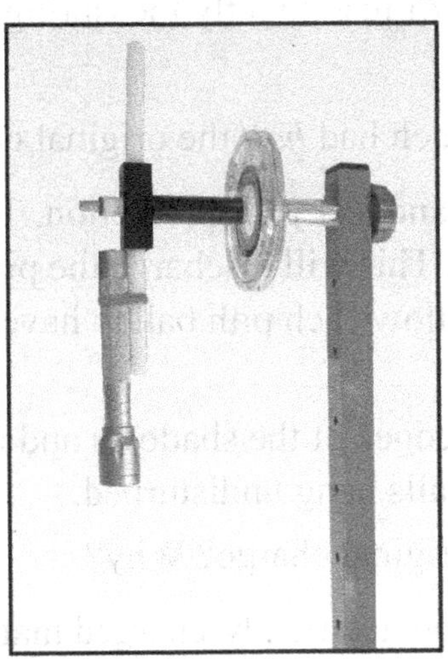

Secure flashlight to spinner arm.

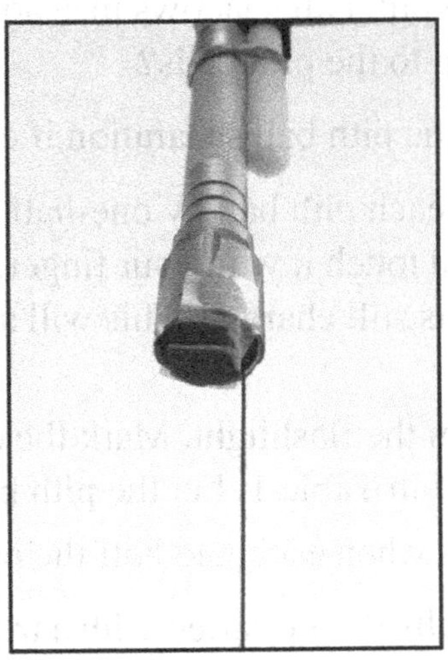

Tape the threads to flashlight.

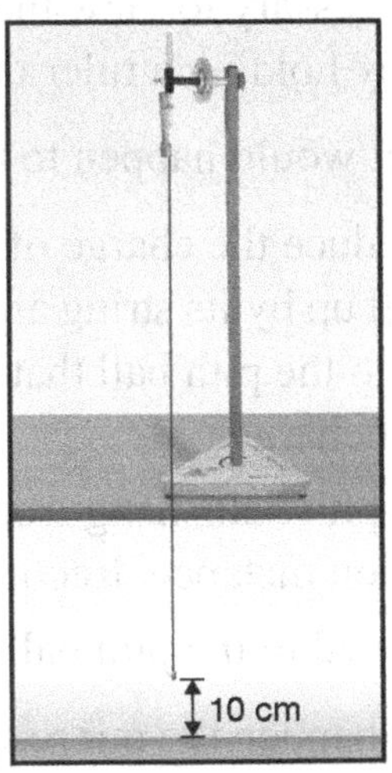

Complete setup.

1. Attach the spinner to the top of the physics stand. Carefully secure the flashlight to one spinner arm with several rubber bands. The flashlight will point down at the floor.
2. Tie the free end of each pith ball string to long pieces of thread. The complete length of string/thread attached to the pith balls should be about 1.5 m.

3. Tape the free ends of both pieces of thread to the flashlight LED cover. The pith balls will hang from the flashlight and will be touching each other.
4. Place the physics stand on a table so the pith balls hang to a point about 10 cm above the floor. Adjust the length of thread if necessary.
5. Put a plain piece of white paper on the floor beneath the pith balls.
6. Turn on the flashlight. You should see clearly-defined shadows of the pith balls on the white paper. If necessary, darken the room somewhat, and adjust the spinner arm so the flashlight beam points straight down.
7. Triboelectrically charge the static electricity box with a piece of wool. Vigorously rub the wool over the box at least 100 times for the best results.

a. What charge does the box now have, based on the triboelectric series table from the previous investigation?

8. Now it is time to transfer the charge to one of the pith balls. Hold a pith ball by its string a few centimeters away from where the pith ball is attached. Touch it to the static electricity box for several moments. You will notice that it is immediately attracted to the box, and then it will fly away and hover.
9. Keep the pith balls separated. Touch the uncharged pith ball to make sure it is neutral. Next, touch the charged pith ball to the uncharged pith ball. This will allow each pith ball to be equally charged.
10. Let the pith balls hang from the flashlight. Mark the edges of the shadows and measure the separation distance. Record in Table 1. Let the pith balls hang undisturbed.

b. Do the pith balls have a positive charge or a negative charge? What happens to the position of the pith balls? Why?

c. Why is it necessary to view the pith ball shadows instead of just directly measuring the pith ball separation by holding a ruler up to the pith balls?

d. Predict what would happen to the pith ball separation if each had *half* the original charge.

11. Let's reduce the charge of each pith ball by one-half and test your prediction. To do this, hold one pith ball up by its string and touch it with your finger. This will discharge the pith ball. Next, touch it once to the pith ball that is still charged. This will allow each pith ball to have about one-half the original charge.
12. Let the pith balls hang from the flashlight. Mark the edges of the shadows and measure the separation distance. Record in Table 1. Let the pith balls hang undisturbed.

e. What happened to the pith balls when each had half the original charge? Why?

f. What might happen to your results if you started with a triboelectrically-charged material other than the static electricity box? Explain your reasoning.

13. Repeat steps 7–12 with at least two different triboelectrically-charged objects other than the box. For example, you could try a balloon, the PVC tube from the wave box, or an old vinyl record album. You can use the wool to triboelectrically charge each one. Record your results in Table 1.

g. Summarize your results in a paragraph. Explain your results using the terms *positive charge*, *negative charge*, *force*, and *distance*.

Table 1: Pith ball results

Triboelectrically charged object	Charged pith ball trials	Pith ball separation (cm)
Static electricity box	1 (full charge)	
	2 (1/2 charge)	
	1 (full charge)	
	2 (1/2 charge)	
	1 (full charge)	
	2 (1/2 charge)	

2 Electrical forces in a penny

COULOMB'S LAW

$$F = K\frac{q_1 q_2}{R^2}$$

Force (N); Constant (9×10^9 N·m²/C²); Charges (C); Distance (m)

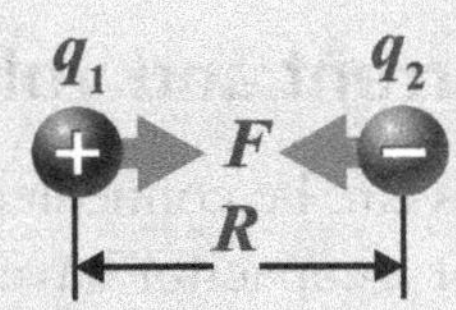

Pennies are made of copper and zinc. A single penny is made of approximately 2×10^{22} atoms that contain 7×10^{23} electrons and an equal number of protons.

a. A single proton or electron has a charge of 1.602×10^{-19} coulombs. How many coulombs of positive charge are in one penny? How many coulombs of negative charge?

b. Imagine you have two pennies, A and B. You remove all of the electrons from penny A and place them on penny B. Is penny A positively or negatively charged? Is penny B positively or negatively charged?

c. Calculate the force of attraction between the two pennies if they are held one meter apart.

d. A diesel locomotive weighs approximately one million newtons. Determine the number of locomotives necessary to have a weight equal to the strength of the force between the pennies.

e. You were probably surprised when you calculated the force between the pennies. What does your answer tell you about the amounts of charge normally transferred in activities such as brushing your socks on carpet or rubbing a balloon on your hair?

21.3 Capacitors

How does a capacitor work?

A capacitor is a device that stores charge. Electric current flowing into a capacitor charges it. Once it is charged, current can flow out until the capacitor is discharged. If you have used a camera with a flash, you have used a capacitor. You might have noticed that after you use the flash once, you have to wait a few seconds before using it again. Today, you will find out how and why that happens.

In this investigation, you will:

- measure the voltage across a capacitor as it charges and discharges.
- explain how the current in a circuit with a capacitor changes with time as it charges and discharges.
- describe the relationship between the circuit resistance and the charging rate of a capacitor.

Materials List

- 2 D-cell batteries
- Electric Circuits components: 2 battery holders, 2 switches, wires
- Circuits RC pack: 1 470-μF capacitor, 1 each of 22-kΩ, 47-kΩ, and 100-kΩ resistors
- DataCollector
- Voltage sensor

1 Current and voltage for a capacitor

Capacitors can be connected in series or parallel circuits, just like resistors. Unlike resistors, the current and voltage for a capacitor have a more complex relationship. The current into or out of a capacitor depends on three things:

- the amount of charge already in the capacitor, measured by its voltage;
- the voltage applied to the capacitor by the circuit; and
- any resistance that limits the current flowing in the circuit.

2 Setting up the investigation

1. Open your switches if they are not already open. Build a circuit with the two open switches, two batteries, a 470-μF capacitor, and a 47-kΩ resistor (diagram at right).
2. Connect the DataCollector to a power source and turn it on. Connect the voltage probe.
3. You will see the Home screen. Select Data Collection mode.

Build this circuit.

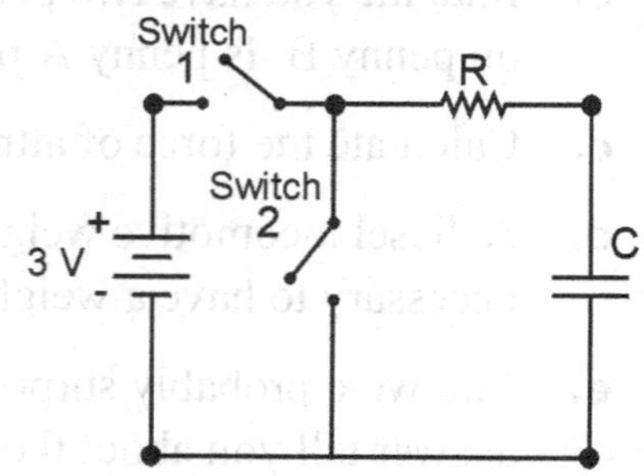

Home screen

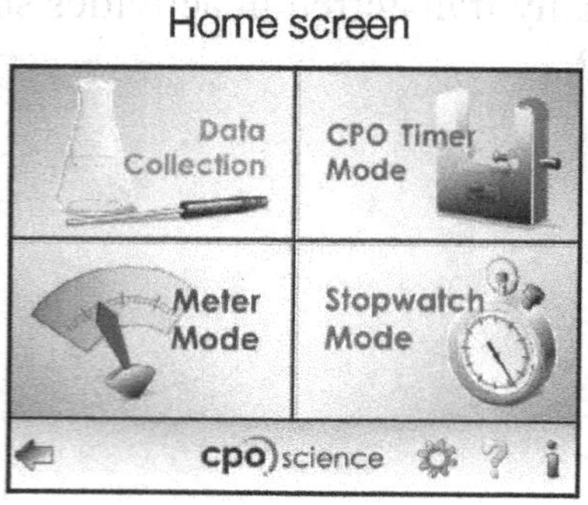

4. Connect the voltage probe across the terminals of the capacitor as shown in the diagram below. You will use the voltage probe to measure the voltage across the terminals of the two batteries.

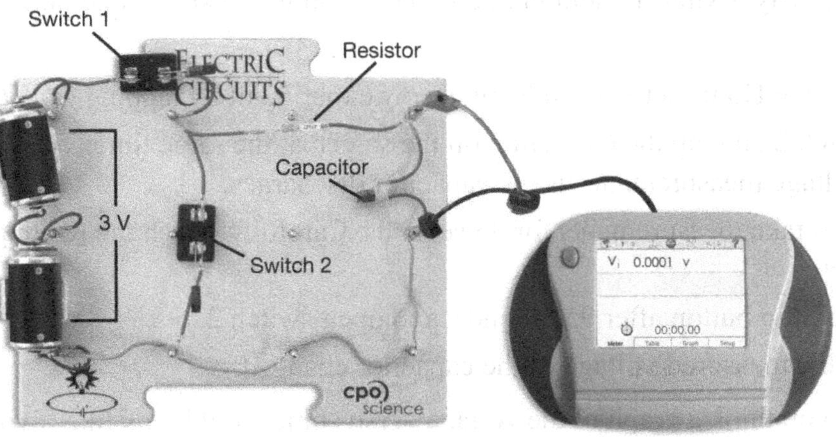

a. Describe the flow of current when switch 1 is closed and switch 2 is open. This is the *charging* circuit for the capacitor.

b. Describe the flow of current when switch 2 is closed and switch 1 is open. This is the *discharging* circuit for the capacitor.

Safety Note: DO NOT CLOSE SWITCH 1 AND 2 AT THE SAME TIME! This will create a short circuit.

3 Charging the capacitor

Collecting data for this investigation requires that you pay close attention and act quickly. Read the instructions below *before* you close switch 1 so you know exactly what you are doing. Switch 2 should remain open.

1. Close switch 1 and tap the Go button on the screen at the same time. This will begin taking and storing voltage measurements as the capacitor charges.
2. Continue to take measurements for 95 seconds. Carefully watch the readings as they are being taken.
3. Hit the Go/Stop button after 95 seconds, and open switch 1.

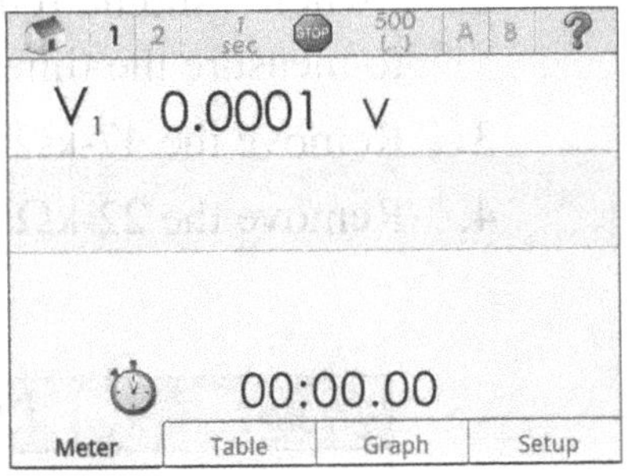

a. Describe how the measured voltage of the capacitor changed.

b. Describe what you think a graph of the voltage versus time would look like if voltage was on the *y*-axis and time was on the *x*-axis. What shape do you think it would have? Would it be linear?

c. Tap the Graph tab on the bottom of the DataCollector's screen and examine the graph. How does your prediction compare to the graph made by the DataCollector?

d. If you were to continue charging the capacitor for a long time, would the voltage reach a maximum value or continue to climb steadily? Your graph should help you answer this question.

e. Suppose you wanted to charge your capacitor to the full battery voltage. Would this take twice as long as the time to reach half the battery voltage? Why or why not?

4 Discharging the capacitor

For this part of the activity, switch 1 should remain open. You will take voltage measurements with switch 2 closed.

1. Go back to the Home screen and begin a new experiment by tapping Data Collection mode.
2. Close switch 2 and tap the Go button on the screen at the same time. This will begin taking and storing voltage measurements as the capacitor discharges.
3. Continue to take measurements for 95 seconds. Carefully watch the readings as they are being taken.
4. Hit the Go/Stop button after 95 seconds, and open switch 2.

a. Describe how the measured voltage of the capacitor changed.

b. Describe what you think a graph of the voltage versus time would look like if voltage was on the *y*-axis and time was on the *x*-axis. What shape do you think it would have? Would it be similar to the graph made during charging? If not, how would it be different?

c. Tap the Graph tab on the bottom of the DataCollector's screen and examine the graph. How does your prediction compare to the graph made by the DataCollector?

d. Imagine you had measured the current in the circuit while charging or discharging the capacitor. Would the current have been constant or would it have changed over time? Sketch the shape of the current vs. time graphs you think you would have measured during charging and discharging.

e. *Challenge!* Both a capacitor and a battery can create voltage differences and drive electric current. Give at least three electrical differences between a capacitor and a battery.

5 The effect of changing the charging or discharging current

1. Touch a wire across the terminals of the capacitor with both switches open. This drains any charge immediately and restores the capacitor voltage to zero. Begin a new experiment by going back to the Home screen and tapping Data Collection again.
2. Switch 2 should remain open. Close switch 1 and use the time display on the bottom of the screen to measure the time it takes the capacitor to charge up to 1.5 volts. Record the time in Table 3.
3. Remove the 47-kΩ resistor and replace it with a 22-kΩ resistor. Repeat steps 1 and 2.
4. Remove the 22-kΩ resistor and replace it with a 100-kΩ resistor. Repeat steps 1 and 2.

Table 3: Charge time data

Resistance (Ω)	Time to reach 1.5V (s)

a. Describe the effect of changing the resistor in the circuit during charging and discharging.

b. How does the current compare for the three different resistors?

c. Explain why the time it took the capacitor to reach 1.5 volts was different for each of the three resistors. Use your knowledge of current, resistance, and capacitors.

22.1 Properties of Magnets

How do magnets interact with each other?

In this investigation, you will experiment with permanent magnets to learn more about magnets, magnetic forces, and magnetic fields. The magnets you will use may not look like other magnets you have seen. The poles are on the opposite faces, as shown in the picture at right.

In this investigation, you will:

- name the properties of a permanent magnet.
- describe and measure the forces that magnets exert on each other.
- sketch magnetic fields.

Materials List

- 10 ceramic magnets (from Electric Motor)
- Navigational compass
- Metric ruler

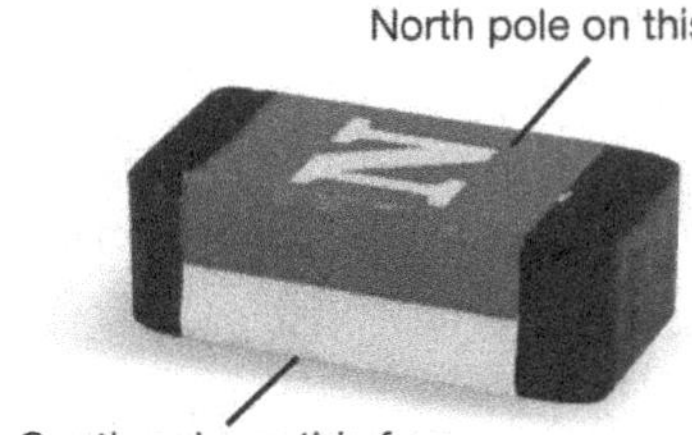

1 Describing the forces that two magnets exert on each other

1. Hold two magnets one centimeter apart with their south poles facing each other. What force do you observe between the magnets—attractive or repulsive?

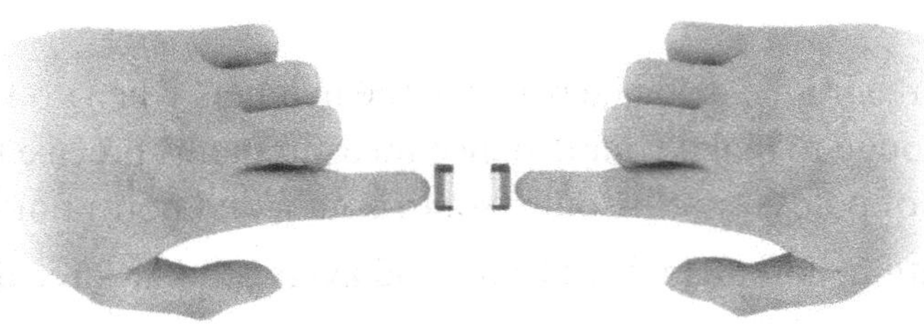

2. Hold two magnets one centimeter apart with their north poles facing each other. What force do you observe between the magnets—attractive or repulsive?

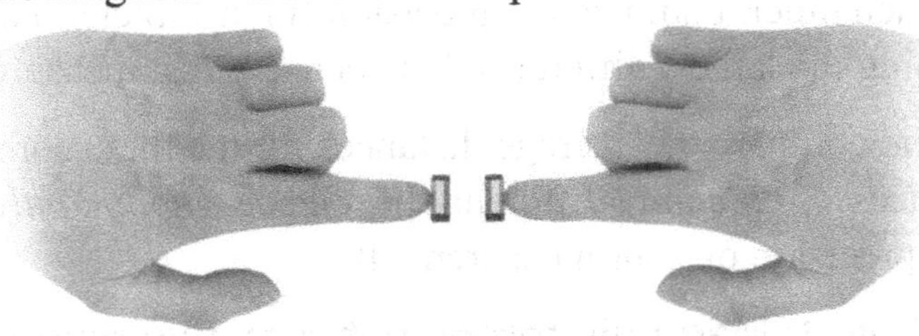

3. Hold two magnets one centimeter apart so one north pole and one south pole face each other. What force do you observe between the magnets—attractive or repulsive?

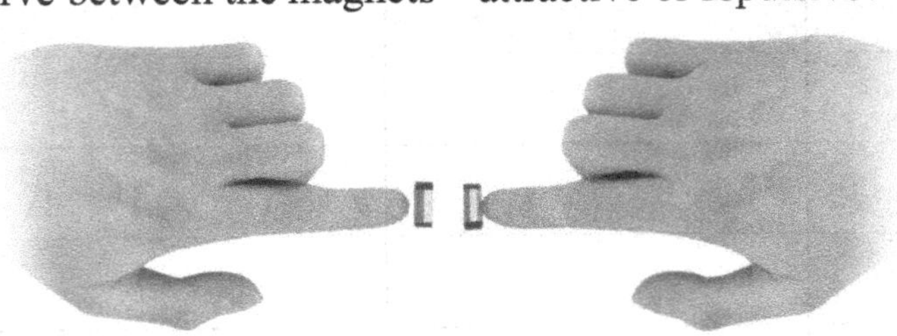

a. Write down a rule that describes how magnets exert forces on each other. Your rule should take into account your observations from steps 1 through 3.

2 Determining how far the magnetic force reaches

How far does the magnetic force of a magnet reach? This is an important question for machines such as motors and generators that use magnets.

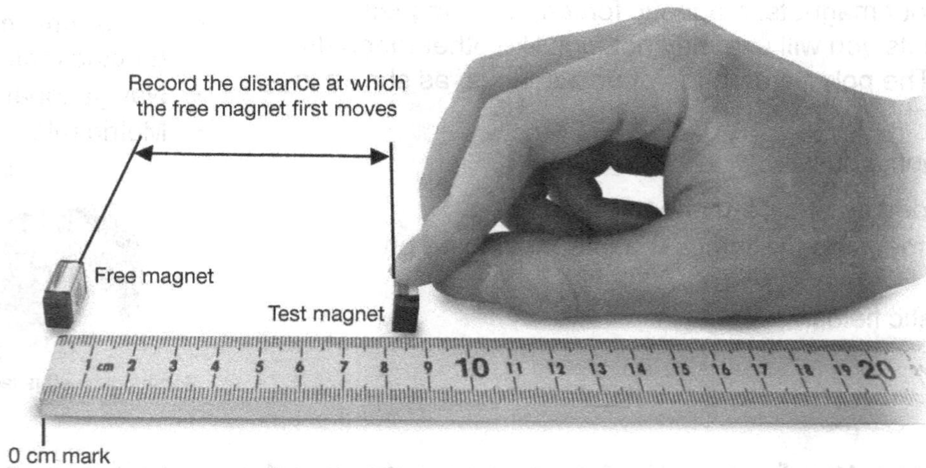

1. Place one magnet at the zero centimeter mark of a metric ruler as shown above. Slide a second magnet closer and closer until the first magnet moves. Practice the technique several times before recording data.
2. In Table 1 below, record the distance between the magnets when you first see movement. Move the magnet slowly enough to make your distance measurements precise to 0.5 millimeters. Try each of the combinations of poles—north-south, south-south, and north-north.
3. For each combination, complete three trials, and average your three distances. Record the averages in Table 1.

a. When referring to many measurements of the same quantity, *precision* describes how close the measurements are to each other. Estimate the precision (in millimeters) of your measured magnet interaction distance using the largest difference between any one measurement and the average.

b. Look at your results and compare the average distances for the three combinations of poles. Are the attract and repel distances *significantly* different? In science, *significantly* means the differences are large compared to the precision of your measurement.

Table 1: Magnetic forces between two magnets

	North-South	South-South	North-North
Distance 1 (mm)			
Distance 2 (mm)			
Distance 3 (mm)			
Average distance (mm)			

3 Magnetic field lines

In this part of the investigation, you will use a compass to discover the direction of the magnetic field around a permanent magnet. A compass is simply a small bar magnet mounted on a support that allows it to rotate. When placed near a magnet, a compass lines up with the magnetic field at that location.

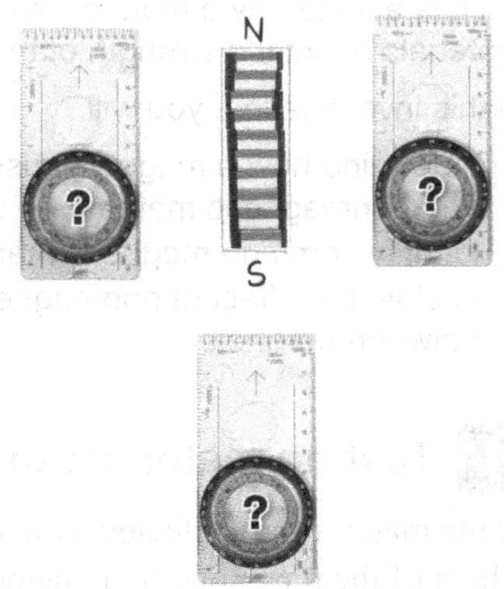

1. Stack 10 magnets on top of each other with alternating poles touching each other. When you join magnets in this way, you create one large magnet with a north pole at one end and a south pole at the other end. A long, narrow magnet of this shape is called a bar magnet.
2. Place the stack of magnets in the center of a blank sheet of paper as shown in the diagram at right. Trace the outline of the magnets and label the north and south poles.
3. Place the compass on the paper beside the magnets as shown. The compass will turn until it is lined up with the magnetic field around the stack of magnets. Remove the compass and then make a small line (1 to 1.5 centimeters long) on the paper in the location where the compass needle had been. Draw a small arrowhead on the end of the line where the red end of the compass needle points. You may want to set your compass on top of the line to check whether it was drawn accurately.
4. Repeat step 3, putting the compass in a different location on the paper. Do this in many locations (at least 50) all around the stack of magnets. You should see a pattern emerge among the lines you draw.

4 Constructing explanations

a. The pattern you drew shows the shape of the magnetic field around the stack of magnets. Connecting the small arrows together forms magnetic field lines. Describe the shape of the magnetic field around the stack of magnets.

b. Do the magnetic field lines point toward or away from the north pole of the stack of magnets?

c. You learned about *electric* field lines when studying positive and negative charges. How are magnetic field lines similar to electric field lines? How are they different?

d. Which of the following has electric field lines shaped similar to the magnetic field lines for a bar magnet?

- one positive charge
- one negative charge
- a pair of positive charges
- a pair of negative charges
- a positive charge next to a negative charge

22.2 Magnetic Properties of Materials

How do magnets interact with different materials?

In this investigation, you will test various materials to determine how each is affected by a magnet. You will also test how nonmagnetic materials affect the strength of the force between permanent magnets.

In this investigation, you will:

- determine how a magnet is used to distinguish between magnetic and nonmagnetic materials.
- identify common magnetic materials.
- explain the effect of nonmagnetic materials on the force between magnets.

Materials List

- 2 Ceramic magnets (from Electric Motor)
- An assortment of materials to test, such as a plastic spoon, wooden pencil, steel paper clip, fabric, aluminum foil, coins, a nail, or chalk
- Metric ruler

1 Testing materials to see if they are affected by magnets

Some materials are affected by magnetic forces and some are not. You will test various materials to determine which of them respond to a magnet. Use one of your permanent magnets to test each object. An object is magnetic if you feel a perceptible force that attracts or repels either pole of your test magnet. An object is nonmagnetic if you do not feel any perceptible force on your test magnet. If the object is magnetic, test it against both poles. Determine whether the object is always attracted, always repelled, or both, depending on whether you use the north or south pole of your test magnet. Record your data in Table 1.

Table 1: How different objects are affected by magnets

Object	Material composition	Attract	Repel	No effect

a. The word *magnetic* is used to describe things that are affected strongly by magnets. What common property do you see in the materials you observed to be magnetic?

b. Do the terms *repulsive* and *nonmagnetic* mean the same thing?

2 Ferromagnetism

The diagram at right shows a microscopic view inside a piece of iron. Iron is an example of a *ferromagnetic* material. The atoms in iron arrange themselves in magnetic domains. Each magnetic domain acts like its own tiny magnet. This occurs because the individual magnetic fields of atoms are aligned within a single domain.

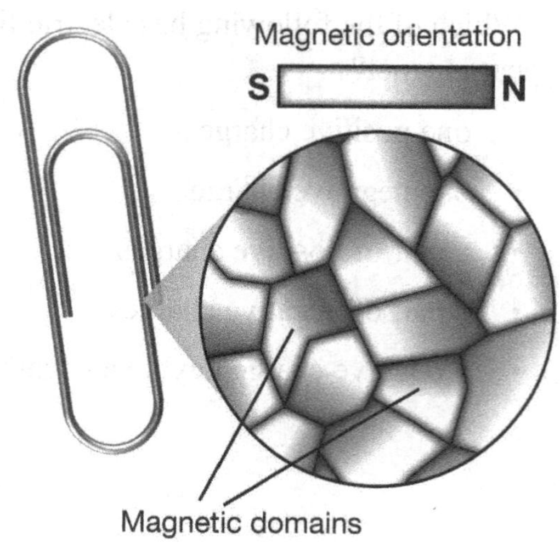

In normal (unmagnetized) iron or steel, the magnetic orientation of domains varies randomly. On average, as many point one way as another way. When a magnet is brought near a ferromagnetic material, the domains close to the external magnet realign themselves as if they were each small magnets.

3 Do nonmagnetic materials affect the magnetic force?

You may know that you can make a magnet move on the top of a table by moving a second magnet underneath the table. Maybe you wondered whether the table affects the strength of the magnetic force. In this part of the investigation, you will find out if magnetic forces get weaker or stronger when they pass through nonmagnetic materials.

1. Choose three nonmagnetic materials from Table 1. Write in these materials in the column headings of Table 2.
2. Place one magnet above the zero centimeter mark on a metric ruler and slide a second magnet closer and closer until the first magnet moves. When you first see movement, record the distance in the first column of Table 2. Repeat for a total of three trials, and calculate the average distance.
3. Place each nonmagnetic material between the two magnets as shown in the diagram below, and repeat the experiment.

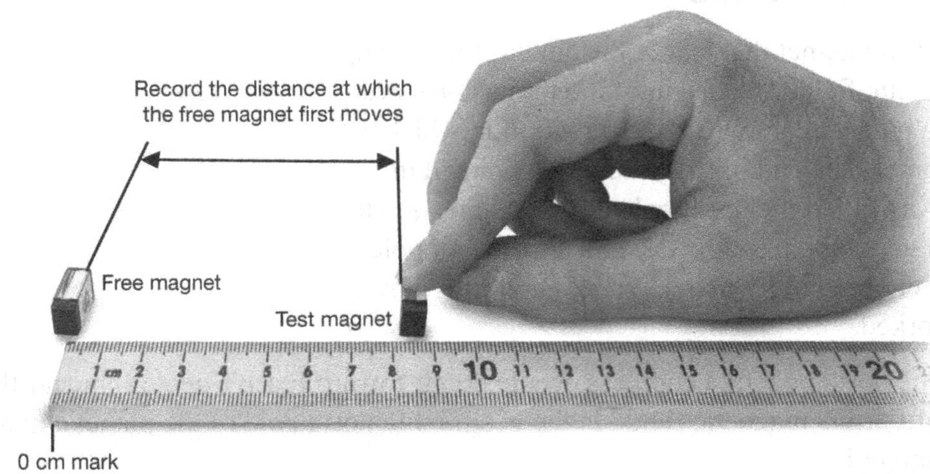

Table 2: Testing nonmagnetic materials

	No material between magnets	material: ________	material: ________	material: ________
Distance 1 (mm)				
Distance 2 (mm)				
Distance 3 (mm)				
Average (mm)				

a. Examine your results in Table 2. Does the strength of the magnetic force diminish by passing through any of the materials you tested?

22.3 The Magnetic Field of Earth

How do we use Earth's magnetic field to tell direction?

Like many other planets, Earth has a magnetic field. Earth behaves as if its core were a huge magnet. The planet's magnetic field is not strong compared with the magnetic field near a small permanent magnet. However, Earth's magnetic field fills a tremendous volume of space around and within the planet. This investigation explores some useful characteristics of Earth's magnetic field.

In this investigation, you will:

- explain how a compass works.
- describe how magnetic declination affects a compass.
- identify materials that will affect a compass magnetically.

Materials List

- Navigational compass
- Magnet (from Electric Motor)
- Steel bucket or baking pan

1 Why a compass works

A compass needle is a magnet that is free to spin until it lines up with the local magnetic field. The origin of the terms *north pole* and *south pole* of a magnet come from the direction in which a compass needle points. The end of the needle that pointed to geographic north was called the north pole of the magnet and the end that pointed to geographic south was called the south pole. These names were chosen before people understood fully how a compass needle worked.

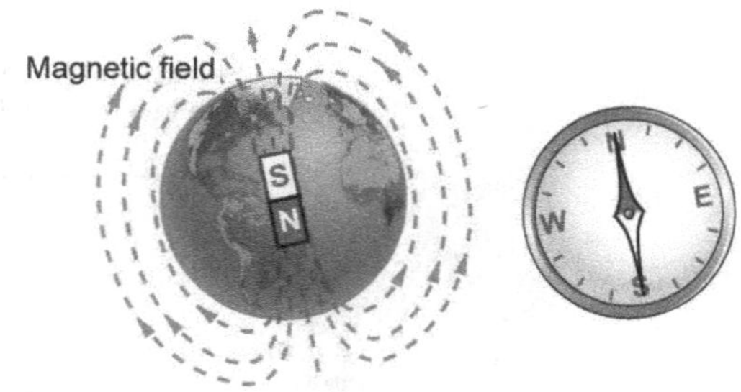

a. If the magnetic north pole of a compass needle points toward Earth's geographic north pole, what is the orientation of Earth's magnetic field? Is the geographic north pole of the planet a magnetic north or a magnetic south pole?

b. The north pole of a compass needle always points toward which pole of a permanent magnet?

2 Reading a compass

Around the needle of the compass is a rotating scale that has degrees marked clockwise from zero to 360. Zero degrees is straight north. Straight south is 180 degrees. East is 90 degrees, and west is 270 degrees. A compass *bearing* means direction in degrees relative to north. For example, a *bearing* of 45 degrees is a direction northeast of where you are.

The arrow on the base of the compass indicates the direction to go when you know the compass bearing. With the compass in your hand, rotate the base until the arrow aligns with the bearing you want to travel. Next, rotate the entire compass until the north end of the needle is aligned with the north arrow *inside* the compass dial. The arrow on the base now points in the direction you wish to go.

Reading a compass

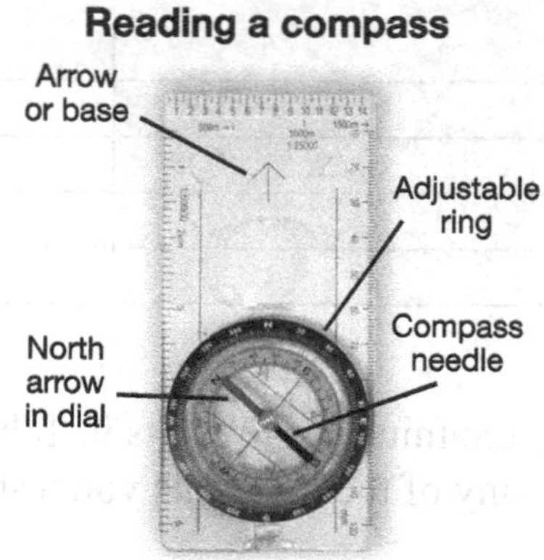

3 Declination

Earth's magnetic field is not lined up perfectly along the north-south axis of the planet. That means a compass does not point directly to the geographic north pole. Depending on where you are, a compass will point slightly east or west of true north. The difference between the direction a compass points and the direction of true north is called *magnetic declination.* Magnetic declination is measured in degrees and indicated on topographical maps.

a. Research the magnetic declination for your area.

b. If the magnetic declination were zero, straight east would be a bearing of 90 degrees on a compass. What compass bearing will you set if you wish to go straight east, but need to correct for the local magnetic declination?

4 The effect of magnetic materials on a compass

The needle of a compass aligns itself with the magnetic field in its vicinity, whether the field comes from a nearby magnet or from Earth. For this reason, magnets and magnetic materials can interfere with navigation when you are using a compass. In the next part of this investigation, you will determine how sensitive a compass is to external magnets and magnetic materials.

1. Place the compass on a large, flat surface such as a floor or table. Rotate the ring on the compass to a bearing of zero degrees (north). Rotate the whole compass until the direction arrow and compass needle are both aligned with north.
2. Obtain one of the small permanent magnets that come with the electric motor experiment. How close can you bring the permanent magnet before the compass needle is deflected from north? You should try the experiment with several different orientations of the magnet to see which one has the largest effect on the compass needle.
3. Obtain a steel bucket or baking pan. The bucket should be magnetic steel (you can test it with your permanent magnet). Orient the compass so that it is aligned with north, and slowly lower it into the bucket. As you lower the compass, rotate it gently back and forth to see whether the needle stays aligned with north. At some point the steel will shield out enough of Earth's magnetic field that the compass will no longer be able to stay in alignment with north.

5 Arguing from evidence

a. How close did the magnet have to be before it affected the compass? There are magnets in many devices, such as television sets, electric motors, and computers. How far away from these devices must you be before you can trust the reading from a compass?

b. The bodies of automobiles are made of magnetic steel, yet you can buy a compass for your car. Some cars even have electronic compasses built in. Based on your observations, would you trust the directions from a compass inside a car?

c. A compass does not work close to Earth's geographic north or south poles. Draw a sketch showing the Earth's magnetic field and use the drawing to explain why compasses have this limitation.

23.1 Electric Current and Magnetism

Can electric current create a magnet?

A magnet that creates its magnetic field from the flow of electric current is called an *electromagnet*. The simplest electromagnet uses a coil of wire often wrapped around a piece of iron or steel. The iron or steel concentrates the magnetic field created by the current in the coil. Stereo speakers, toasters, doorbells, car alternators, and power plant electrical generators are just a few of the many devices that use electromagnets.

In this investigation, you will:

- build an electromagnet.
- explain how electric current affects the strength of the magnetic field in an electromagnet.
- compare permanent magnets and electromagnets.

Materials List

- 2 Ceramic magnets (Electric Motor)
- 1–2 m Magnetic wire, 24 gauge
- Sandpaper
- 1 Galvanized nail (7 cm long)
- 3 D-cells
- Electric Circuits
- Ruler or straightedge
- Multimeter with red and black leads
- 100 Steel paper clips
- Graph paper
- Pencils

Safety Tip: Disconnect your electromagnet when not in use as the wire and batteries get hot quickly.

1 Build an electromagnet

1. Wrap a nail tightly with wire as shown at right. Leave 20 centimeters of uncoiled wire on each end of the nail. Count the number of turns you wrap and write the result here.

   Turns: ____________________

   Make sure your wire turns are neat, tight, and evenly spaced or the electromagnet may not work efficiently. This is your electromagnet coil.
2. Remove any insulation or varnish from the two free ends of the wire.
3. Do not connect the wires to the battery yet.

2 Compare electromagnets and permanent magnets

a. This electromagnet certainly does not look much like the permanent magnets you are familiar with. Using what you know about magnets, think of at least two tests to show that your electromagnet acts like a permanent magnet. Describe your two proposed tests.

b. Connect your electromagnet to the battery and perform the two tests you described in a above. Record your observations.

Safety Notes: Do not leave the battery connected for an extended period of time because the wire will get hot.

c. Does the electromagnet act like a permanent magnet? Explain.

3 The right-hand rule

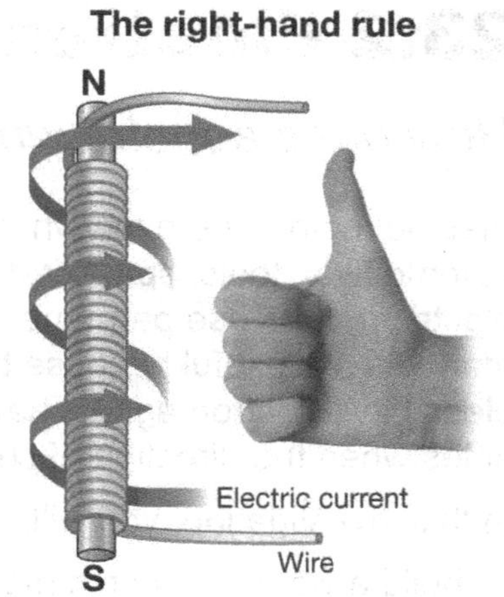

a. Use a permanent magnet or compass to determine the location of the north and south poles of your electromagnet. Explain your method and make a diagram showing the direction of the current from the positive battery terminal, around the nail, and into the negative terminal.

b. Reverse the direction of the current through your electromagnet by switching where the electromagnet wires' ends connect to the battery. Test for the locations of the north and south poles. What effect did reversing the current have on the locations of the poles?

c. The polarity of an electromagnet can be predicted easily if you know the direction of the current in the coil. When the fingers of your right hand curl in the direction of the current, your thumb points toward the electromagnet's north pole. This method of finding the magnetic poles is called the *right-hand rule*. Use the right-hand rule to find the locations of the poles of your electromagnet.

4 Increasing the current through an electromagnet

1. Measure the current through your electromagnet when it is connected to one battery. Test how many paper clips it can pick up and record your results in Table 1.
2. Increase the current through the electromagnet by connecting it to two batteries. Again measure the current and test how many paper clips it can pick up. Record your results.
3. Repeat step 2 using three batteries.
4. Remove half the turns by unwrapping the wire from the nail. Don't change the total length of wire.
5. Repeat steps 1–3 with the "new" electromagnet, which has only half as many turns of wire.

Safety Note: Keep batteries connected for only a short time or the electromagnet will get hot. Check the maximum current your meter can measure; it will be written on the front of the meter near the jacks where the leads plug in. You may need to use different jacks on the front of your meter. The current will be above 0.2 A (200 mA).

Table 1: Current and magnet strength

Turns	No. of batteries	Current (A)	No. of paper clips picked up

5 Constructing explanations

a. Draw a graph showing how the number of paper clips picked up by the magnet varies as the current increases. Label your axes and title your graph.

b. How did changing the number of turns affect the strength of the electromagnet?

c. What is your conclusion about the relationship between current and the strength of the electromagnet?

d. Look at other groups' electromagnets, data tables, and graphs. Did some electromagnets work better than others? Write down anything you notice that might explain differences in performance.

23.2 Electromagnets and the Electric Motor

How does a motor work?

Materials List

- Electric Motor with 4 D-cell batteries
- Multimeter with red and black leads

Electric motors are everywhere. You find them in fans, washing machines, automobiles, tools, spacecraft—anywhere that we use powered machines. All electric motors use permanent magnets and/or electromagnets. Permanent magnets are useful because they create a magnetic field without needing any electricity. Electromagnets have the advantage of being able to reverse their poles when the direction of current is changed.

In this investigation, you will:

- build a working electric motor.
- explain how permanent magnets and electromagnets interact and cause a motor to spin.
- measure the current and voltage of a motor.
- calculate the power used by various motor designs.

1 Getting the rotor to spin

Electric motors use magnetism to spin. Try to get the rotor to spin by manipulating permanent magnets as follows.

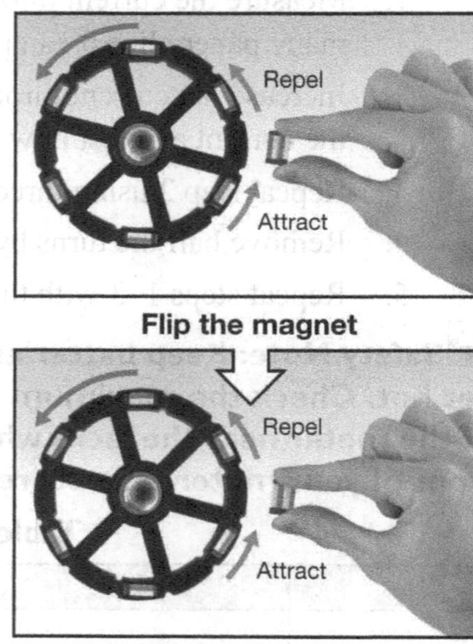

1. Take the motor apart and put six magnets in the rotor so they are evenly spaced and alternate north-south poles facing outward.
2. Bring a stack of two or three magnets close and try to repel one of the magnets in the rotor. The rotor should spin a little.
3. As soon as the rotor spins a little, flip the magnet in your fingers to attract and then repel the next magnet on the rotor.
4. By sequentially flipping the magnet in your fingers, try to get the motor to spin.

You will see that reversing the poles of the magnet in your fingers is the key to making the rotor spin.

a. When is the right time to reverse the magnet in your fingers? Think about where the magnets are in the rotor.

b. How could you make the rotor spin the other way?

2 Making a four-pole electric motor

All electric motors use some kind of switch to change the orientation of the poles of the magnet at the right time. You accomplished this by flipping the permanent magnet. An easier way to do this is with an electromagnet. The locations of the north and south poles of an electromagnet depend on the direction of the current through it. Reversing the current through an electromagnet has the same effect as flipping a permanent magnet. A device called a *commutator* is used to switch the orientation of the north and south poles of the electromagnet. In the electric motor you are building, the commutator is a plastic disk and it uses a beam of light to tell it when to switch the location of the poles.

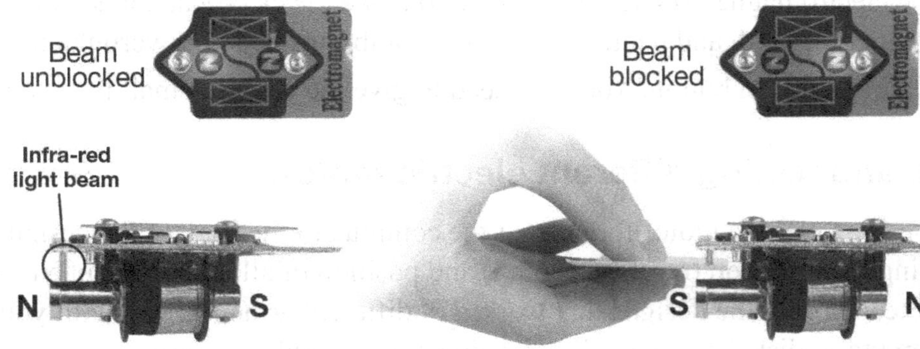

When the light is not blocked (see diagram above), the current flows in a direction to make the north pole at the front of the electromagnet. When the light is blocked, the current flows in the opposite direction, making the north pole at the back of the electromagnet. There are two green lights on the electromagnet. The illuminated light shows which end is the north pole.

The commutator disks have alternating black and clear sections around the edge that switch the direction of current in the electromagnets by blocking the light beam. Align the clear and black edges with the centers of the magnets.

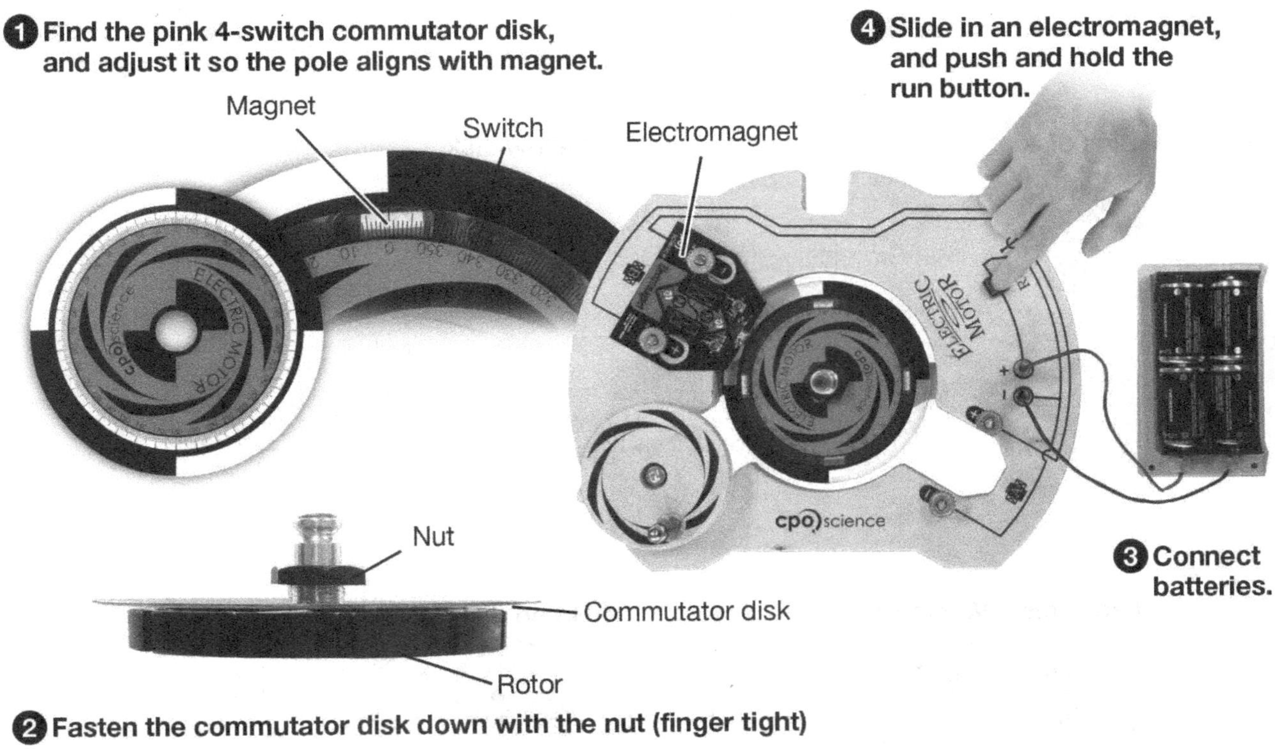

The following steps explain how to set up the electronic motor.

1. Find the pink four-pole commutator disk shown in the picture.
2. Arrange four magnets so the north or south poles alternate. Be sure the disk is aligned so the border between clear and black is centered on each of the four magnets. The border is where the electromagnet will switch from north to south. Finger-tighten the big nut to secure the disk once you have it aligned with the magnets.
3. Connect the battery pack matching the red (+) and black (–) wires to their corresponding terminals on the motor.

4. Attach the electromagnet to position A or B. To make the electrical connections, push the electromagnet forward, and gently tighten the thumb-nuts. Do not overtighten them.
5. Push and hold the Run button. You may need to give the motor a small push to start it.

3 Designing and testing different electric motors

Design and test a working electric motor for each of the commutator disks. Use the design charts to record each design, including the direction (north and south) and position of all magnets. Put an *x* where you placed the electromagnet. Record only the designs that work. Try different arrangements until you get a design that works for each commutator disk. The green disk requires you to think creatively.

Pink disk

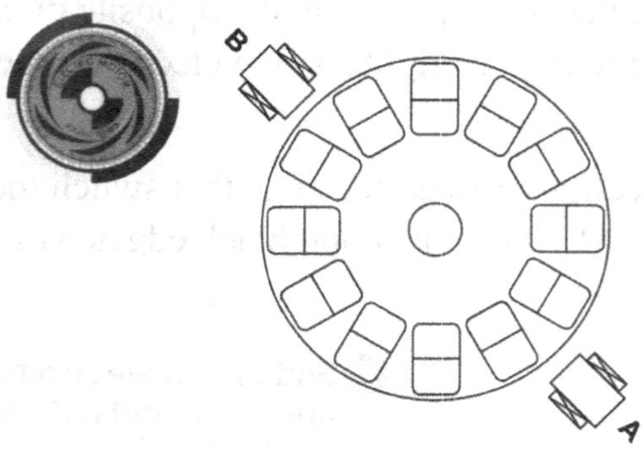

Orange disk

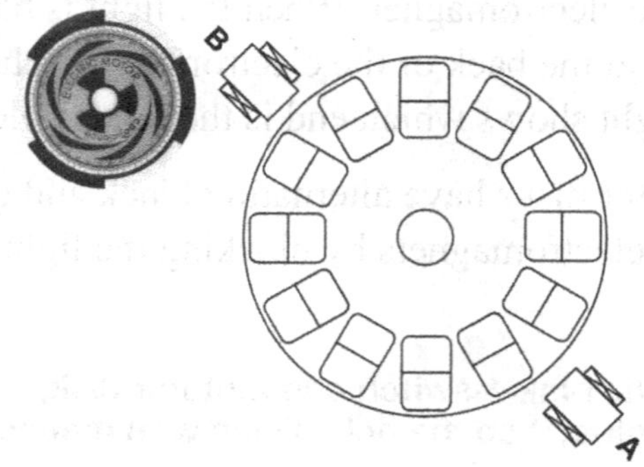

Blue disk

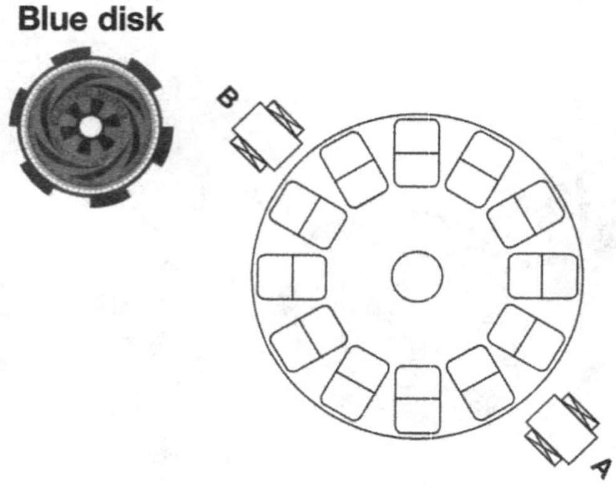

Yellow disk

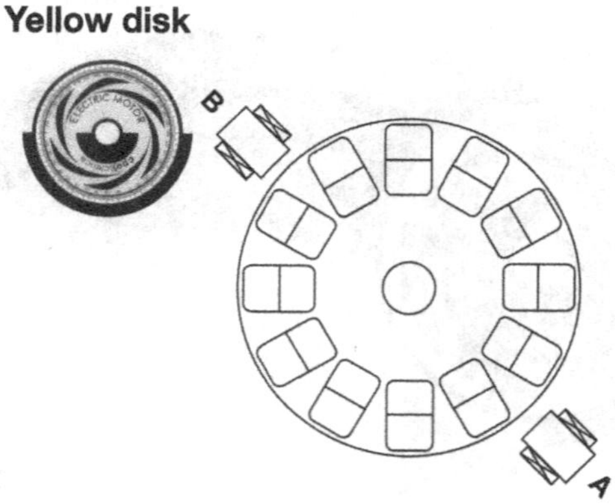

Green disk: Version 1

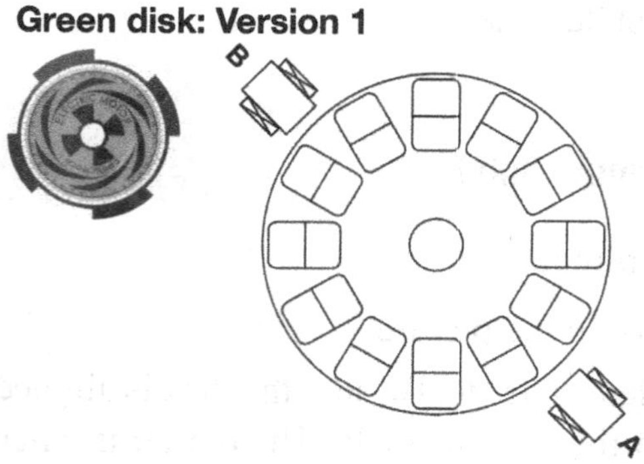

Green disk: Version 2

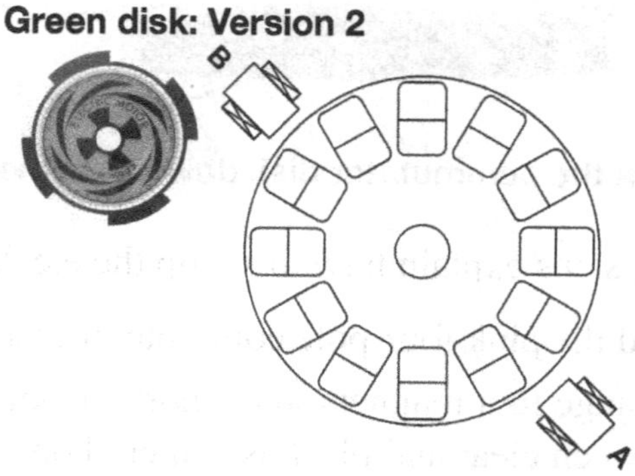

4 Measuring current

When you push the Run button on the motor, it connects an electric circuit that allows current to flow through the electromagnet. The current flows from the red (+) terminal to the black (–) terminal of the battery pack. For you to measure the current in the motor, it must travel through the meter on its path from the batteries, through the motor, and back to the batteries again.

1. Make a working motor with the orange disk, six magnets, and the electromagnet. Remember to alternate the north-south orientation of the magnets.
2. Connect the battery pack so the current flows through the meter before (or after) flowing through the electric motor. Look at the diagram below to make sure you are connecting the wires properly.
3. Set the DataCollector to measure DC current.

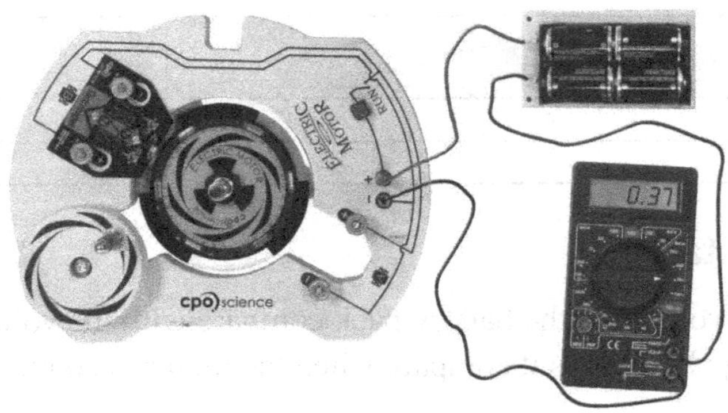

d. How much current is the motor using when it is spinning? Record your answer in Table 1.

e. Now, measure the current with the rotor stopped (use your finger to stop it). How much current is the motor using when the rotor is stopped? Record your answer in Table 1.

Table 1: Motor current data

Current with rotor stopped (amps)	Current with rotor spinning (amps)

5 Measuring voltage

Measure and record the battery pack's voltage when the motor is off. This is called the *no-load voltage*, meaning nothing is drawing current from the battery pack. The battery voltage changes when current is flowing. In the next part of the investigation, you will measure and compare the no-load voltage of the battery pack with its voltage under load while the motor is running.

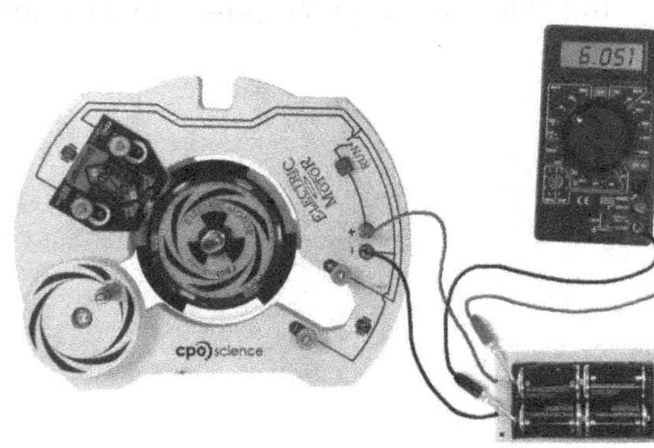

6 Collecting the data

Measure the current and voltage of the motor for four different motor designs using different numbers of permanent magnets. Record the current and voltage both with the rotor spinning and with the rotor held in place while the Run button is held down.

Table 2: Current and voltage for different motor designs

Number of permanent magnets	Current with rotor spinning (amps)	Current with rotor stopped (amps)	Voltage with rotor spinning (volts)	Voltage with rotor stopped (volts)

Use the data from Table 2 to calculate the power consumed by the motor.

Table 3: Power used by the motor

Number of magnets	Power with rotor spinning (W)	Power with rotor stopped (W)

7 Analyzing the data

a. How does the no-load voltage of the battery pack compare with the voltage when the motor is connected and running? How does it compare when the motor is connected but stopped?

b. How does the motor's voltage when it is on but stopped compare to the voltage when it is running?

c. How does the current when the motor is on but stopped compare to the current when it is running?

d. How is the number of permanent magnets used related to the power consumed by the motor?

e. Does the motor use more power when it is running or when it is stopped? Use your observations to explain why electric motors in machines often burn out if the machine jams and the motor is prevented from turning when the electricity is still on.

f. When the motor is running, most of the energy from the battery goes to overcoming friction and adding kinetic energy to the rotor. Where does the energy go when the motor is stopped from turning?

g. How much power does your motor use compared with a 100-watt light bulb? Your answer should show a calculation of how many motors you could run using the electricity used by the 100-watt bulb. Base the calculation on the power used when the rotor is spinning (use any motor configuration).

23.3 Induction and the Electric Generator

How does a generator produce electricity?

Electric generators convert mechanical energy into electrical energy. This Investigation explores the design, operation, and principles of an electric generator that works by rotating magnets past a coil of wire. The passing of the magnets causes a magnetic field to change in the coil of wire. The changing magnetic field produces electricity.

In this investigation, you will:

- explain how an electric generator works.
- build and test several electric generator designs.
- use Faraday's law of induction to explain why the amount of electricity generated depends on the speed and number of magnets in the generator.

Materials List

- Electric Motor
- DataCollector with 1 photogate
- Multimeter with red and black leads
- Graph paper
- Rubber band

1 Faraday's law of induction

A changing magnetic field near a coil of wire can induce current to flow in the wire. Current flows because the changing field creates a *potential difference* (also called *voltage*) across the ends of the wire. The effect of this voltage is similar to what happens when a battery is connected to the ends of the wire. This process is called *electromagnetic induction.*

Faraday's law of induction states that the voltage is proportional to the number of turns of the wire and to the rate at which the magnetic field through the coil changes. The faster the magnetic field changes, the greater the voltage induced in the coil. The polarity of the voltage depends on the direction of the magnet's motion relative to the coil.

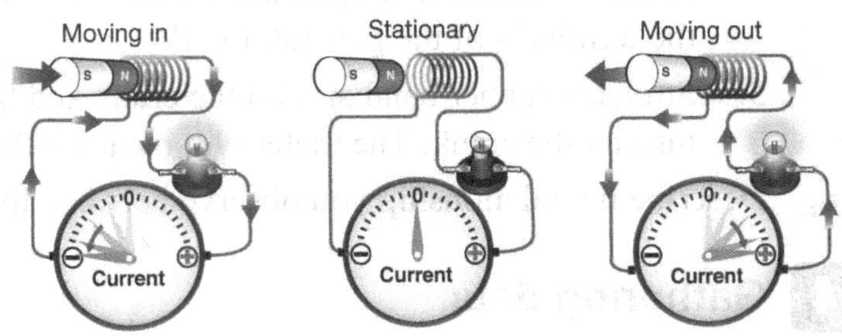

2 Observing the induced current in a coil

1. Remove the coil from the generator (or motor). Attach the leads of the digital multimeter to the positive and negative terminals of the coil. Set the multimeter to measure AC volts.
2. Move a permanent magnet back and forth very close to the coil. Watch the multimeter. What do you notice?
3. Move the magnet at a faster rate. What happens to the voltage?
4. Slowly move the permanent magnet away from the coil while continuing to shake it at a fast rate. What happens to the voltage?
5. Place the magnet on the coil. What voltage is induced when the magnet is not moving?

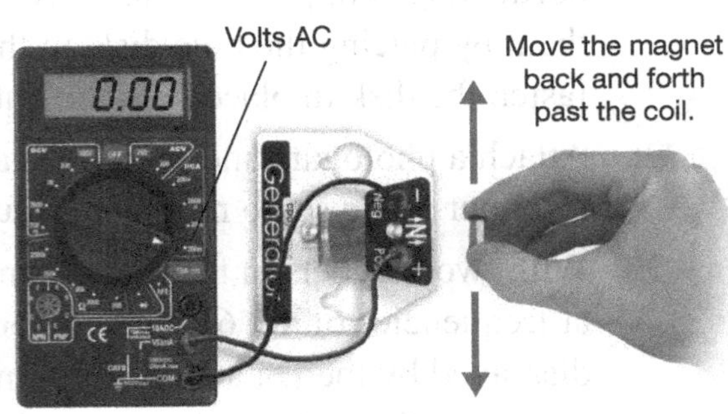

3 Building the generator

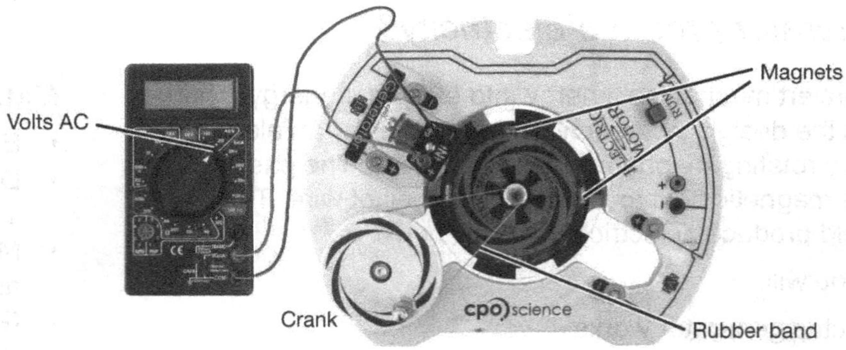

A generator uses induction to convert the mechanical energy of the spinning rotor into electrical energy. When the rotor spins, the permanent magnets pass by the coil. This causes current to flow in the coil. When a north pole passes the coil, current flows in one direction; when a south pole passes it, current flows in the opposite direction. If the magnets are arranged so alternating poles face outward, the current in the coil moves back and forth as the rotor spins, creating alternating current, or AC.

1. Place four magnets on opposite sides of the rotor with alternating poles facing out. Place the blue disk on the rotor and fasten it with the nut. This secures the magnets in place while the rotor spins. Always attach a disk to the rotor before spinning it with magnets inside.
2. Slide the generator coil over the thumbscrews in either the A or B position. Slide the generator coil toward the rotor and tighten the screws to hold it in place. Connect the leads of the multimeter to the terminals of the generator coil.
3. Stretch a rubber band around the crank and the spindle of the motor. You can spin the rotor by turning the crank. The meter will measure the AC voltage produced.

a. Describe the relationship you observe between the speed of the rotor and the voltage produced.

4 Gathering data

There are two main variables that affect the performance of the generator: the number of magnets and the rate at which the rotor spins. The photogate and DataCollector can be used to measure the spin rate.

1. Place two magnets in the rotor with alternating north and south poles facing outward. Secure them by placing the blue disk on the rotor and fasten the disk in place with the nut.
2. Attach a photogate and set the DataCollector to Timer mode set to measure frequency.
3. With two magnets in the rotor, spin the motor at frequencies of 40, 60, and 80 hertz as displayed by the DataCollector. It may take some practice to keep the rotor spinning at the correct frequency long enough to read the voltage. Measure and record the voltage you generate in Table 1.

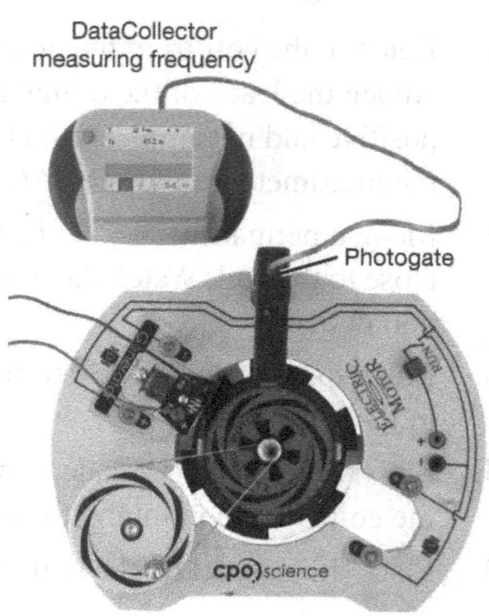

4. Repeat the last step with 4, 6, and then 12 magnets in the rotor. The magnets should always have alternating north and south poles facing outward. If you loosen the generator coil, make sure you slide it back toward the rotor and tighten the thumbscrews before collecting data.

5 Analyzing the data for the alternating poles generator

Table 1: Generator data with alternating magnetic poles facing out

Rotation frequency	Voltage with 2 magnets	Voltage with 4 magnets	Voltage with 6 magnets	Voltage with 12 magnets
40 Hz				
60 Hz				
80 Hz				

a. Use the data from Table 1 to make a graph of voltage versus number of magnets for each frequency. Plot all three sets of data on the same graph using three separate lines. Clearly label each one to indicate the frequency.

b. How does changing the number of magnets affect the voltage generated? If you double the number of magnets, how much does the voltage change?

c. If you doubled both the number of magnets and the speed, what change would you expect in the voltage?

d. Suppose you had a rotor with a different magnet slot spacing. If you used 16 magnets, what would you expect the voltage to be for each frequency? You should use your graph to answer this question.

e. Why do increasing the rotor speed and changing the number of magnets have a similar effect on the voltage generated? You should mention Faraday's law in your answer.

f. Why is it important for the generator coil to be positioned close to the rotor? If you loosen the generator coil screws and slide the coil away from the motor, what happens to the voltage generated?

6 Changing the orientation of the magnets

Flip the magnets so they all have a north pole facing outward. Repeat the experiment measuring speed and voltage for a range of rotation frequencies. Record your measurements in Table 2.

Table 2: Generator data with north poles facing out

Rotation frequency	Voltage with 2 magnets	Voltage with 4 magnets	Voltage with 6 magnets	Voltage with 12 magnets
40 Hz				
60 Hz				
80 Hz				

a. Make a graph of voltage versus number of magnets for each frequency.

b. Compare the voltage generated with all north poles facing out to the voltage of the generator with alternating poles facing out. Which design produced the greater voltage overall?

c. Is the difference between the voltage of the two motor designs more noticeable at low or high speeds? Why do you think this is?

d. Compare the voltage created by each design using 12 magnets. Why do you think the generator with all north poles facing out did not work well with 12 magnets?

e. How could you improve the design of the generator? Discuss three improvements you could make that would increase the voltage generated.

24.1 Semiconductors

What are some useful properties of semiconductors?

It is almost impossible to do anything today without being affected by electronics. Electronic devices are in telephones, computers, video games, cars, and watches—a virtually endless list. Diodes, a basic element of electronics, are used to control the direction of current flowing in a circuit. Diodes are also the building blocks for transistors. This investigation explores the properties of diodes.

In this investigation, you will:

- describe the electrical behavior of a diode.
- describe how a diode is different from a resistor in terms of its current and voltage characteristics.

Materials List

- 2 D-cell batteries
- Electric Circuits
- Several diodes to test
- Multimeter with red and black leads

1 Testing a diode

A common diode looks like a small cylinder with a stripe on one end. A diode is a one-way valve for electric current. Current can only flow one way through a diode and not the other way.

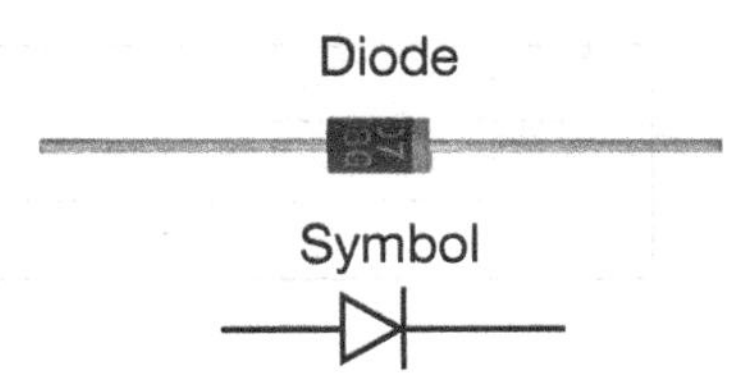

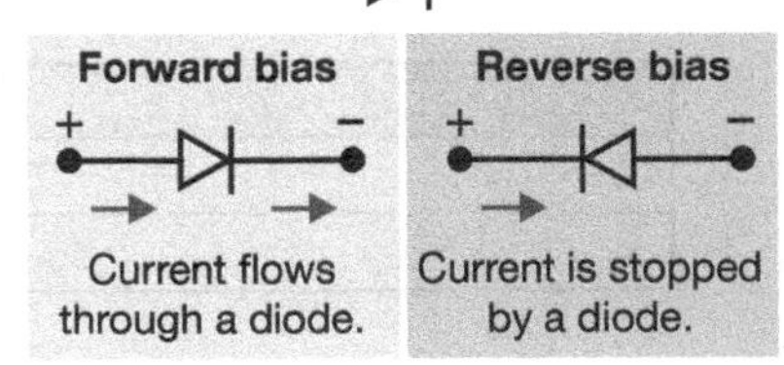

1. Connect the circuit below using two batteries, a switch, the potentiometer, and a diode.

Build this circuit.

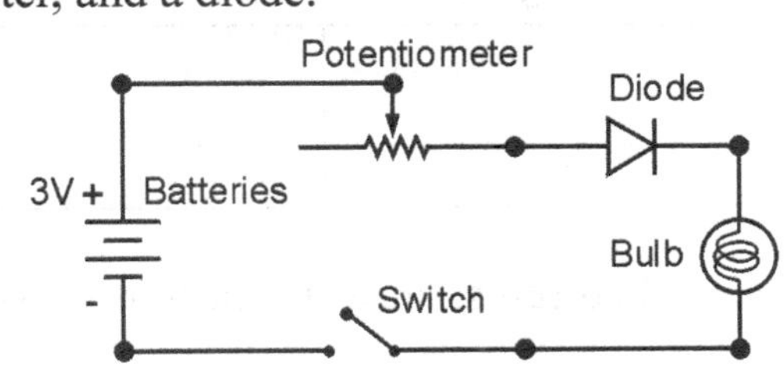

2. Close the switch and turn the potentiometer so the bulb lights. Next, open the switch and remove and reverse the diode in the circuit. Close the switch and try to light the bulb.

2 The current versus voltage curve for a diode

The next part of the experiment measures the current versus voltage curve for a diode. You will first measure positive voltage and current with the diode forward biased. You will then reverse the battery connections to measure negative voltage and current.

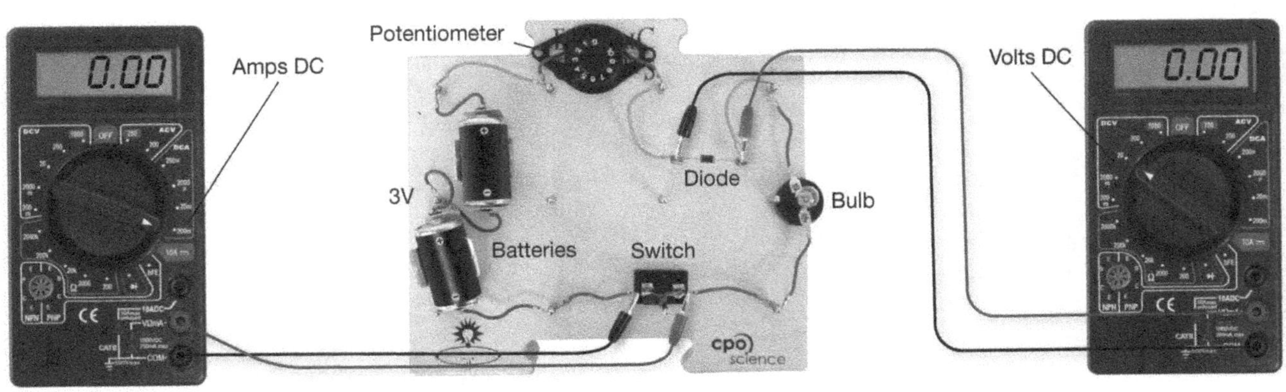

1. Replace the diode in the proper direction in your circuit so current flows through the diode toward the striped end.
2. Turn the potentiometer all the way to the left to make the bulb as bright as possible.
3. Set the meter to DC volts. Close the switch and measure the voltage across the diode by touching the leads of the meter to the terminals of the diode. Record the value in Table 1.
4. Set the meter to DC amps. Measure the current by opening the switch and inserting the leads of the meter between the terminals of the switch.
5. Turn the potentiometer down and repeat the voltage and current measurements for at least five settings of the potentiometer across its full range.
6. Reverse the battery connections so the diode is now biased in the reverse direction. Relative to the diode, all voltage and currents are now negative.
7. Measure voltage across the diode and current in the circuit for at least five settings of the potentiometer. The bulb will not light because the diode is blocking all the current. Record the values in Table 2.

Table 1: Positive voltage vs. current

Potentiometer setting	Current (A)	Voltage (V)

Table 2: Negative voltage vs. current

Potentiometer setting	Current (A)	Voltage (V)

a. Showing both positive and negative voltages and currents, draw the current vs. voltage graph for the diode.

b. Explain how the graph shows that the resistance of the diode changes with the direction of current.

3 LEDs

The acronym LED stands for *light emitting diode*. Repeat the experiment in Part 3 except use an LED instead of an ordinary diode and replace the bulb with a 100Ω resistor.

24.2 Circuits with Transistors

What are some useful properties of transistors?

A transistor allows you to control current, not just block it in one direction like a diode. A transistor is like a variable flow valve for current. A good analogy for a transistor is a pipe with an adjustable gate. When the gate is closed, the pipe has very high resistance and not much water flows. When the gate is open, the pipe has low resistance and water flows easily. The adjustable gate is a control for the flow through the pipe.

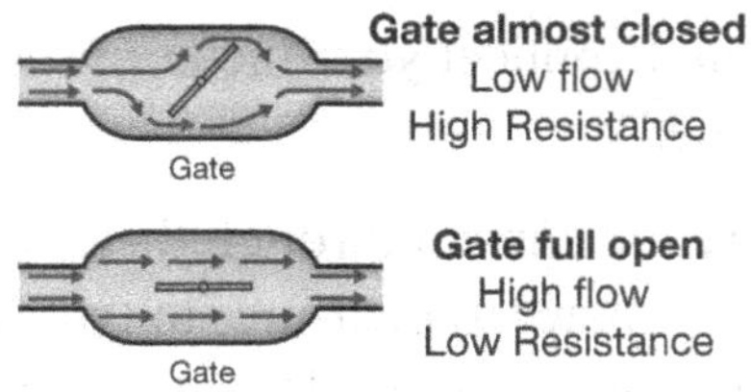

In this investigation, you will:

- describe the three terminals of a transistor and their functions.
- describe how to make a transistor work as a switch.
- connect a circuit using a transistor as a switch.

Materials List

- 2 D-cells
- Electric Circuits
- 2 jumper wires with alligator clips
- 1.5 V to 3V DC electric motor
- Solar cell
- Several npn transistors to test
- An assortment of resistors with values from 100Ω to 100 kΩ
- Multimeter

1 Connections and circuits for a transistor

A transistor has three terminals. The main path for current is between the *collector* and *emitter*. The *base* controls how much current flows. You can think about a transistor like a potentiometer that can be electrically controlled. The effective resistance of a transistor is changed by altering the current flowing into the base.

When used as a switch, the current needed to "turn on" a transistor is very small: a few millionths of an amp is typical. In a circuit, the current into the base is limited by a resistor with a resistance that is large, often 100 kΩ or more.

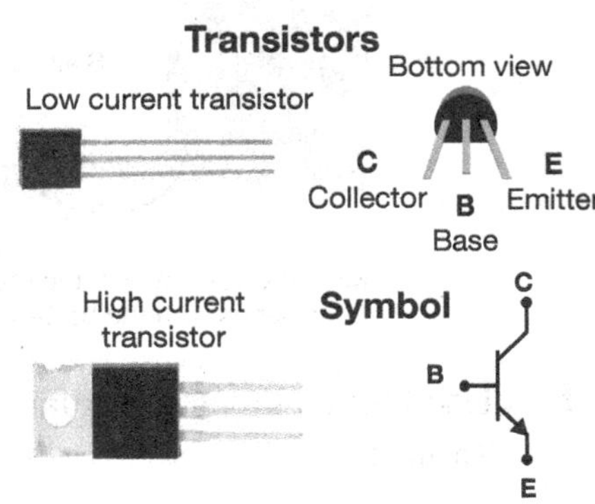

2 A transistor switch circuit

Transistor switches are often used when small input signals control larger output currents used to drive motors, lights, or other electrical devices. One example is a light-controlled fan. In this circuit, a fan goes on when there is light. The solar cell does not produce enough electricity to drive the fan motor directly, but it does produce enough current to turn the transistor on. The transistor switches the motor on by allowing current to flow from the battery to the motor.

Build this circuit.

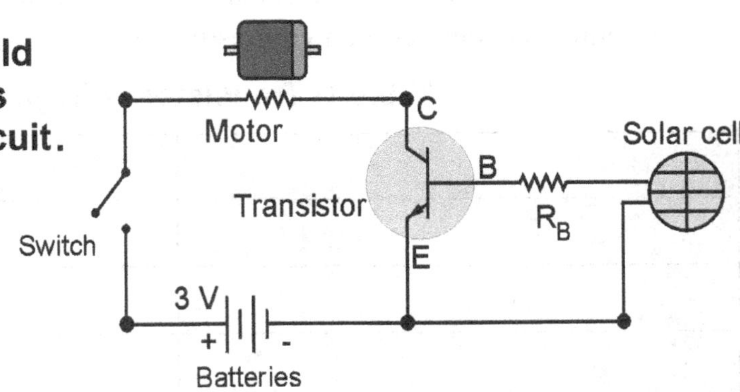

1. Connect the circuit shown in the diagram.
2. You will need to try different values for the base resistor R_B until the motor switches on and off when light reaches or is blocked from the solar cell.

3 Analyzing the circuit

a. What size resistor did you use for the base of the transistor?

b. Measure the voltage across the base resistor when the transistor is switched on (low resistance). Use Ohm's law to calculate how much current flows into the base of the transistor when it is turned on.

c. Break the circuit at the battery and connect your meter in series to measure the current through the motor when the transistor is turned on.

d. Cover the solar cell and measure the current through the circuit when the transistor is turned off. This current may be too small for your meter to measure; in that case, you can only estimate what the maximum current through the circuit could be. The meter rounds measurements to the nearest 0.001 amps. For example, if the meter reads 0.000 A, the maximum current that can be flowing is 0.0005 amps.

e. Reconnect the circuit and measure the voltage across the transistor when it is turned on. Also, measure the voltage across the transistor when it is turned off.

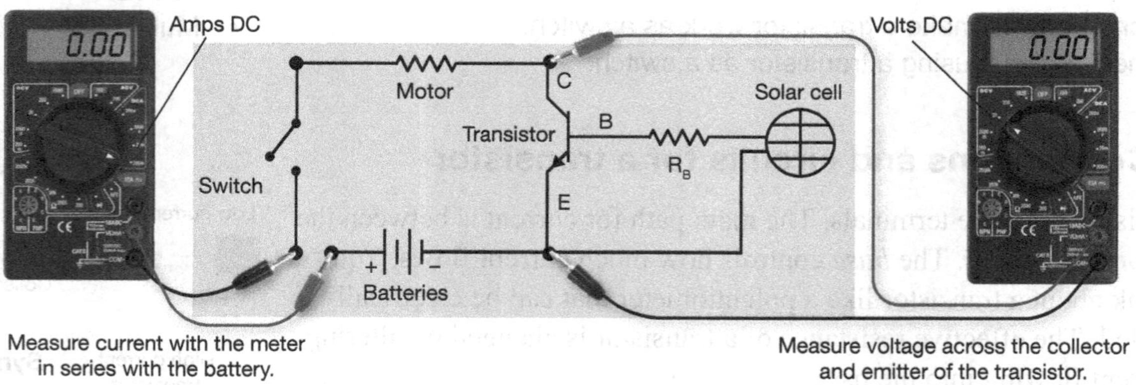

f. Use Ohm's law to calculate the resistance of the transistor in its "on" state, and record your answer in Table 1.

g. Use Ohm's law to calculate the resistance of the transistor in its "off" state and record your answer in Table 1. If your meter read 0 A, your calculation is really only telling you the *minimum* resistance the transistor could have. The actual resistance could be much higher, making the current through the circuit even smaller than your estimate.

Table I: Transistor voltage and current measurements

	Motor on	Motor off
Current (A)		
Collector - emitter voltage (V)		
Transistor effective resistance (Ω)		

24.3 Digital Electronics

How do you construct electronic logic circuits?

Digital electronics use voltages to represent numbers and decisions. The word *digital* describes a system where there are only two possibilities: there is no in-between or "maybe" allowed. For example, in a decision-based scenario, 5 V can mean yes, and 0 V can mean no. When representing a number, 5 V can mean one and 0 V can mean zero. At first glance, it might seem like you could not do much when the only possibilities are yes/no or 0/1. However, computers, CD players, and many other complex electronic devices use millions of yes/no decisions or 1/0 numbers as instructions to perform their various functions.

In this investigation, you will:

- build circuits using electronic logic to evaluate inputs and make a decision (output).
- practice using integrated circuits ("chips").

Materials List

- Solderless breadboard
- 3 D-cell batteries
- An assortment of small 22-gauge, solid-core wires
- Small flat bladed screwdriver
- Small pair of needlenose pliers
- Wire stripping tool (or use pre-stripped wire)
- Electric Circuits
- 2 Jumper wires with alligator clips
- 6 Small LEDs
- Several logic integrated circuits
- 6 330Ω resistors
- Multimeter

1 Digital logic

Circuits called *logic gates* are the building blocks of almost all digital systems. The fundamental logic gates are called AND, OR, NAND, and NOR. As their names imply, these gates compare two input voltages and produce an output voltage based on the inputs. The diagram below shows the output of each of the four logic gates for every combination of inputs. The NAND gate is named for the combination Not AND, which is the opposite of AND. The NOR gate is the opposite of OR.

Basic logic circuits

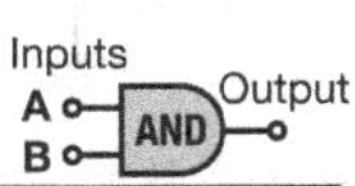

Inputs A	Inputs B	Output
0	0	0
0	1	0
1	0	0
1	1	1

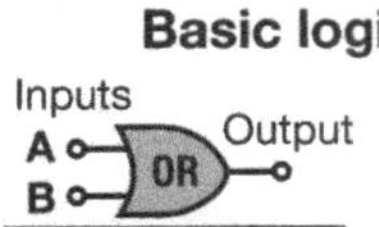

Inputs A	Inputs B	Output
0	0	0
0	1	1
1	0	1
1	1	1

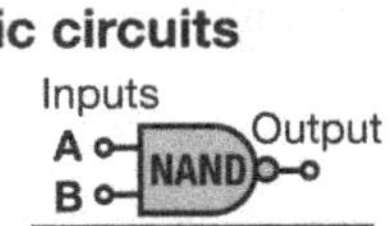

Inputs A	Inputs B	Output
0	0	1
0	1	1
1	0	1
1	1	0

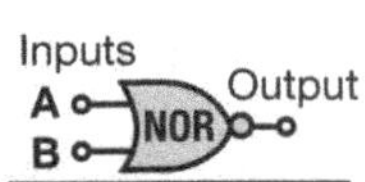

Inputs A	Inputs B	Output
0	0	1
0	1	0
1	0	0
1	1	0

Combination circuit

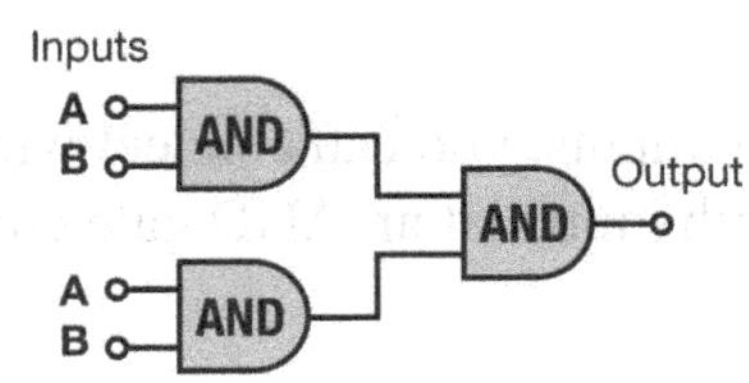

The basic logic gates can be combined to make more complex decisions. For example, the outputs of two AND gates can be connected to the inputs of a third AND gate. The output of this circuit is one *only* if all four of the inputs are also one. Electrically, the output is 5 V only when all four inputs are also 5 V. If *any one* of the inputs goes to 0 V, the output also goes to 0 V.

2 Integrated circuits

Logic gates are built from many transistors in *integrated circuits* commonly known as "chips." The diagram at right shows a picture of a chip that has four AND circuits. This chip operates with 5-V signals and must be supplied with 5 V to pin 14 and 0 V to pin 7. Inputs and outputs are connected to pins 1–6 and 8–13.

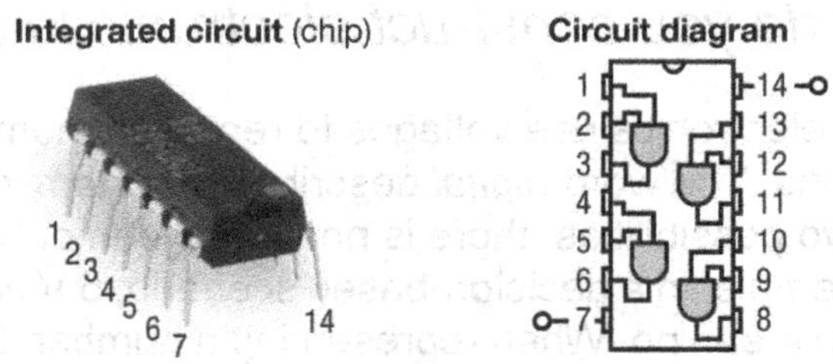

3 Building a logic circuit

The hole spacing on a breadboard is designed to accept standard integrated circuits. The long rows are connected to 5 V and 0 V (ground). The diagram at right shows the AND chip on the breadboard with power and ground connected. The notch on one end is how you find pin 1. Pin 1 is to the left of the notch and the pins are numbered counterclockwise around the chip.

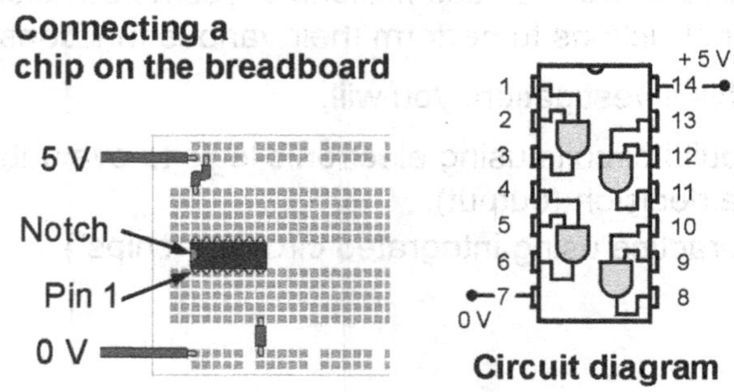

How will you know if your circuit works? When testing circuits, it is useful to attach LEDs to the outputs of the logic gates. The LEDs light up when the output is 5 V. The 330Ω resistors limit the current that must be supplied by the chip to drive the LED. The diagram below shows LEDs and resistors attached to the outputs of an AND gate. When building the circuit, remember the *diode* in light-emitting diode: diodes allow current to flow in only one direction. If you connect the LED backward, it will never light!

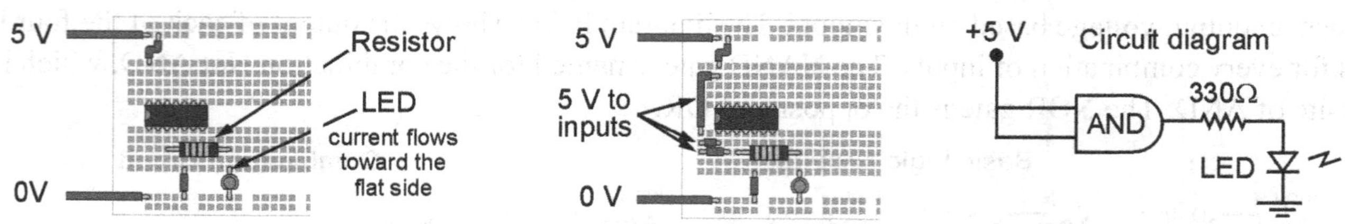

To supply inputs, you can connect wires from the 5-V supply line to the inputs of the chip. The diagram shows both inputs of an AND gate connected to 5 V so the LED lights up.

4 Designing and building some logic circuits

1. Construct a circuit with two inputs (A and B) and one output. The output should be 5 V when either A or B input is at 5 V. Use an LED to indicate the output.
2. Construct the four-input AND circuit shown on the previous page. The output should be 5 V only when all four inputs are 5 V. Use an LED to indicate the output.
3. Design and construct a circuit that has an output described by Table 1. This kind of table is called a *truth table*. This truth table describes the function OUTPUT = (A AND B) OR C. Your circuit should use an LED to indicate that the output is 5 V.

Table 1: Truth table for (A AND B) OR C

Inputs			Output
A	B	C	
0	0	0	0
0	0	1	1
0	1	0	0
0	1	1	1
1	0	0	0
1	0	1	1
1	1	0	1
1	1	1	1

25.1 Matter and Atoms

What are the properties of different elements?

All elements are made of atoms containing the same three basic particles: protons, electrons, and neutrons. Although they are made of the same particles, elements can have very different properties. In this investigation, you will research one element.

In this investigation, you will:

- research an element to learn about its properties.
- construct a poster summarizing what you learned about the element.
- share your poster with your class.

Materials List

- Copy of the periodic table
- Access to the library, Internet, and/or science resource materials
- Poster board
- Colored markers
- Glue and tape
- Supplies for creating the poster, such as magazines, rulers, etc.

1 Atoms and elements

Nearly all the matter in the world is made from 92 different elements. Water is made from the elements hydrogen and oxygen. Air contains mostly nitrogen and oxygen. Steel is mostly iron and carbon with a few exotic elements like vanadium and chromium mixed in. Rocks are mostly silicon and oxygen. We get different kinds of matter by using different combinations of the 92 basic elements.

Each of the 92 elements is made of a unique type of atom. All atoms of a given element are similar to each other. If you could examine a million atoms of carbon, you would find them all to be similar. However, carbon atoms are different from iron atoms or oxygen atoms. The atoms of an element are similar to atoms of the same element but different from atoms of other elements. One single atom is the smallest particle of an element. Since atoms of the same element are similar, every atom of carbon has the properties of carbon. Similarly, every atom of gold has all the properties of gold.

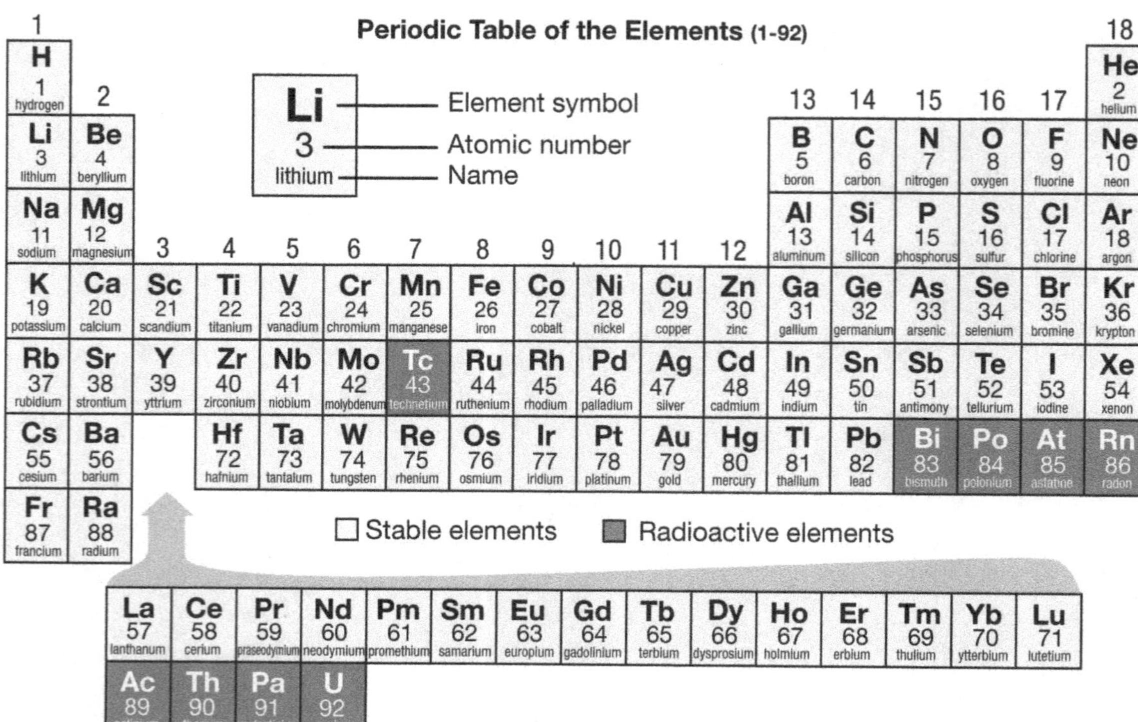

2 Research your element

Each group will select or be assigned a different element. Research your element to find the information below. Record the bibliographic data for all of the sources you use in your research.

- symbol
- atomic number
- average atomic mass
- boiling point
- density
- color
- origin of the element's name
- one compound that contains the element (state its molecular formula)
- one mixture that contains the element
- one use for the element
- three interesting facts about the element

3 Communicating your results

Make a one-page poster that neatly displays the information you found for your element. Include a bibliography on the back of the poster.

4 Characterizing the element and comparing it with other elements

Answer the following questions regarding the element you researched.

a. Name three other elements with similar chemical properties. Where on the periodic table are elements with similar properties located in relation to each other?

b. At 500°C, is your element a solid, liquid, or gas?

c. At 1,500°C, is your element a solid, liquid, or gas?

d. Does your element play a role in living organisms, such as in your body's chemistry, or in food?

25.2 Temperature and the Phases of Matter

How do you determine the melting/freezing point of cetyl alcohol?

A heating/cooling curve shows what happens to the temperature of a substance as it changes phase. In this investigation, you will study a pure substance, cetyl alcohol, as it melts and freezes. Based on your observations, measurements, and graphs, you will be able to determine the melting and freezing point of cetyl alcohol.

In this investigation, you will:

- melt and freeze cetyl alcohol.
- create and analyze the heating and cooling curves for cetyl alcohol, and determine the melting/freezing point.
- explain what happens to the temperature of a substance as it changes phase.

Materials List

- DataCollector
- Temperature probe
- Test tube, 22 × 150 mm
- Cetyl alcohol
- Hot water
- Safety goggles
- Lab apron
- Hot mitts
- 250 mL Beaker
- 400–600 mL Beaker
- Electronic scale

WARNING — This lab contains chemicals that may be harmful if misused. Read cautions on individual containers carefully. Not to be used by children except under adult supervision.

1 Making a prediction

a. Cetyl alcohol can be produced from vegetable oils, and is used in the manufacturing of soap, shampoo, skin creams, and lotions, among other things. The cetyl alcohol you will be using is a pure substance.What do you expect to observe in terms of the temperature range for the freezing/melting point?

b. Suppose your cetyl alcohol sample became contaminated with another substance. How would this affect the freezing/melting point?

2 Using a temperature probe with the DataCollector

1. Turn on the DataCollector and choose Data Collection mode. Plug in a temperature probe. Push the Go button to begin a new experiment.
2. Once an experiment begins in Data Collection mode, the Meter window appears (see example at right). The Meter window displays the current readings for active sensors. The clock at the bottom of the window shows elapsed time.

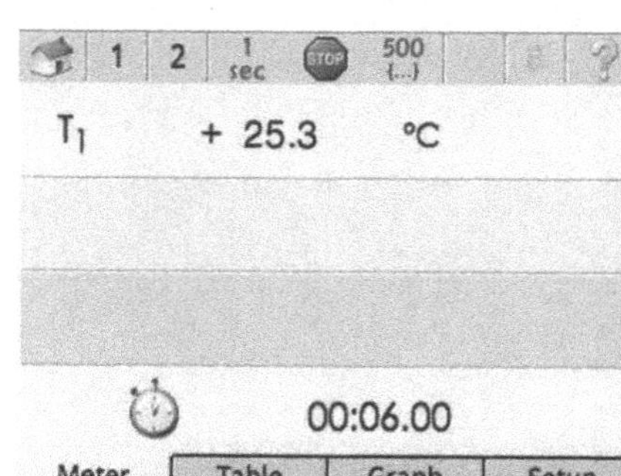

3. To stop an experiment, push the Go button on the DataCollector enclosure. The Meter window will continue to show current sensor readings, although they are not being stored in the data table.
4. To resume the experiment, push the go button again. You can switch to Table or Graph view while the experiment is running by tapping on those selections at the bottom of the screen.
5. If you want to start a new experiment, tap on the Home icon at the top of the screen and choose Data Collection mode again.

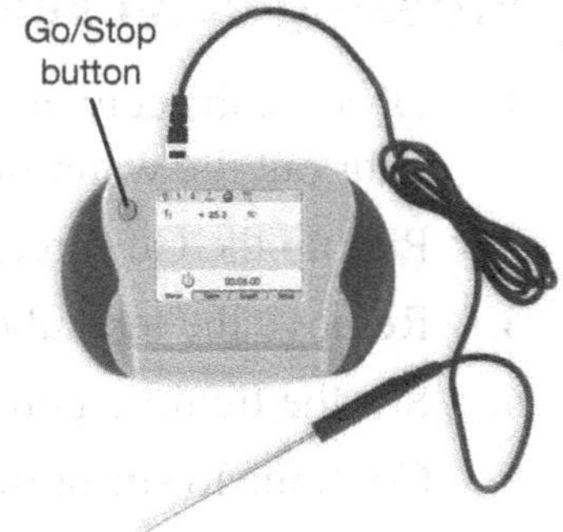

3 Melting cetyl alcohol

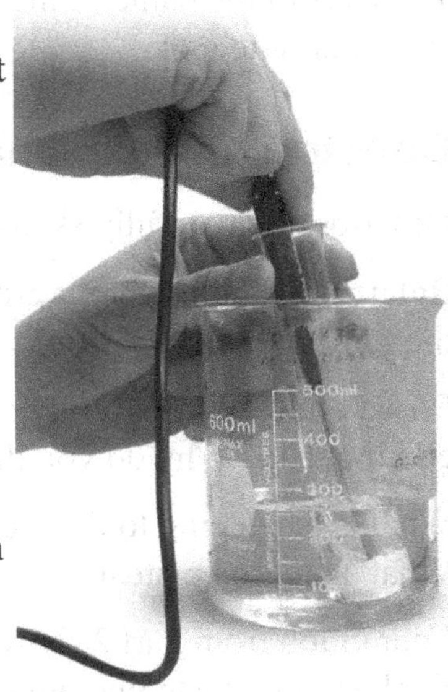

1. Fill a 250-mL beaker with cold water.
2. Fill a 400- to 600-mL beaker with approximately 250 mL of hot water (75–80°C).
3. Measure out approximately 4–5 grams of cetyl alcohol in a test tube.
4. Measure the temperature of the hot water to make sure it is within range (75–80°C).
5. Quickly immerse the temperature probe into the cold water. Then remove and dry it, making sure it is cooled to room temperature.
6. Pull up the Go screen on the DataCollector (tap the Home icon and then Data Collection mode to get to the Go screen).
7. Place the temperature probe into the test tube, put the test tube containing the solid into the water bath, and press go on the DataCollector. View the experiment from the graph window. You will see time on the x-axis and temperature on the y-axis.
8. Stir the solid vigorously with the temperature probe.
9. Continue to stir as long as some solid remains.
10. Once all of the material is liquid, press the Stop/Go button to end Data Collection. The experiment is automatically named with the word *experiment* and a number. To see the name of the experiment for which you just collected and stored data, tap on the setup choice at the bottom of the screen. Make a note of the experiment name. Leave the sample in the warm water (with the temperature probe still in it) for use in Part 5.

4 Analyzing and interpreting data

a. Predict the effect of sample size on the melting/freezing point.

b. Describe and carefully sketch the shape of the graph from Part 3.

5 Freezing cetyl alcohol

1. Observe the cetyl alcohol in the test tube to make sure it is still all liquid. If some solid has formed, pour out the water in the warm beaker and add more warm water.
2. Pull up the Go screen in Data Collection mode.
3. Remove the test tube from the warm beaker and press go to start the DataCollector experiment.
4. Stir the liquid vigorously with the temperature probe as the solid begins to form.
5. Continue to stir until all of the liquid turns into a solid. (*Note*: If you need to repeat any part of the experiment, you may reuse your sample.)
6. When you are finished recording data, press the Stop/Go button on the DataCollector and return your sample of cetyl alcohol to your teacher.
7. Record the name of the experiment from the DataCollector (found in the Setup window) in your notebook.

6 Constructing explanations

a. Describe and carefully sketch the graph from Part 5.

b. Referring to both of your graphs, determine the melting and freezing point of cetyl alcohol. Are they the same? Should they be?

c. Based on the shapes of your graphs, which data do you think is more reliable, the heating data or the cooling data? Why do you think this is so?

d. What is happening to the molecules of cetyl alcohol during the diagonal portions of the heating curve? What about the plateau?

e. Your graph from Part 2 shows that during a change from solid to liquid, the temperature stays the same. Explain why the temperature does not increase, even though energy is being added.

25.3 The Specific Heat of a Metal

How can you use specific heat to identify an unknown metal sample?

If you have ever walked barefoot on a concrete walkway during a hot day, you have felt its warmth on your feet. In fact, the thermal energy transferred to your feet may send you retreating to the grass or even a swimming pool. Why is the temperature of the concrete so different from the temperature of the soil or the swimming pool water? Although the sunlight shines on all three surfaces, it is easier to raise the temperature of concrete than the water in a swimming pool. A hot summer day may only raise the temperature in a pool by one or two degrees.

In this investigation, you will:

- use a calorimeter, the specific heat of water, and the law of conservation of energy to determine the specific heat of an aluminum sample and two unknown metal samples.

Materials List

- DataCollector
- Temperature probe
- 3 Foam cups (with or without lids)
- Electronic scale or triple beam balance
- Safety goggles
- Lab apron
- Aluminum and zinc samples
- Unknown metal sample
- Hot water
- Tongs

1 Doing the investigation

1. Find the mass of the aluminum sample and record it in Table 1.
2. Make a calorimeter by nesting two foam cups.
3. Pour about 150 mL of room temperature water into the calorimeter. Measure and record the mass of the water you add to the calorimeter (density of water = 1g/mL). (*Note*: Make sure there is sufficient water in the calorimeter to submerge your metal samples completely.)
4. Fill a styrofoam cup two-thirds full with hot water.
5. Place your aluminum sample into the cup with hot water. Allow the sample to sit in the water for about a minute so it will get warmed by the hot water.
6. Read Part 2 from the previous investigation, "Using a temperature probe with the DataCollector," if you need to review how to run experiments with the DataCollector and temperature probe.
7. Pull up the Go screen on the DataCollector (tap the Home icon and then Data Collection mode to get to the Go screen).
8. Tap the go button and use the temperature probe to measure the temperature of the hot water. Record this temperature, once it stabilizes, as the initial temperature of the metal in Table 1.
9. Remove the temperature probe from the hot water and place it in the calorimeter. When the temperature stabilizes, record it as the initial temperature of water. Leave the temperature probe in the calorimeter.
10. Use the tongs to remove the aluminum from the hot water and place it into the calorimeter. Try to transfer as little hot water as possible into the calorimeter when moving the sample. Go as quickly as possible but do not spill any hot water.
11. Once the temperature stabilizes, record the final temperature in Table 1.
12. Repeat steps 1–11 with both the zinc sample and the unknown metal sample provided by your teacher. You may be asked to conduct two trials with your unknown samples, if time allows.

Table I: Mass and temperature data

	Aluminum	Zinc	Unknown metal
Mass of metal (g)			
Mass of water in calorimeter (g)			
Initial temperature of water (°C)			
Initial temperature of metal (°C)			
Final temperature of mixture (°C)			

2 Analyzing the data

a. Calculate the temperature change of the water.

b. Calculate the temperature change of the metal.

c. Calculate the heat gained by the water using the equation below.

HEAT EQUATION

$$E = mC_p(T_2 - T_1)$$

Where: E = thermal energy (Joules) lost or gained by the water in the calorimeter
m = original mass of the measured water in the calorimeter (grams)
C_p = specific heat of water (4.184 J/g °C)
$T_2 - T_1$ = change in temperature of the water (also referred to as ΔT)

d. The amount of energy released by the metal is equal to the energy absorbed by the water. Knowing the value of E for the metal, calculate the specific heat (C_p) of the metal. (Remember, since the energy is being released by the metal, change the sign of E from answer 2c to use for answer 2d.)

e. Identify your unknown metal by comparing your calculated value of its specific heat to known specific heat values of common metals provided by your teacher.

f. Calculate the percent error for your unknown sample.

3 Constructing explanations

a. Compare and contrast the specific heats of aluminum and zinc. How do their positions on the periodic table help explain their specific heat differences? If you add the same amount of heat energy to both, which metal would experience the greatest change in temperature?

b. How close were your experimental specific heat values to the known values for aluminum and zinc? Were your experimental values too high or too low? Based on the procedure, give an explanation for why your experimental values might have been too high or too low.

c. What did you determine was the identity of your unknown sample? How did you come to this conclusion?

d. How does the law of conservation of energy allow us to make the calculations needed to determine the specific heat of the unknown sample?

e. Suppose you had to design a small metal container to keep soup warm for several hours. Would you rather use a metal that his a relatively high specific heat, or one that has a relatively low specific heat? Why?

f. Water has a high specific heat. How does the fact that humans are largely made up of water help us regulate our body temperature?

26.1 Heat Conduction

How does heat pass through different materials?

Conduction is the transfer of heat through materials by the direct contact of matter. All materials conduct heat at some rate. Solids usually are better heat conductors than liquids, and liquids are better conductors than gases. The *thermal conductivity* of a material describes how well the material conducts heat. Materials with high thermal conductivity, such as copper, aluminum, and other metals, are good thermal *conductors.* Materials with low thermal conductivity, like fiberglass and foam, are *insulators*. This investigation explores how different materials conduct heat.

In this investigation, you will:

- compare the thermal insulating or conducting properties of materials.
- measure the heat flowing between two different temperatures.

Materials List

- Shallow baking pan
- Thin plastic cup
- 2 Foam cups
- 4 Copper nails
- Metric ruler
- Permanent marker
- Temperature probe
- Small pot
- Hot plate
- DataCollector
- Crushed ice
- Water
- Digital scale or triple beam balance (accurate to at least 0.1 gram)

1 Heat flowing from a cup

This experiment will investigate how quickly heat moves from hot water in a cup to cold water in a pan through different substances. Fill a shallow baking pan with two centimeters of water and crushed ice. You will also need a container of water that has a temperature of at least 60°C. Record your data in the top section of Table 1.

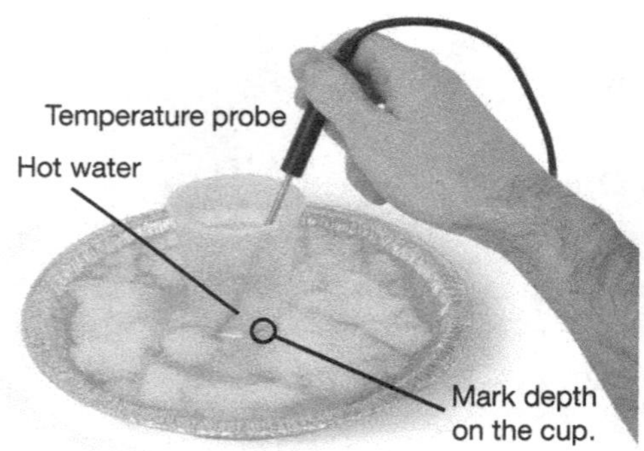

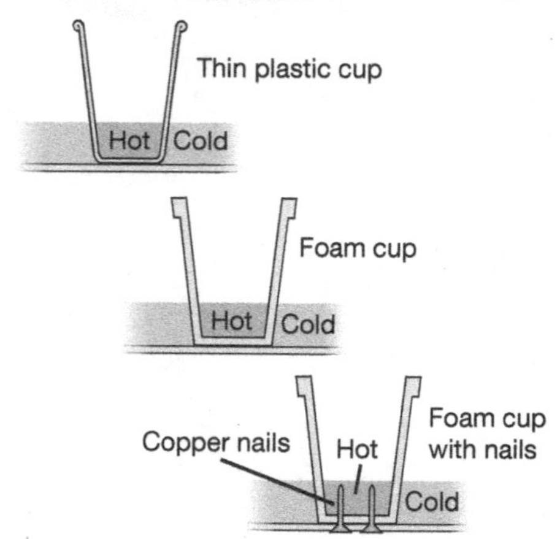

1. Use the DataCollector in Meter mode and the temperature probe to measure the temperature of the cold water bath. Record this and all other measurements in Table 1.
2. Measure the depth of water in the cold pan. Mark the depth on a thin plastic disposable cup with a marker. Fill the cup with hot water to the mark.
3. Hold the cup of hot water in the pan of cold water and swirl it gently with the temperature probe. Using the running clock at the bottom of the Meter screen, measure the time it takes for the temperature to drop from 50°C to 40°C, and from 30°C to 20°C. Use the same starting and final temperatures for all variations of the experiment.
4. Repeat the same experiment, but use a foam cup to hold the hot water. Use the plastic cup to transfer the new hot water to the foam cup so the volume of water stays constant.
5. Prepare another foam cup by pushing four copper nails through the bottom as shown in the diagram. Repeat the experiment again with this setup. Remember to swirl the cup constantly with the temperature probe as you are monitoring the temperature.

6. Use the plastic cup to measure the mass of water used in the experiment. Enter the mass in all three rows of Table 1 since it is the same for each trial.

2 Heat flowing into a cup

For this part of the investigation, set up the opposite circumstance. Heat will flow from the hot water in the pan to the cold water in the cup. Fill the pan with two centimeters of hot water that is at least 60°C. You may need to add hot water from time to time to keep the water temperature within 3–4 degrees of 60°C. Fill a second container with crushed ice and enough water to refill the cup several times.

1. Repeat the same three experimental variations you did in Part 1. This time, the heat will flow *into* the cup. Measure the time it takes for the temperature of the water in the cup to go from 20°C to 30°C, and from 40°C to 50°C. Again, be sure to gently swirl the cup around the bottom of the pan to mix the water and even out the temperature. Keep the temperature probe in the cup.

2. Record the temperature of the water in the pan and the mass of water in the cup in the appropriate rows along with the times, using the bottom part of Table 1.

Table 1: Temperature and mass data

Heat flowing from cup to pan	Cold temp (T_c, °C)	Mass (kg)	50°C to 40°C (time in s)	30°C to 20°C (time in s)
Plastic cup				
Foam cup				
Foam cup, nails				

Heat flowing from pan to cup	Hot temp (T_h, °C)	Mass (kg)	20°C to 30°C (time in s)	40°C to 50°C (time in s)
Plastic cup				
Foam cup				
Foam cup, nails				

3 Analyzing the data

a. Use the concept of thermal conductivity to explain the differences between the measurements for the plastic cup, the foam cup, and the foam cup with the nails pushed through the bottom.

b. Based on your observations, is the copper in the nails a thermal insulator or a thermal conductor?

c. Compare the time it took the water in the cup to cool down from 50°C to 40°C with the time it took to cool down from 30°C to 20°C. Explain the difference using what you know about how heat flows.

d. Compare the time it took the water in the cup to heat up from 20°C to 30°C with the time it took to heat up from 40°C to 50°C. Explain the difference using what you know about how heat flows.

e. The specific heat of water is 4,184 J/kg°C. Use the heat equation to calculate how much energy was used to change the temperature of the water from 20°C to 30°C.

f. Compare the amount of energy used to heat the water from 20°C to 30°C with the amount of energy used to heat the water from 40°C to 50°C. Is the energy the same? Do the same comparison for the energy removed to cool the water from 50°C to 40°C, and from 30°C to 20°C.

26.2 Convection

Can moving matter carry thermal energy?

Convection is the transfer of heat by the motion of liquids and gases. Free (or natural) convection occurs whenever there is a temperature difference in a gas or liquid. Free convection in a gas occurs because gas expands when heated. Hot gas with lower density is lighter than surrounding cooler gas and floats upward, carrying heat along with it. Convection in liquids also occurs because of differences in density. Hot liquid is less dense than cold liquid. Therefore, hot liquid rises and cold liquid sinks. The resulting current transfers heat from hotter regions of liquid to cooler regions.

In this investigation, you will:

- observe the currents created by free convection in water.
- observe the rapid exchange of thermal energy in forced convection.

Materials List

- 2 Identical clear glasses
- 1 Colored ice cube
- Uncolored ice cubes
- Tongs
- Stirring rod
- Hot tap water

1 Observing free convection in water

1. This experiment requires preparation a day ahead of time. Make a colored ice cube by mixing five drops of food coloring in enough water to fill 1 pocket of an ice cube tray. Fill the rest of the tray with clear water and place the tray in the freezer overnight.
2. Obtain two clear glasses that are the same size and shape. Fill one with hot water from the faucet.
3. Fill the other glass with water and a few ice cubes that are not colored. Stir the water and ice well.
4. Remove the colored ice cube from the tray. **Handle the colored ice cube with tongs because the color may stain your clothes or fingers.**
5. Using the tongs, dip about 2 centimeters of the colored ice cube into the hot water and observe what happens. What do you see?
6. Remove the ice cubes from the glass of cold water so they do not interfere with observing. Using the tongs, dip about 2 centimeters of the colored ice cube into the cold water and observe what happens. Remove the colored ice cube and put it back in the freezer.

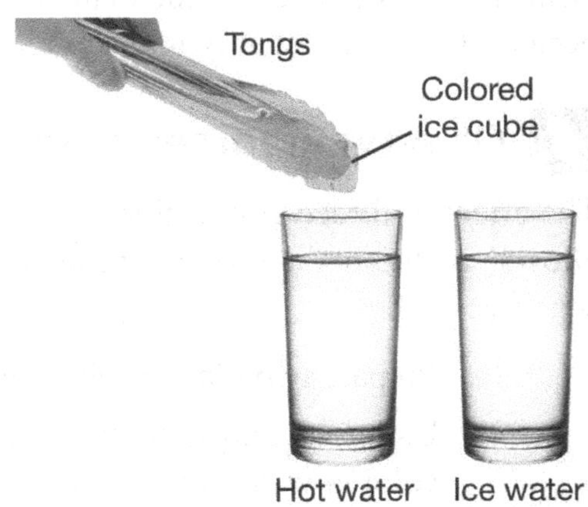

2 Constructing explanations

a. Describe the motion of the cold water melting off the colored ice cube when it was partially immersed in hot water. Did the melting colored water move upward, downward, or to the side? Explain what you observed using the concept of density of hot and cold liquids.

b. Compare the motion of the colored water melting off the colored ice cube when it was partially immersed in cold water. Did the melting colored water move upward, downward, or to the side? Was the motion faster or slower than it was in the hot water? Explain what you observed using the concept of density of hot and cold liquids.

3 Forced convection

When the flow of gas or liquid is circulated by pumps, fans, or anything other than natural buoyancy forces, the heat transfer is called *forced convection*. Using a fan on a hot day is an example of forced convection. The breeze blowing over your skin carries away heat much faster than natural convection would in still air. Forced convection is used in many human technologies.

To see why forced convection is so effective, consider a warm surface in the air. Heat flows from the surface to the air by conduction because the molecules in the air make contact with the molecules of the surface. The air near the surface gets slightly warmer and rises, carrying the heat with it by natural convection.

The rate at which heat flows depends on the difference in temperature between the surface *and the air right near the surface*. Natural convection circulates cooler air against the surface slowly. After a while, the rate of heat flow drops because the air near the surface is not much cooler than the surface itself. The diagrams below show how the temperature changes with distance away from the surface.

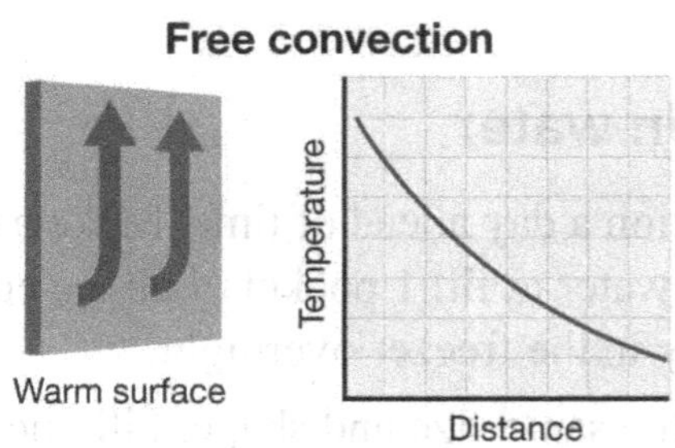

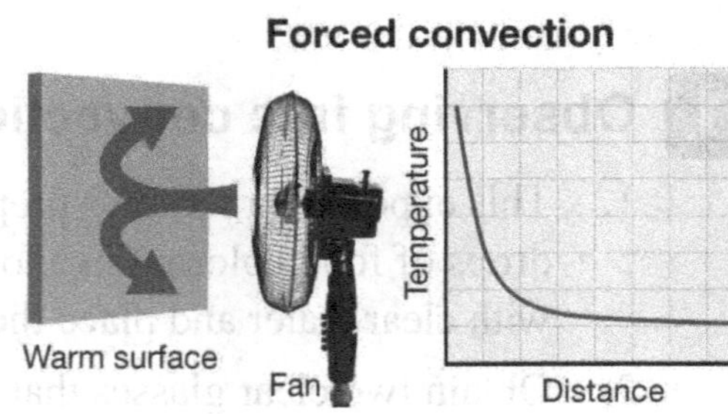

Now consider the situation with a fan. Cool air is blown against the surface, displacing the warmer air right near the surface. The difference between the air temperature and the surface temperature is much greater. Heat flows faster because the temperature difference is greater. The temperature difference remains great because the fan constantly replaces air that has been warmed by the surface with cooler air from farther away.

4 Observing forced convection

The effect of forced convection can be demonstrated using the colored ice cube. With the ice cube, the color allows you to see the mixing of low and high temperature that would normally be detectable only with sensitive probes.

1. Prepare two glasses with hot water from the faucet.
2. Using the tongs, dip the colored ice cube into one cup of water and observe the rate at which the colored water mixes with the clear water. The rate at which the liquids mix reflects the rate at which heat is being carried by convection from the water farther from the ice cube to the ice cube itself. Remove the colored ice cube before it melts to half its original size.
3. Have a partner stir the hot water in the other glass. While the water in the glass is still spinning, dip the colored ice cube in the glass with the stirred water (and keep stirring). Observe the rate at which the colored water mixes with the clear water while stirring.

Stirring is a form of forced convection because the water is circulated by forces other than natural convection.

a. Which situation mixed the colored water more rapidly into the clear water, free convection or forced convection?

b. In which situation would you expect more rapid flow of heat, free convection or forced convection?

26.3 Radiant Heat

How does heat get from the Sun to Earth?

Thermal radiation, or heat, is electromagnetic waves (including light) produced by objects, like the Sun or a sunflower, because of their temperature. All objects with a temperature above absolute zero give off thermal radiation. Thermal radiation comes from the thermal energy of atoms. Power given off as thermal radiation increases with higher temperatures because the thermal energy of atoms increases with higher temperatures. This investigation explores some of the properties of thermal radiation both from the Sun and from an electric light.

In this investigation, you will:

- demonstrate that sunlight carries heat.
- observe how the light spectrum emitted by hot objects changes with their temperature.
- recognize that all objects emit and absorb electromagnetic radiation.
- identify objects that are blackbody radiators and those that are not.

Materials List

- Cardboard
- Double-sided tape
- Incandescent lamp/dimmer switch
- Spectrometer
- Scissors
- Glue
- Aluminum foil
- DataCollector
- Temperature probe
- Metric ruler
- Pencil

1 Creating a focusing mirror for collecting solar radiation

Radiation carries energy through the vacuum of space—93 million miles between the Sun and Earth. One way to collect solar power is to use a mirror that concentrates the radiation from the Sun on a small area. A parabolic mirror reflects parallel rays of light to a point at the focus. Solar collectors use parabolic mirrors because they are the most efficient shape for concentrating radiation from the Sun.

1. To create a parabolic mirror, start with one 20 × 28-centimeter piece of cardboard. Draw lines two centimeters apart parallel to the short side across the cardboard. Using a ruler, plot the points in Table 1 on the cardboard, working left to right. The points describe the surface of a parabola with two flat sections, one on either side.
2. Connect the points with a smooth curve and cut out the shape as shown in the diagram at right. Also cut 2-centimeter-square notches out of the corners.

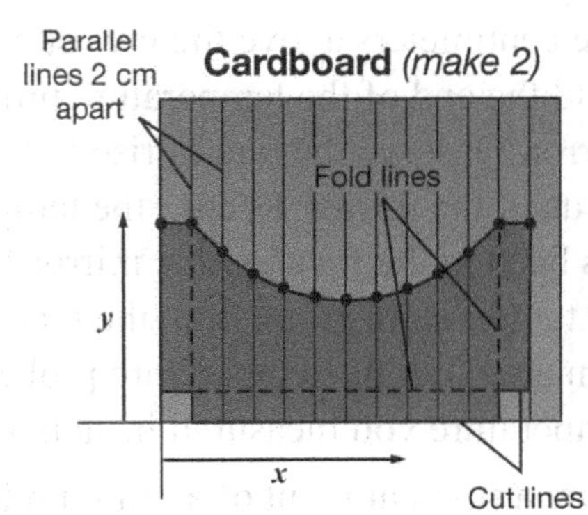

Table 1

x (cm)	y (cm)
0	13.0
2	13.0
4	11.2
6	9.8
8	8.8
10	8.2
12	8.0
14	8.2
16	8.8
18	9.8
20	11.2
22	13.0
24	13.0

3. Make a second, identical cardboard shape and fasten the two shapes together with tape or glue by folding and overlapping the tabs on both sides and on the bottom, as shown at right.
4. Cut two more pieces of cardboard that measure 3 × 14 centimeters. Apply three strips of double-sided tape to each of the cardboard rectangles. Carefully lay aluminum foil shiny side up on the cardboard so it sticks to the tape. Cut the foil a little wider than the cardboard and fold it around the edges. Be careful not to wrinkle the foil because it is the reflective surface for your parabolic mirror.
5. Carefully lay the two strips of foil-covered cardboard on the parabolic shape and glue them in place. (A hot glue works well for this purpose. Safety precautions must be taken!) The completed mirror should look like the diagram at right.

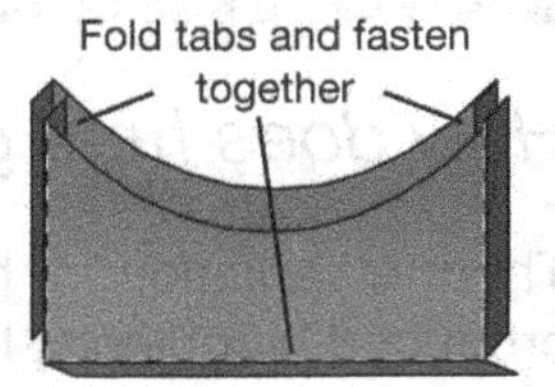

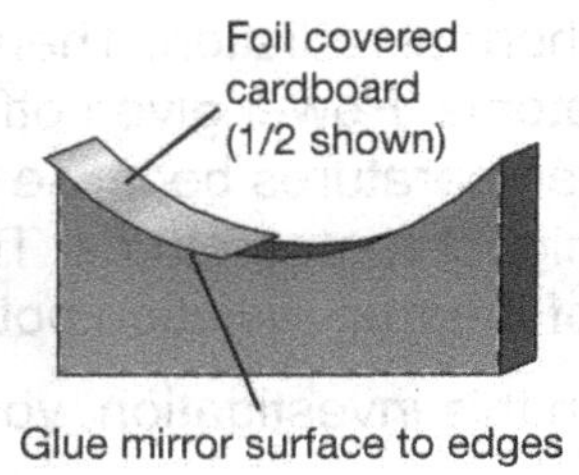

2 Observing radiant energy in action

1. Read Part 2 from investigation 25.2, "Using a temperature probe with the DataCollector," if you need to review how to run experiments with the DataCollector and temperature probe.
2. Tap the Home icon and then select Data Collection mode on the Data Collector.
3. Tap Go and hold the temperature probe in open air and sunlight until it comes to thermal equilibrium (the temperature reading holds steady). Record the temperature in Table 2.
4. Hold the mirror so its axis faces the Sun. Hold the temperature probe in the focus of the mirror. The focus is five centimeters above the lowest point and along the axis. Hold the end of the temperature probe at the focus of the mirror. Observe the rate of rise in the temperature at the focus of the mirror. Record the temperature after the probe has been in the focus of the mirror for 30 seconds. Remove the temperature probe from the focus to prevent it from being damaged. Let the temperature probe cool down to the temperature you measured in step 3.
5. Wrap a small amount of aluminum foil around the end of the temperature probe. Repeat step 4, holding the foil-wrapped end of the probe at the mirror's focus. Again, hold the temperature probe at the focus for 30 seconds and record the temperature in Table 2. Remove the thermometer from the focus and let it cool down again.

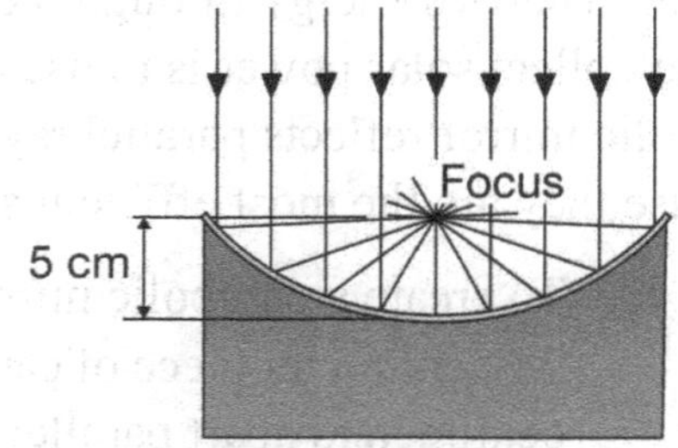

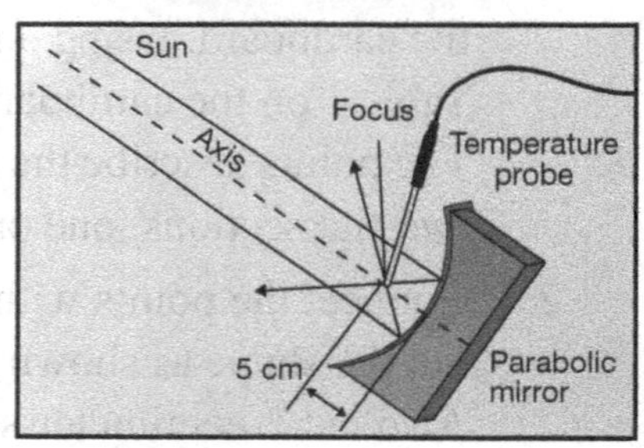

Safety Tip: Never look directly at any bright light source, especially the Sun!

Table 2: Temperature data

	Temp (°C)
In sunlight, outside of mirror	
After 30 seconds at focus	
After 30 seconds at focus, covered with aluminum	

3 Using your model

a. Explain why the thermometer at the focus of the mirror measured a much higher temperature than the same thermometer in sunlight outside the mirror. Your explanation should refer to the concepts of radiant heat and intensity.

b. Explain why the thermometer wrapped in aluminum foil measured a lower temperature rise than when the thermometer was not wrapped. Your answer should use the concepts of absorbed and reflected radiant energy.

4 The blackbody spectrum

To a physicist, an object is perfectly black when it absorbs all light that falls on its surface. If all light is absorbed, any light coming off the surface can only be from thermal radiation. A perfect blackbody radiator is a surface that reflects nothing and emits only thermal radiation. To a physicist, the white-hot filament of a bulb is a good blackbody. All light from the filament is thermal radiation; almost no light is reflected from other sources.

The power vs. wavelength graph for a perfect blackbody is called the blackbody spectrum. The blackbody spectrum changes with the temperature of the surface emitting the radiation. Near room temperature, the spectrum does not extend into visible light. Therefore, we cannot "see" objects by their own light. We see ordinary objects by reflected light. Up to a few thousand degrees Celsius, most of the energy in thermal radiation is in invisible infrared light. As objects heat up, they start to give off visible light, or glow. The curve for 600°C at right shows some power emitted in red and orange visible light. At 600°C, objects glow dull red like a burner on an electric stove. As the temperature rises, thermal radiation produces shorter-wavelength, higher-energy light. At 1,000°C, the color is yellow-orange and turns to white at 1,500°C.

The blackbody spectrum
(Power vs. wavelength)

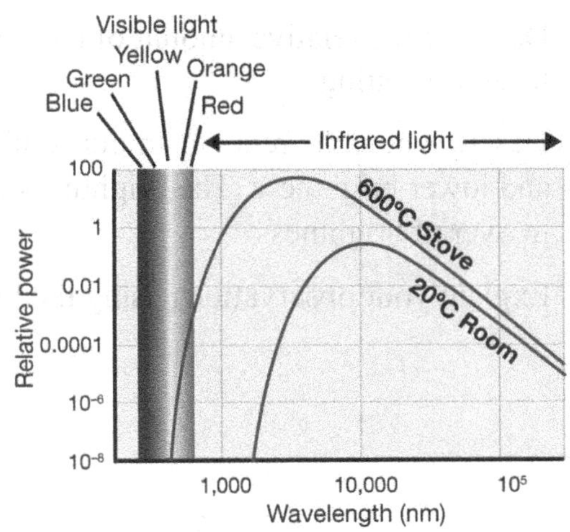

a. Once it is turned on, a heating element on a stove will change from black to orange-red. Describe what's happening in terms of thermal radiation.

b. Identify whether these objects are blackbody radiators or not: an orange, a cup of hot coffee, a green tee-shirt, a star in the night sky.

c. For each of the objects in question 4b, describe which colors of visible light are involved in seeing these objects. Which colors do you see that are related to thermal radiation?

5 Observing the blackbody spectrum

If you carefully watch a bulb on a dimmer switch, you can see the color of the bulb change as the filament gets hotter. The bright white light from an incandescent bulb is thermal radiation from a very hot filament—near 2,600°C.

1. Obtain a halogen lamp that has a dimmer switch. Halogen bulbs operate at very high temperatures (*be careful*). An ordinary incandescent lamp with a dimmer switch also will work but a fluorescent light will not work.
2. Look at the light through the spectrometer as you adjust the dimmer switch. Remember, to use the spectrometer, point the slit at the light and look left to see the spectrum.

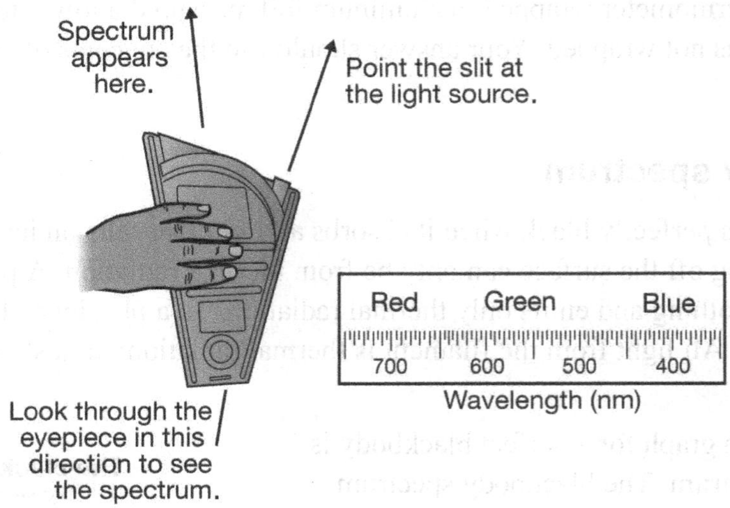

a. Describe the relative amount of red, green, and blue light in the spectrum when the bulb is at its brightest setting.

b. Describe how the relative amounts of red, green, and blue light change as the bulb is turned to lower and lower brightness. The brightness diminishes, but try to separate the color changes from the change in overall brightness.

c. Explain your observations using the blackbody spectrum.

27.1 Properties of Solids

How do you measure the strength of a solid material?

Physical strength refers to an object's ability to hold its form even when force is applied. Solid materials vary widely in strength. We often choose a material for a given application based on the strength of the material. For example, steel is good for building bridges because it is solid and strong. You would not build a bridge from wax, even though wax is a solid.

The strength of an object (such as a bridge) depends on the answers to two questions.

1. How much does the object bend or deform under applied force?
2. How much force can the object take before it breaks?

In this investigation, you will:

- measure the breaking strength of a material.
- calculate the tensile stress at fracture.
- observe the deformation of a material as it breaks in tension.

Materials List

- Modeling clay (1 cup)
- Metric ruler
- 5-Newton spring scale
- Graph paper
- Packaging tape
- Calculator

1 Force, strength, and stress

The strength of an object can be broken down into considerations of *design* and *materials*. Design means the size and shape of an object. Even a cable made from a weak material like wax can support a car if the cable is thick enough. A thin steel cable can support the same car but with much less material. To properly assess the strength of a material itself, the effects of size and shape must be separated out.

The *stress* in a material is the ratio of the force acting through the material divided by the *cross-section* area through which the force is carried. The cross-section area is the area perpendicular to the direction of the force. Dividing force by cross-section area (mostly) separates out the effects of size and shape from the strength properties of the material itself. The Greek letter sigma (σ) is used as the symbol for stress. Stress (σ) is force (F) divided by cross-section area (A).

Force (N)

Stress (N/m²) $$\sigma = \frac{F}{A}$$

Area (m²)

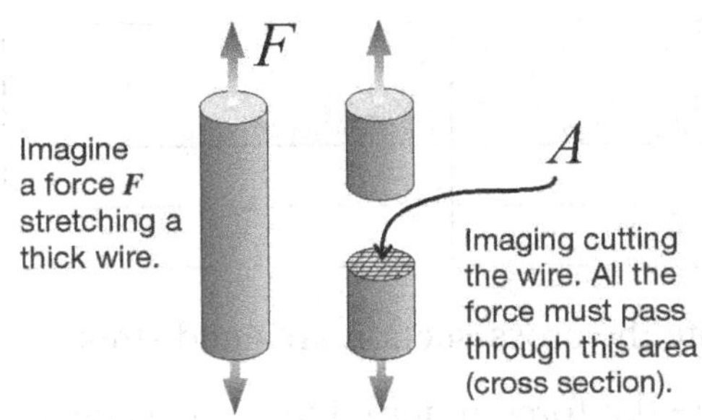

Materials break when the stress within them reaches the limit the material can take. Soft solids like wax break at a low value of stress. Materials like steel break at a much higher stress level. The *tensile strength* is the stress at which a material breaks under a tension force. Tension forces are "stretching" forces. Strong materials like steel have high tensile strength. Weak materials like wax and rubber have low tensile strength. Materials like wood and plastic have tensile strength between rubber and steel.

2 Measuring the tensile strength

Although clay breaks at a relatively low value of stress, the technique used for measuring its tensile strength is exactly the same as used for stronger materials such as steel and aluminum.

Preparing the test samples

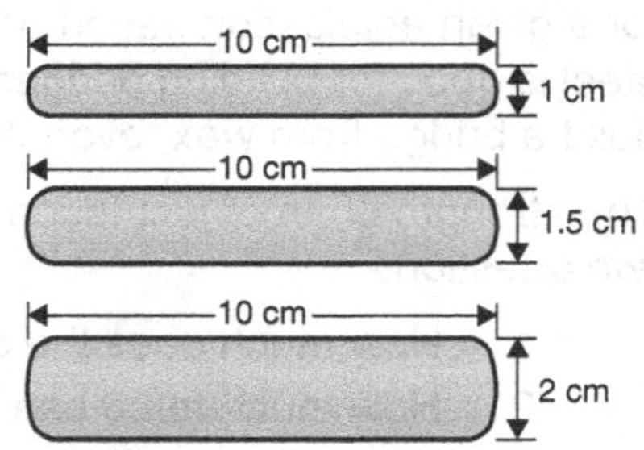

1. Get about one cup of modeling clay. Work the clay with your hands until it is smooth and uniform with no voids or cracks inside. Any cracks or voids in the clay will cause it to fail at a much lower value of stress.
2. Roll the clay into three cylinders as shown in the diagram at right. The cylinders should be 10 centimeters long and have diameters of 1, 1.5, and 2 centimeters.
3. The breaking force for each of the cylinders will be measured with a 5-newton spring scale. Adjust the calibration of the scale so that the pointer reads 0 N when held horizontal.
4. Use a small length of tape to make a loop on one end of each clay cylinder. Attach the spring scale to the tape and slowly pull the clay until it stretches and breaks. Be sure to watch the scale carefully. The breaking force is the scale's reading *just as the clay starts to stretch and break*. Once the clay starts to stretch, it will keep stretching rapidly if you keep a constant force applied. A material is considered to have failed as soon as it starts to stretch rapidly.

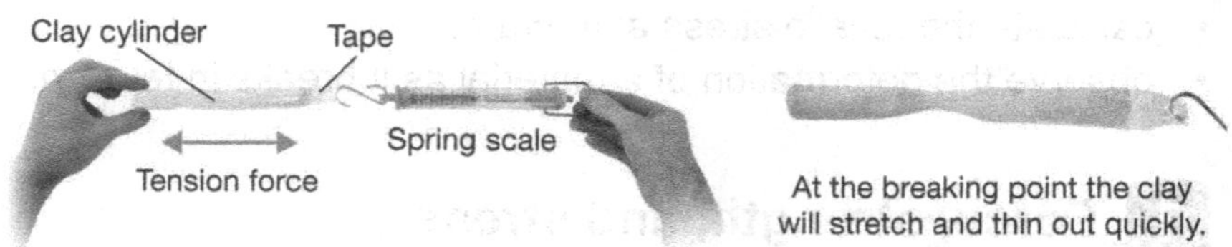

At the breaking point the clay will stretch and thin out quickly.

5. Repeat the breaking force measurement for all three samples of clay. You will have to remake each clay sample several times to get consistent measurements. Practice your technique and try to be consistent about identifying the force at which the clay just begins to stretch toward failure. Record your most consistent results in Table 1.

Table 1: Breaking force data

Diameter (cm)	Breaking force (N)	Cross-section area (cm^2)	Stress (N/cm^2)

a. Calculate the cross-section area and stress for each of the clay cylinder samples.

b. Compare the force required to break each sample with the stress in each sample. Which is a better measure of the strength of the clay as a material, the breaking force or the stress?

c. Plot a graph showing the breaking force versus the cross-section area. What mathematical property of the graph is the tensile strength? Which is the maximum stress the clay can take before breaking?

d. Use your data to estimate the force required to break a cylinder of clay that is three centimeters in diameter.

e. Suppose you could keep a constant force applied as the clay stretches. What happens to the stress in the clay once it starts stretching?

27.2 Properties of Liquids and Fluids

What are some implications of Bernoulli's equation?

When forces are applied to fluids, they create *pressure* instead of stress. Like stress, pressure is a ratio of force per unit area. Unlike stress, however, pressure acts in all directions and not just the direction of the applied force. Pressure is as important to understanding fluids as force is in understanding motion. Fluids flow because of differences in pressure. For example, water flows away from high pressure in a pipe to low pressure at an open faucet. This investigation explores Bernoulli's equation, which relates the pressure, height, and speed of a moving fluid.

In this investigation, you will:

- demonstrate how the movement of a fluid creates pressure differences.
- build a device for measuring the speed of moving air.
- use Bernoulli's equation to determine the speed of moving air from a pressure difference.

Materials List

- Plastic straw with flexible "elbow"
- 3x3" sticky note
- 8" length of 3/4" pvc pipe with pre-drilled hole
- 8" length of 1/2" pvc pipe with pre-drilled hole
- Coupling to join the two pipes
- 0.5-m length of 1/4" clear flexible plastic tubing
- Red food coloring
- Water
- Paper cup
- Clear tape
- Meter stick

1 Bernoulli's equation

Pressure and energy are related. Energy represents the stored ability to exert force and do work. Fluid in a container has energy: any pressure created by the fluid pushes on the sides of the container with forces that can do work. The law of conservation of energy applies to fluids just like anything else. When written for a fluid, the law of conservation of energy is called *Bernoulli's equation.* Bernoulli's equation relates the pressure (P) to the speed (v), the mass density (ρ), and the height (h). The variable involving height is the potential energy, written in variables convenient to fluids. The variable involving speed is the kinetic energy. The variable involving pressure is the energy that is present because of pressure differences in the fluid.

BERNOULLI'S EQUATION

$$\rho g h + P + \frac{1}{2}\rho v^2 = constant$$

(along any streamline in a fluid)

ρ	Density (kg/m³)
g	Strength of gravity (N/kg)
h	Height (m)
P	Pressure (Pa)
v	Speed (m/s)

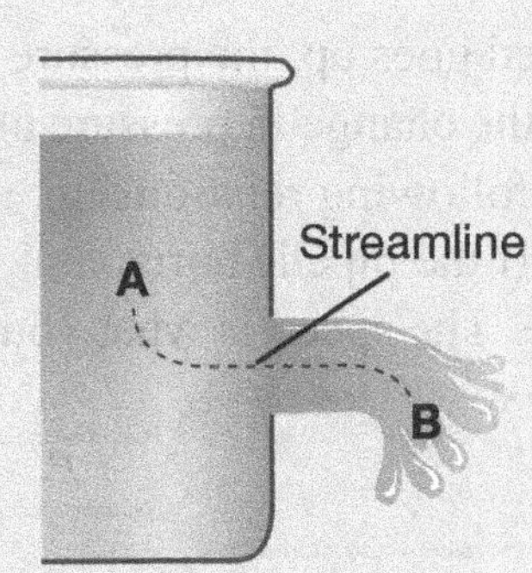

Streamlines are imaginary lines that show the flow of fluid. We draw streamlines so they are always parallel to the direction of flow. For example, a streamline connects point A and point B in the diagram above, meaning fluid from A moves to B. Bernoulli's equation tells us that if one or two of the variables increases along a streamline, at least one of the others must decrease to compensate. For example, if the speed goes up, the pressure must go down. This is how aircraft wings generate lift forces. The shape of a wing causes the air to flow faster on the upper surface than on the lower surface.

2 Demonstrating Bernoulli's equation

For this experiment, you will need a plastic straw and a sticky note that is about three inches square.

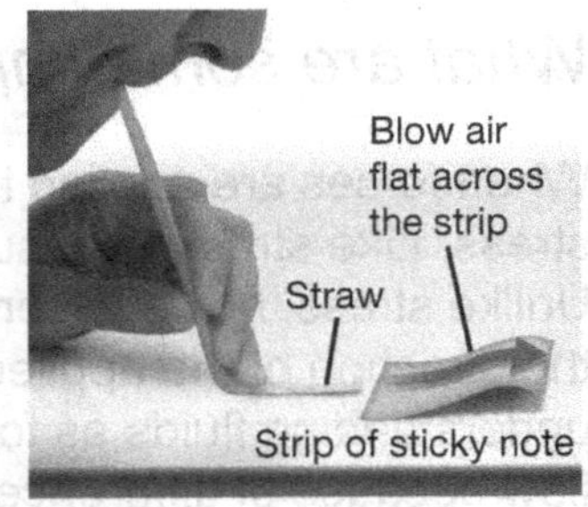

1. Cut a strip of the sticky note about two centimeters wide. Stick the note on a desk as shown at right.
2. With the straw, blow air across the sticky note from the sticky side toward the free side. Position the straw so the air blows parallel to the paper. The end of the straw should not overlap the paper by more than one centimeter.
3. Observe what happens to the paper as the moving air blows across its upper surface.

a. Does the paper curl up into the moving air or down away from the moving air?

b. Use the relationship between speed and pressure to explain how the paper moves.

3 Using Bernoulli's equation to measure the speed of moving air

Bernoulli's equation gives us a simple way to measure the speed of air moving in a pipe. Consider air flowing in a pipe that has a reduction in its diameter. When air flows through the pipe, the speed is greater in the smaller pipe than in the larger pipe. The difference in speed creates a difference in pressure that you can measure.

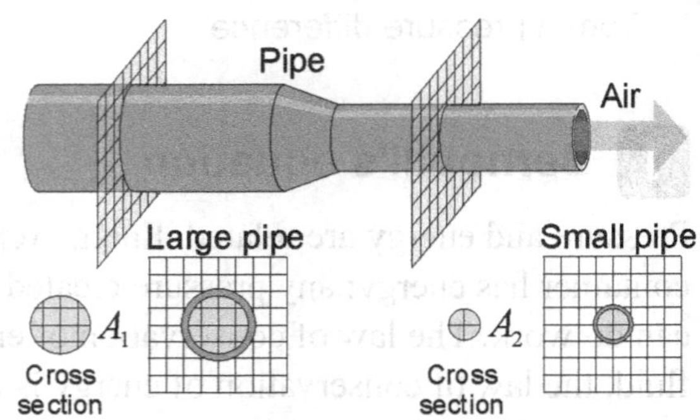

Relationship between speeds

$$\frac{v_1}{v_2} = \frac{A_2}{A_1}$$

Imagine cutting a cross-section of the pipe in two places. The same amount of air must cross each place in equal amounts of time. That means the speed of the air increases by the inverse of the ratio of the areas. If the area of the larger pipe is twice the area of the smaller pipe, the speed of the air in the smaller pipe is twice the speed of the air in the larger pipe.

If the speed goes up, the pressure must go down. Because you know the ratio by which the speed must increase, the change in pressure allows us to measure the air speed. Applying Bernoulli's equation to the pipe gives the following relationship: since the variation in pressure is very small, the density of the air is nearly constant. If the pipe is horizontal, the height is also constant. The only unknowns are the speed and pressure difference. The speed is what you want to know, and the pressure difference you can measure.

Bernoulli's equation

Large pipe Small pipe

$$\rho gh + P_1 + \tfrac{1}{2}\rho v_1^2 = \rho gh + P_2 + \tfrac{1}{2}\rho v_2^2$$

Relationship between areas and speeds

$$v_2^2 - v_1^2 = v_2^2 - v_2^2\left(\frac{A_2}{A_1}\right)^2 = v_2^2\left(1 - \frac{A_2^2}{A_1^2}\right)$$

$$P_1 - P_2 = \tfrac{1}{2}\rho(v_2^2 - v_1^2) \quad\rightarrow\quad P_1 - P_2 = \tfrac{1}{2}\rho v_2^2\left(1 - \frac{A_2^2}{A_1^2}\right)$$

Final result

$$v_2 = \frac{\sqrt{P_1 - P_2}}{\sqrt{\frac{\rho}{2}\left(1 - \frac{A_2^2}{A_1^2}\right)}}$$

4 Building an air-speed tester

Construct an air-speed tester from PVC pipes. For the large pipe, use an eight-inch length of 3/4" PVC pipe that has an inner diameter of 0.75 inches (.019 meters). Use a similar length of 1/2-inch PVC pipe for the smaller pipe (with an inner diameter of 0.5 inches, or 0.012 meters). Drill two 31/32-inch holes in the centers of both pipes for plastic tubing. Connect the two pipes with a coupling. You do not need to glue the coupling; the pipes will fit tightly enough without it.

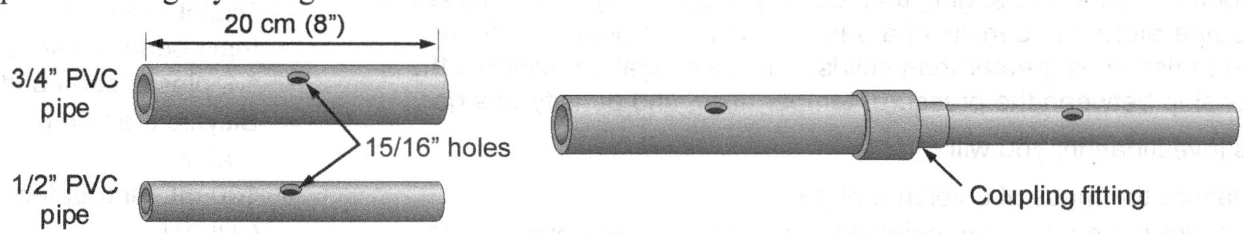

To measure the pressure difference, you will use the difference in height of a water-filled tube. A water-filled tube used to measure pressure differences is called a *manometer*. Make your manometer from a half-meter length of 1/4" clear flexible plastic tubing. Mix some food coloring in 1/4 cup of water. Slowly fill the tube about 1/4 full of colored water so that there are no air bubbles in the filled section of the tube. Squeeze the tube into the holes so that the tube ends are nearly flush with the inner wall of the pipe. Use tape to fasten the tubes as shown. The completed air speed tester should look like the diagram at right.

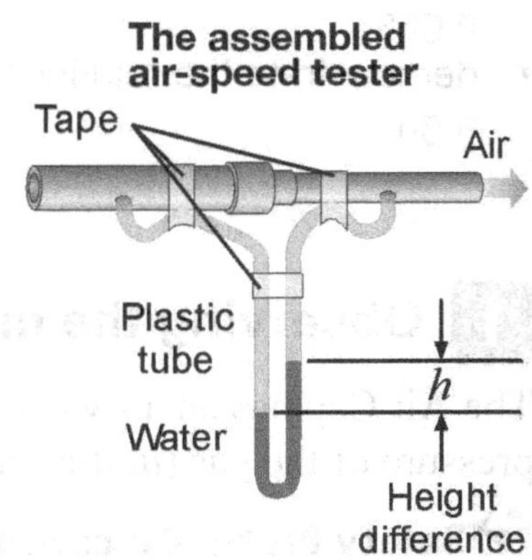

When you blow through the larger pipe, you will see the water level in the tube change as the pressure changes. You can measure the difference in height with a meter stick. The greater the height difference, the greater the difference in pressure.

5 Calculating the air speed

At standard atmospheric pressure and temperature, the density of air is 1.22 kg/m^3. At low speeds, the density of air remains constant as it passes through the pipe. The ratio of areas between the large and small pipe is 2.25, meaning the speed in the smaller pipe is 2.25 times the speed in the larger pipe. That means the complicated factor under the square root reduces to a simple value:

$$\sqrt{\frac{\rho}{2}\left(1-\frac{A_2^2}{A_1^2}\right)} = \sqrt{\frac{(1.22\ \text{kg/m}^3)}{2}\left(1-\frac{1}{2.25^2}\right)} = 0.6996$$

One further simplification is needed. The difference in pressure is related to the height difference in the water in the tube by the equation $P_1 - P_2 = \rho gh$ where ρ is the density of water (1,000 kg/m^3).

$$v_2 = 1.43\sqrt{P_1 - P_2} \qquad P_1 - P_2 = \rho gh = 9{,}800h$$

If we substitute the values for ρ and g into the formula for the speed, we get a nice, simple result. The speed of the air in the small pipe (v_2) is 142 × the square root of the height difference in meters!

$$v_2 = 142\sqrt{h}\ \text{m/s}$$

Blow through the tube, measure the water's height difference, and use the relationship above to calculate the speed of the air in the large and small pipes.

27.3 Properties of Gases

How much matter is in a gas?

Gases are fluids because they can change shape and flow when forces are applied. Gases are different from liquids because they can expand and contract, greatly changing their density. Depending on the pressure and temperature, the density of a gas can vary from near zero (outer space) to densities greater than solids. This investigation explores the relationship between the pressure, temperature, and density of a gas.

In this investigation, you will:

- measure the mass of a volume of gas.
- measure the relationship between mass, density, and pressure for a gas.
- demonstrate the relationship between pressure and temperature for a gas.

Materials List

- Air Cap
- Digital scale or triple beam balance
- 1 L plastic soda bottle
- Bicycle tire pump
- Water
- 100 mL Graduated cylinder
- Tire pressure gauge
- Bucket of ice water
- Graph paper
- Calculator

1 Observing the mass of a gas

The Air Cap has an air valve to which a bicycle pump may be attached. The pump can be used to raise the pressure of the gas (in this case, air) in the bottle. Pressurized gas stores energy.

Safety Note: Be careful with the bottle, and DO NOT exceed 80 pounds per square inch (psi) of pressure.

1. Use the balance to measure and record the mass of the bottle (and air) at ordinary atmospheric pressure. Record the mass and all other data for this part of the investigation in Table 1.
2. Fill the bottle to the very top with water. Empty the water into a graduated cylinder to measure the volume of the bottle. Record the volume in Table 1.
3. Screw the Air Cap onto the bottle. Use the bicycle tire pump to increase the pressure of gas in the bottle to 10 psi. You should take the pressure a little higher than 10 psi and use the valve to let some air out, adjusting the pressure until it equals 10 psi. Check the pressure with your gauge, as the gauge on most bicycle pumps is not accurate enough for this experiment.
4. Use the balance to measure and record the mass of the bottle (and air) at 10 psi pressure. Record the mass in Table 1.
5. Repeat the pumping and mass measurement for pressures between 10 and 80 psi. *Do not exceed 80 psi!*

Air Cap

Plastic soda bottle

Table 1: Pressure and mass data

Gauge pressure (psi)	Mass (g)	Volume (ml)

2 Analyzing the data

The pressure gauge you use measures pressure relative to the pressure of the atmosphere outside the gauge. This type of pressure is called *gauge* pressure. The *absolute* (total) pressure inside the bottle is equal to the gauge pressure *plus* the pressure of the atmosphere. On average, the atmosphere has a pressure of 14.7 psi. To calculate the absolute pressure, you need to add 14.7 psi to each pressure you measured on the gauge.

In metric units, pressure is measured in pascals (Pa). One Pa equals 1 newton of force per square meter of area (1 Pa = 1 N/m2). To convert between psi and Pa, use the relationship 1 psi = 6,895 Pa.

a. Calculate the first column of Table 2 by adding 14.7 psi to the gauge pressures from Table 1.

b. Calculate the second column of Table 2 by converting the absolute pressures in psi to Pa.

c. To calculate the mass of air, you need the mass of the bottle so that you can subtract it from the total mass measured in Table 1. However, when the gauge reads zero pressure, there is still 14.7 psi of air in the bottle. *The balance measures only the mass of the bottle itself* and is not sensitive to the mass of air in the bottle as long as the pressure inside and outside the bottle is the same. Explain why this is true.

d. Calculate the third column of Table 2 by subtracting the mass of the bottle from the total mass.

e. Estimate the mass of air in the bottle at an absolute pressure of 14.7 psi (when the gauge measured zero pressure). Use your measurements to make the estimate.

f. Calculate the fourth column of Table 2 by adding the result from e above to the masses in column 3. The result is the total mass of air in the bottle as a function of the absolute (total) pressure in the bottle.

g. Make a graph showing the pressure of air versus the mass of air in the bottle. Describe a mathematical rule that the graph shows. For example, is the graph a straight line? What would that demonstrate?

Table 2: Calculated pressure and mass data

Absolute pressure (psi)	Absolute pressure (Pa)	Mass increase (g)	Total mass (g)

3 The effect of temperature on pressure

1. Pump the bottle up to a pressure of exactly 60 psi (using your gauge).
2. Immerse the pressurized bottle in a bucket of ice water. Hold the bottle under water for at least five minutes (using something other than your hand) so the temperature of the air in the bottle has time to reach the same temperature as the ice water. Measure the pressure of the cold air.

a. What happens to the pressure in the bottle when the temperature is reduced?

b. Does reducing the temperature change the total mass of air in the bottle?

c. Give a physical reason that the pressure should depend on the temperature.

28.1 Atomic Structure

How does the size of the nucleus compare with the size of an entire atom?

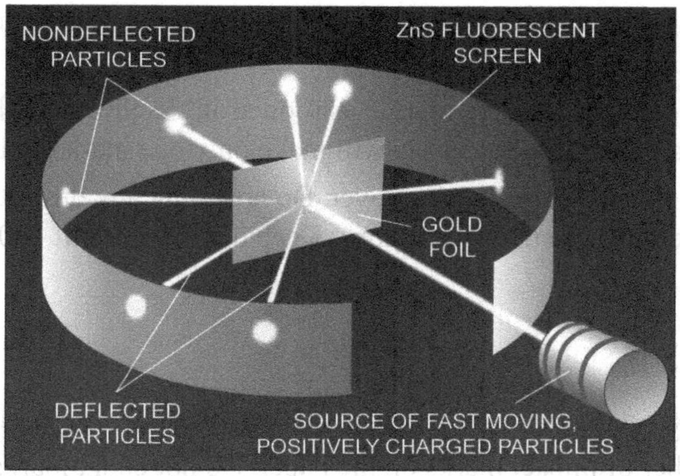

In 1911, three British scientists—Ernest Rutherford, Hans Geiger, and Ernest Marsden—conducted one of the most important experiments of the 20th century. They aimed a lot of tiny, fast-moving, positively-charged particles (called alpha particles) at a thin sheet of gold foil. Most of the alpha particles went right through the foil and hit a screen behind it. But the men were surprised when some of the particles bounced back. The particles must have hit areas of the foil that were more dense! From this experiment, Rutherford thought that an atom must be made up mostly of empty space. This would explain how most of the alpha particles could pass through the foil, they just went through the empty space. He thought that there was a tiny core called a nucleus in the center of each atom. He believed that most of the atom's mass was located in this tiny core. He figured that the alpha particles that bounced back from the gold foil must have hit the nuclei of atoms.

Materials List

- Carbon paper
- Circle sheet (duplicated from this investigation)
- Plastic marble (from Marble launcher)
- Calculator
- Metric ruler
- Tape

The three scientists compared the number of alpha particles that bounced back to the total number of alpha particles that they aimed at the gold foil. By doing this, they could estimate how much space was taken up by the nucleus in a gold atom. As you work, you will use a similar technique to estimate the radius of a circle.

In this investigation, you will:

- use indirect measurement to find the radius of a circle.
- compare and contrast your work with Rutherford's classic experiment.

1 Bombarding the hidden circle sheet with a plastic marble

1. Place a sheet of carbon paper on the floor, carbon side facing up.
2. Place the circle sheet (duplicated from the last page of this investigation) on top of the carbon paper. Make sure the printed circle side is facing down. You should not be able to see the circles during this activity.
3. Tape the circle sheet to the floor so that it does not move.
4. Choose one person to be the "dropper." The other person will be the "catcher." You will switch roles after 25 drops.

5. The dropper should hold a marble about 20 cm above the paper and drop the marble onto the paper. When the marble hits the paper, the carbon paper will make a mark on the other side of the circle sheet. Each dropper should cover the paper evenly with marble marks. Drop the marbles all over the paper, not just in one spot. Later you will count the marks on the circle side to see how many marbles hit inside the circles. You will compare that number with the total number of marks all over the paper.
6. The catcher must be sure to catch the marble as soon as it hits the paper. Do not let the marble bounce. This way, each drop of the marble will make only one mark on the paper.
7. If the marble misses the paper, do the drop over. Do not count any drops that miss the paper. After 25 drops, switch roles and do another 25 drops. Keep doing this until 100 successful drops have been completed.

2 Recording the data

1. Pick up your circle sheet. Save the carbon paper (it can be reused by another group sometime).
2. Count the total number of impact marks on the paper. Do not count marks that go off the edge of the paper. Record your data in Table 1.
3. Count the number of marks that are completely inside a circle. Do not count any marks that touch the edge of a circle. Record your data in the table below.
4. Measure the length and width of the paper in millimeters. Find the area of the paper. Do this by multiplying the length times the width. Record the area in the table below.
5. We know that the circles are all the same size. We also know that the placement of the marks was random. Therefore, we can set up a proportion:

$$\frac{\text{number of marks in circles}}{\text{number of marks on paper}} = \frac{\text{total area of circles}}{\text{total area on paper}}$$

6. Solve this equation. The answer will tell you the total area of all the circles. Record your answer in Table 1.

Table 1: Marble drop data

Number of marks on paper	Number of marks in circles	Area of paper (mm^2)	Total area of circles (mm^2)

3 Analyzing the data

a. Figure out a way to use a piece of data from Table 1 and the total number of circles on your paper to calculate the *area of one circle* on your paper. This area represents an indirect measurement, since you did not directly measure a circle to find the area.

b. Use direct measurement (with a metric ruler) to find the radius of one circle in millimeters. Next, use the formula for the area of a circle to calculate the area of one circle on the paper.

c. You now have two values for the area of one circle; the first area you found in 3a using indirect measurement, and the second area you found in 3b using direct measurement. Find the percent error for your indirect measurement.

d. Compare your percent error with the percent error of several other groups. Discuss your findings.

4 Arguing from evidence

a. Write a paragraph to explain how this activity is like the experiment with the gold foil. Be sure to talk about each piece of equipment that was used. For example, what does the marble represent?

b. In step 3c, you used a formula to find the percent of error of your results. Could Rutherford's group of scientists find their percent error in the same way that you did? Why or why not?

c. What could Rutherford's group have done to prove that their findings were accurate?

d. Would your percent error have been lower if you had dropped the marble 200 times? Or 1,000 times?

e. Name two possible sources of error in your experiment. What could you change to decrease these errors?

f. *Challenge!* How many marks should fall inside the circles on your paper if you dropped the marble 500 times?

28.2 Electrons and Quantum States

How do atoms create and interact with light?

Quantum physics is the branch of science that deals with extremely tiny systems, such as how electrons behave inside an atom. When scientists look at light emitted by an atom, they see a few very specific colors. The colors are different for atoms of different elements. The energy of light depends on the color. The fact that atoms emit only certain colors of light means that the process of emitting light from an atom must only work for certain special amounts of energy.

In this investigation, you will:

- simulate the absorption and emission of light from an atom using the atomic structure model.

Materials List

- Photons and Lasers card deck
- Laminated atom model sheet
- Water soluble marker
- Paper towel

The colors emitted by the element hydrogen

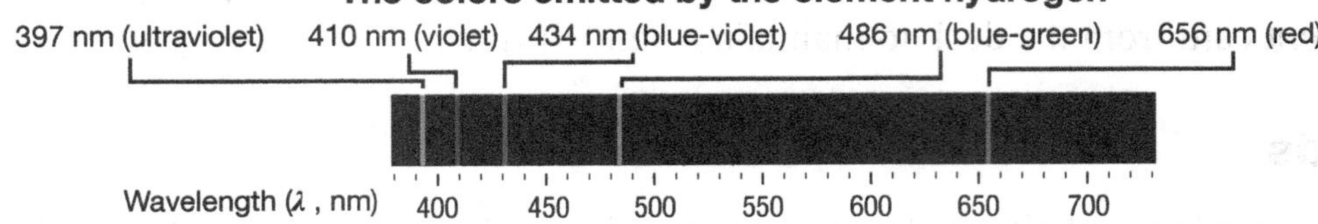

1 Modeling the absorption and emission of light

This investigation uses colored and numbered cards, called Photons and Lasers cards, to represent the absorption and emission of different colors of light by the atom. For example, a red "Pump 1" card represents a photon of red light absorbed by the atom. When the atom absorbs one red photon, one electron moves up one energy level. There are yellow (2), green (3), and blue (4) cards that raise an electron 2, 3, and 4 levels. The word *pump* comes from the technology of lasers. In the first lasers, the process of injecting energy into the atoms in the laser was compared to pumping water into a bucket with a hand pump. Each stroke of the pump puts an equal amount of water into the bucket as each photon puts an equal amount of energy into the atom.

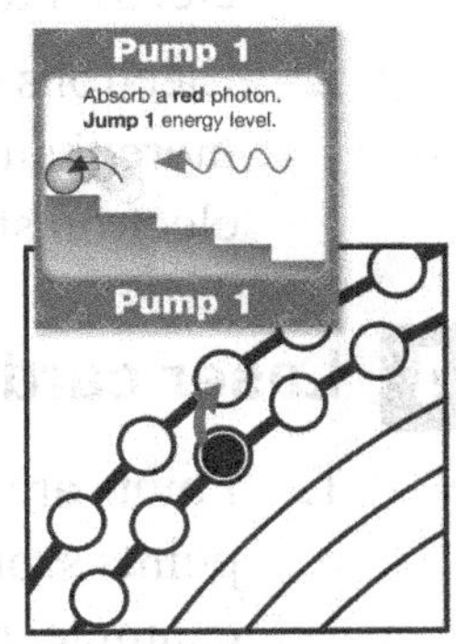

Electron moves up one level

An atom with one or more electrons above the ground state is said to be in an *excited state*. Atoms in an excited state can be triggered to release the energy of the elevated electrons. The trigger is a photon of light that is that same color as the light the atom would release if the electron were to come down from its higher level. This process is called *stimulated emission* because photons are used to stimulate the excited atom to emit more photons of matching color. The word *laser* is really an acronym standing for *light amplification by stimulated emission of radiation.*

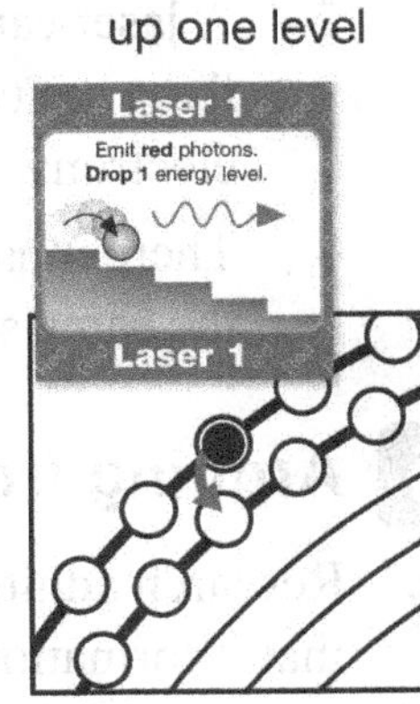

Electron moves down one level

Laser cards represent the trigger photons that stimulate the atom to emit matching light. The emitted photons are emitted by the atom as electrons fall to lower energy levels. For example, if an atom has an excited electron that *can* move down, a red Laser 1 card represents a photon of light that triggers the electron to move down. Laser cards allow all the electrons on a particular level to move down, but only if there are any empty levels for them to move down to.

2 Playing the game

The game of Photons and Lasers is a simulation of the processes that go on inside atoms of a material that is absorbing and emitting light. Although the game is played using one model, real atoms only absorb a single photon at a time. Think of the single model as representing a collection of many atoms.

The objective of the game is to win by scoring 10 points. Points are scored by playing laser cards when the atom has electrons that are above the ground state (excited). One point is scored for each electron for each level it moves down. Of course, an atom in the ground state has no excited electrons. Players must play pump cards to excite electrons before points can be scored. The strategy of playing involves balancing the use of pump and laser cards to make points.

1. Set up the atomic structure model to represent neon-20 (10 protons, 10 neutrons, and 10 electrons). Do this by coloring in the appropriate electron spaces on the laminated atom model sheet. The electrons should be in the lowest possible energy levels. When electrons are in the lowest energy levels, we say the atom is in its *ground state*.
2. Each player is dealt a hand of five cards. A player chooses one card to play per turn, and draws a replacement card from the deck to maintain a hand of five.

3 Pump cards

1. Each pump card played allows the player to move *any single electron* up the number of levels corresponding to the card. Do this by erasing electrons from their current state and coloring the appropriate circles in at a higher energy level. For example, a pump 2 (yellow) card allows one electron to go up two levels. A move cannot be divided among more than one electron.
2. Electrons cannot be moved to places where there are already electrons. The quantum rules forbid more than one electron in any given quantum state. The open circles in each level represent the electron states that exist in all atoms.

4 Laser cards and scoring

1. Points are scored by playing laser cards. The following rules apply to laser cards: The number of points scored equals the number of electrons moved multiplied by the number of levels moved. For example, a laser 2 card can score six points by moving three electrons down two levels each.
2. A laser card may only move electrons from *one* level during any given turn. For example, suppose there are two excited electrons on level 5 and two more on level 4. A laser card can move only two electrons, from either level 5 or level 4, but not electrons from both levels in the same turn.
3. There must be open states for an electron to be moved into. Electrons are not allowed to pile up in the same state or be placed into the nucleus.

5 Arguing from evidence

a. Research additional sources of information about lasers to find out more about how they work. Using that information, think of an application for a laser that has not been thought of before. Describe your new application of a laser.

b. Write a short paragraph that describes how fireworks work. Use what you learned from playing the Photons and Lasers game and research additional sources of information to write your paragraph.

28.3 The Quantum Theory

How can a system be quantized?

To a physicist, if something is quantized, it can only exist in whole units, not fractions of units. A *quantum* of something is the smallest amount that can exist. For example, when light is quantized, one photon is the smallest unit, or quantum, of light. In the quantum theory, all matter and energy have both wave-like properties and particle-like properties. Light acts like a wave from far away. However, up close, light acts "particle-like" because the wave is made of individual photons. An electron acts like a particle when it is free to move and far from other electrons. However, if an electron is confined in a small space (like an atom), it behaves like a wave. This investigation uses a vibrating string to model different aspects of the quantum theory.

Materials List

- Sound and Waves
- DataCollector used with the AC adapter
- Physics Stand
- Calculator

In this investigation, you will:

- show how a vibrating string has similar properties to a quantum system.
- identify the "quantum states" of a vibrating string.

1 Setting up the investigation

Connect the DataCollector to the Sound and Waves Generator as shown in the diagram. The telephone cord connects the DataCollector and wave generator. The black wire goes between the wave generator and the wiggler.

1. Attach the fiddlehead to the top of the stand, as high as it goes.
2. Attach the wiggler to the bottom of the stand, as low as it goes.
3. Stretch the elastic string a little (5–10 cm) and attach the free end to the fiddlehead. Loosen the knob until you can slide the string between any two of the washers. *Gently* tighten the knob just enough to hold the string.
4. Turn on the DataCollector and select Timer mode.
5. Set the wave generator to Waves using the button. The wiggler should start to wiggle back and forth, shaking the string.
6. Set the DataCollector to measure frequency. You should read a frequency of about 10 Hz. Ten hertz means the wiggler is oscillating back and forth 10 times per second.

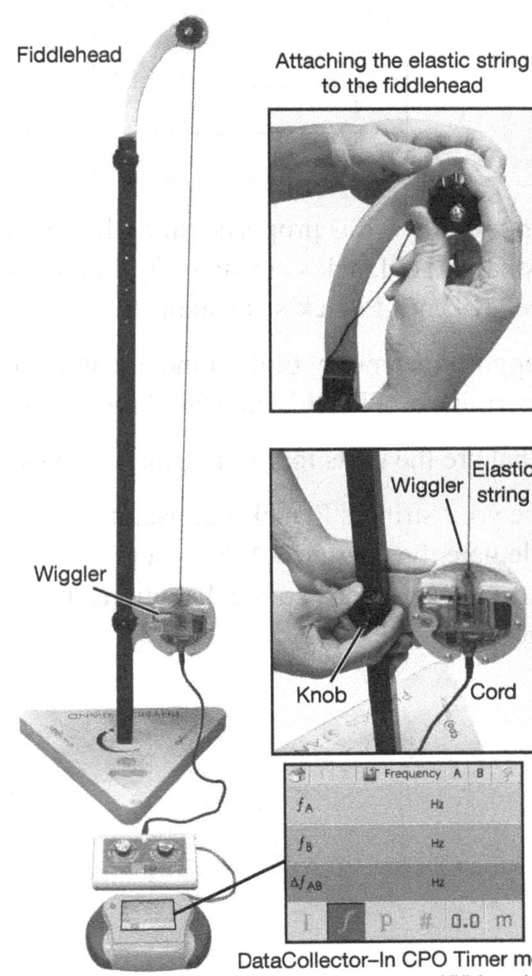

DataCollector–In CPO Timer mode, choose "Frequency(f)" function.

7. Try adjusting the frequency of the wiggler with the frequency control on the wave generator. If you watch the string, you will find that interesting patterns form at certain frequencies.

2 Waves on the vibrating string

The first five harmonics of a vibrating string

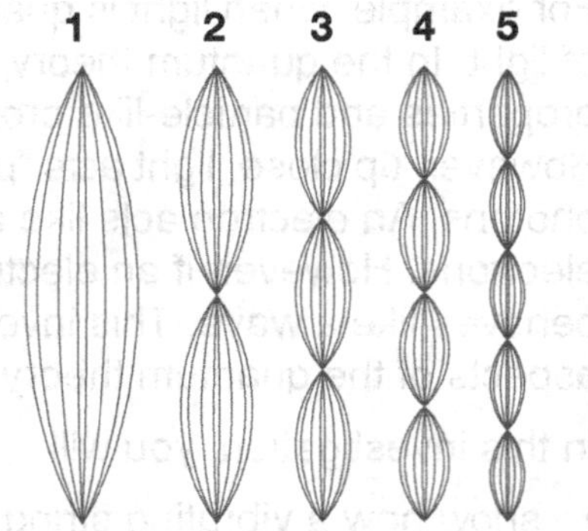

At certain frequencies, the vibrating string will form wave patterns called *harmonics*. The first harmonic has one bump, the second harmonic has two bumps, and so on. The harmonic number is like the *quantum number* for the electron wave in an atom. The first harmonic (fundamental) has quantum number 1. The second harmonic has quantum number 2 and so forth.

Adjust the frequency to obtain the first 8 "quantum numbers" for the string and record the frequency for each one in Table 1. You should fine-tune the frequency to obtain the largest amplitude before recording the data for each harmonic.

Table 1: Frequency and energy data

Quantum # (harmonic #)	Frequency (Hz)	Energy (joules)
1		
2		

The energy of a wave is proportional to the frequency of the wave. In quantum physics, the constant of proportionality is Planck's constant, h, which has the value $h = 6.626 \times 10^{-34}$ J-s. The energy of the wave is its frequency times Planck's constant, $E = hf$.

a. Suppose the energy of the fundamental wave on the string (quantum 1) is one joule. Use your data to determine a "Planck's constant" for the vibrating string.

b. What are the units for your string's Planck's constant?

c. Use your string's Planck's constant to calculate the energies of the other harmonics. Record your calculations in Table 1.

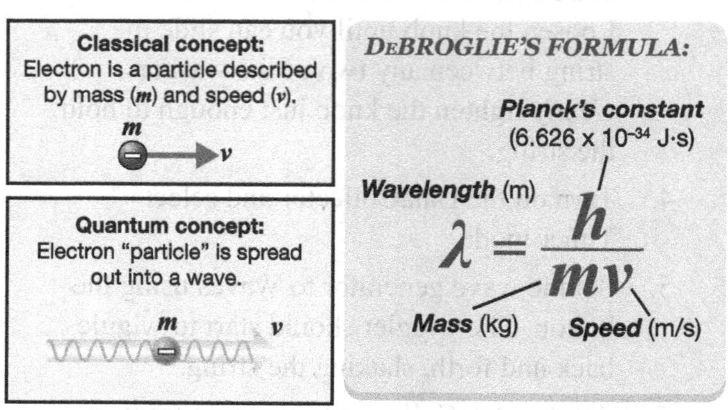

3 The wavelength of a particle

The wavelength of a particle (λ) depends on its mass (m) and speed (v) according to DeBroglie's formula. The wavelengths of particles tend to be extremely small. An electron moving at a million meters per second has a wavelength of only 7×10^{-10} meters. The short wavelength is why an electron looks like a particle most of the time. Its wave properties only become apparent when it is confined to a space near the size of its wavelength, such as an atom.

a. Calculate the wavelength of a steel ball with a mass of 28 grams that is moving at 45 m/s (100 mph).

b. Is the wavelength you calculated a large or small number compared to the size of the ball?

c. Suppose the ball is confined to a box that has a width of 10 cm. Do you think you need quantum mechanics to describe the motion of the ball in the box? Base your answer on the comparison between the wavelength of the ball and the size of the box.

29.1 Radioactivity

How do we model radioactivity?

Radioactivity is the release of energy and/or particles from the nucleus of an atom. If the nucleus has too many neutrons or is unstable for any other reason, the atom becomes radioactive and may spontaneously change into another kind of atom. For example, if you leave a carbon-14 (C^{14}) atom alone, it eventually changes into a nitrogen-14 (N^{14}) atom. The most common form of radioactivity is called *radioactive decay*. The word *decay* means to break down. In radioactive decay, the nucleus breaks down and forms a different nucleus. This investigation simulates some of the properties of radioactive decay.

In this investigation, you will:

- learn about chance and probability.
- simulate the radioactive decay and half-life of a radioactive isotope using coin tosses.
- explore the statistical nature of random processes such as coin tosses and radioactive decay.

Materials List

- 100 Pennies in a jar or paper cup
- Tray or box to collect pennies when they are poured out
- Graph paper

1 Probability and chance

We cannot predict when a specific atom will decay and give off radioactivity. What you can know is the *probability* that any atom will decay in a certain interval of time. For example, atoms of the isotope sodium-25 (Na^{25}) have a 50 percent probability of decaying into magnesium-25 (Mg^{25}) every minute. If you had a sample of Na^{25} and waited one minute, half of it would have changed into Mg^{25}.

Radioactive Decay of Na^{25} into Mg^{25}

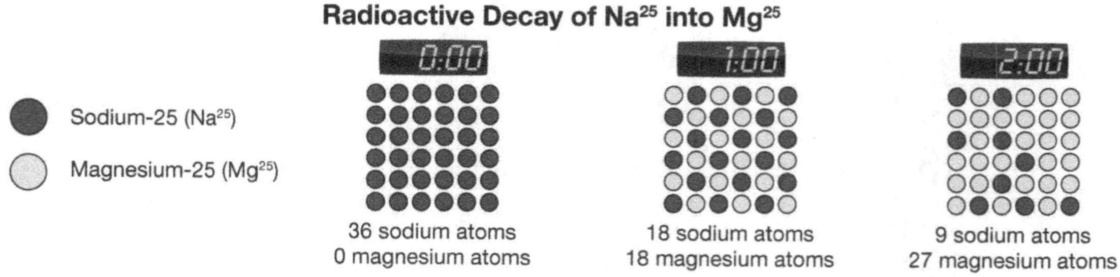

In mathematics, the word *chance* means the likelihood that a particular outcome will occur. For example, a 40 percent chance of rain on Friday means that it rains on about 40 out of every 100 days with weather conditions similar to Friday's. Probability and chance are related. Probability describes the likelihood of a single event occurring. Chance describes the outcome of a whole system, which may be dependent on many single events.

Radioactive decay depends on chance. It is possible to accurately predict the average rate of decay of lots of atoms, although the decay of a single atom is impossible to predict. One very useful prediction is the *half-life*. The half-life is the time it takes for one half of the atoms in any sample to decay. For example, the half-life of sodium-25 is one minute. If you start out with 36 grams of Na^{25}, one minute later, only 18 grams will still be Na^{25}. The rest will have decayed to Mg^{25}. Wait another minute, and half of the remaining 18 grams of Na^{25} will decay, leaving 9 grams of Na^{25} and 27 grams of Mg^{25}. Wait a third interval of a minute, and you will be down to 4.5 grams of Na^{25}. *One half of the radioactive atoms decay during every time interval of one half-life.*

2 Simulating radioactive decay

Tossing a coin is a good example of a process that is described by probability. There is a 50 percent probability of any single toss coming up heads or tails. You cannot predict whether a specific toss will come up heads or tails. However, you can make a good prediction of the average outcome of many tosses. For example, the graph on the right shows the chance of getting a certain number of heads out of 100 tosses. According to the graph, if you repeated the experiment of tossing 100 coins, you could expect 55 out of every 1,000 experiments to come up exactly 50 heads. Most experiments would come up with 40 to 60 heads out of every 100 tosses. The graph also shows that very few experiments would have less than 30 heads or greater than 70 heads per 100 tosses.

Outcome of 100 coin tosses

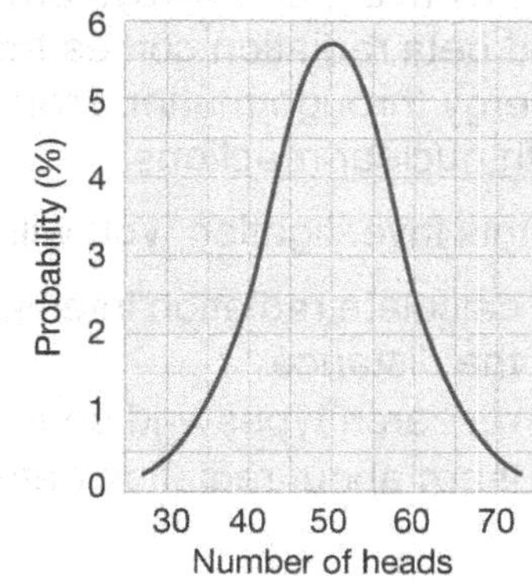

Since coin tosses are based on chance, an experiment of tossing 100 coins can simulate the radioactive decay of 100 atoms. Let a heads-up coin represent one atom of Na^{25}. If the coin lands tails-up, the atom it represents has decayed and become Mg^{25}. Each toss of the coins represents one half-life, since 50 percent of the coins can be expected to "decay" with every toss.

1. Place 100 pennies in a large paper cup so they can be shaken and dumped on a table.
2. Shake the cup and dump the pennies on the table. Separate and count the number of pennies that land heads-up and the number that land tails-up. Record the counts in Table 1 in the row for the first toss, which represents one half-life. The number of heads represents the number of remaining Na^{25} atoms, and the number of tails represents the number of Mg^{25} atoms.
3. Place all the pennies that landed heads-up back in the cup. Do not reuse the pennies that landed tails-up since they represent atoms that have already decayed. Shake the cup again and dump the pennies on the table. Record the number of heads and tails in the row for the second toss, which represents two half-lives. The number of heads represents the number of Na^{25} atoms remaining. The number of Mg^{25} atoms is the number of (new) tails plus the previous number of Mg^{25} atoms.
4. Repeat the experiment using only the pennies that landed heads-up. Keep going until you have one or no pennies left.

Table 1: Coin toss decay simulation

	Heads	Tails	Na^{25} Atoms	Mg^{25} Atoms
Start			100	0
First toss (1 half-life)				
Second toss (2 half-lives)				

a. Make a graph showing the number of Na^{25} atoms and the number of Mg^{25} atoms on the vertical axis. Put time on the x-axis, with each toss of the coins representing one minute. Provide a key to identify the data for each type of atom.

b. Add the data from your entire lab or class for each toss. For example, if there are six groups, the class data set would represent the decay of 600 atoms. Make the same graph with the class data.

c. How does the class graph compare with any single group graph for 100 atoms? Sketch what the graph would look like for 10^{20} atoms, a typical number of atoms in a sample of matter.

29.2 Radiation

What are some types and sources of radiation?

Materials List

- Scientific calculator
- Access to the Internet or library for research

The word *radiation* refers to the flow of energy through space. Light, radio waves, microwaves, and X-rays are forms of electromagnetic radiation. The energy in alpha and beta radiation comes from moving particles. Ultrasound is radiation of sound energy through matter. Many people mistakenly think of radiation as only associated with nuclear reactions.

In this investigation, you will:

- calculate radiation intensity and show how it diminishes with the inverse square of the distance.
- research types and sources of radiation.
- learn about radiation detectors and dosimetry through research.

1 Intensity and energy

The intensity of radiation measures how much energy flows per second per unit of area. In metric units, intensity is power per square meter or watts per square meter (W/m^2).

When radiation comes from a single point, the intensity decreases by one over the square of the distance. This is called an *inverse square law.* If you get two times farther away from a radiation source, the intensity goes down by one-fourth, which is one divided by the square of two ($1 \div 2^2$). If you get 10 times farther away, the intensity goes down by a factor of 100, which is $1 \div 10^2$.

a. A bright lamp emits a power of 100 watts of light and infrared radiation. Calculate the intensity of radiation from the lamp at a distance of one meter and at a distance of 10 meters.

b. How far would you have to be from a source of radiation for the intensity to decrease by a factor of 1,000 compared with the intensity at a distance of one meter?

c. The intensity of radiation from the sun is about 1,000 W/m^2 at Earth's surface. Use the inverse square law to estimate the intensity of solar radiation on the planet Mercury.

2 Types of radiation

Radiation becomes harmful when it has enough energy to remove electrons from atoms. The process of removing an electron from an atom is called *ionization.* Ultraviolet (UV) light is harmful because an ultraviolet photon has enough energy to eject an electron from an atom. UV light is an example of *ionizing radiation.* The diagram below shows a few forms of ionizing radiation. The energy in the radiation can come from matter particles or from waves such as electromagnetic waves.

Three typical forms of ionizing radiation

	Charged particles	Neutral particles	Energy
Example	Alpha radiation Beta radiation 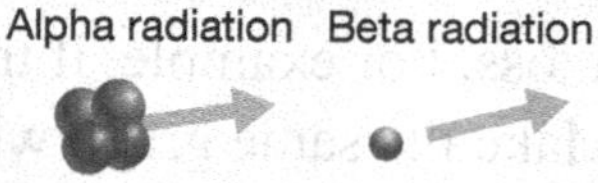		 Gamma ray
Description	Helium 4 nuclei or electron moving at a high speed	Neutrons moving at high speed	Electromagnetic waves (photons)

3 Sources and risks from ionizing radiation

a. Name one form of nonionizing radiation you experience every day.

b. Identify one source of ionizing radiation that is charged particles.

c. Identify one source of ionizing radiation that is neutral particles.

d. Identify one source of ionizing radiation that is electromagnetic radiation.

e. If all three forms of ionizing radiation had the same energy and intensity, which would be most difficult to stop: charged particles, neutral particles, or electromagnetic radiation?

4 Cosmic rays

Earth is constantly bathed by radiation from space. The term *cosmic rays* describes all forms of radiation that come from beyond our solar system. Research cosmic rays and answer the following questions.

a. List three types of radiation present in cosmic rays.

b. Give the intensity of at least one form of cosmic ray radiation in W/m^2.

c. Identify two sources of cosmic rays.

d. Are cosmic rays ionizing or nonionizing radiation, or do cosmic rays include both types?

5 Radiation detectors

Since we cannot see most forms of ionizing radiation, we need to use detectors. Detectors are instruments or materials that respond to radiation in a way that we can measure.

a. Research and identify a detector of radiation.

b. What physical principle does the detector use to sense the radiation?

c. What forms of radiation can the detector sense?

d. What forms of radiation cannot be sensed by the detector?

e. Where is the detector used?

6 Dosimetry

Ionizing radiation absorbed by people is measured in a unit called the *rem*. The total amount of radiation received by a person is called a *dose*, just like a dose of medicine. Radiation doses in people are measured in rems instead of watts because different kinds of radiation are absorbed differently by body tissues.

a. Which would have the highest dose in rems to a person: neutrons, alpha particles, or gamma rays?

b. What is the average annual dose in rems you receive from background radiation in the environment?

c. What is the average dose limit allowed for people who work in nuclear reactors?

29.3 Nuclear Reactions and Energy

Where does nuclear energy come from?

Nuclear energy comes from the organization of protons and neutrons into the nuclei of atoms. When protons and neutrons are rearranged into the nuclei of different elements, vast amount of energy are either absorbed or released.

In this investigation you will:

- interpret the nuclear energy curve.
- calculate the energy released or absorbed in nuclear reactions.

Materials List

- Scientific calculator
- Access to the Internet or library for research

1 Setting the scale

When working with nuclear energy, the numbers become very large, so it is important to set up a way to understand them. Consider heating up one kilogram of water from ice cold, 0°C, to boiling temperature, 100 °C. You might do this by pouring a one-liter bottle of ice water into a pot on the stove. Water has a specific heat of 4,184 J/kg°C, so heating 1 kg up by 100 degrees takes 4,184 × 100, or 418,400 joules. Two liters of water would take twice this, or 840,000 J. Therefore, 1 megajoule (1 MJ) is approximately the energy you use to cook a single meal.

A power plant must provide energy for thousands or millions of people. If an entire city of a million people cooked the same meal at the same time, the energy requirement would be one million megajoules, which is 10^{12} J, or one terajoule (1 TJ). The following discussion of nuclear energy uses units of terajoules because the amounts of energy exchanged are so large.

Imagine you could choose to assemble 1 kilogram of protons, neutrons, or a combination of the two into any element you choose. For example, you could group them into nuclei with six protons and six neutrons and make carbon. You could also group a kilogram of protons and neutrons into nuclei with 92 protons and 146 neutrons and make uranium. The amount of energy stored in the nucleus would be very different in either case. The graph below shows the energy binding the nucleus together as a function of atomic number.

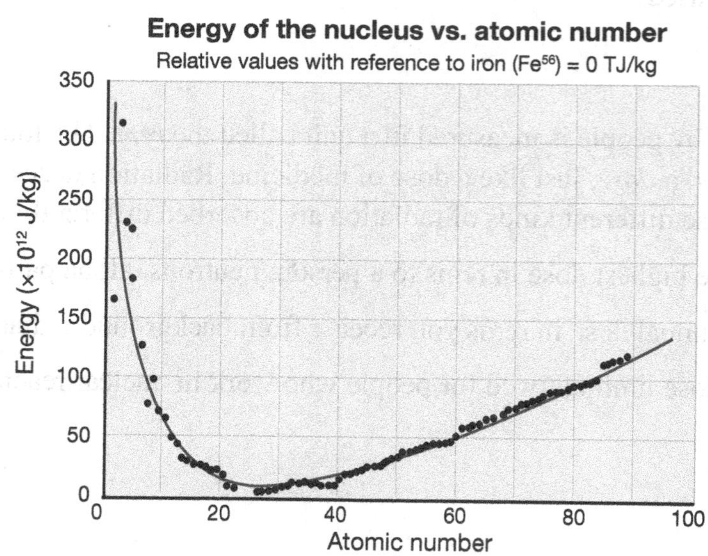

2 Using the binding energy graph

a. What element is at the lowest point of the graph? The graph has been scaled to make this element zero energy. (*Hint*: the atomic number is 26.)

b. Helium has an atomic number of 2. How much nuclear energy is contained in one kilogram of helium nuclei? How does it compare to one kilogram of iron nuclei?

c. How much energy is contained in one kilogram of uranium nuclei? Uranium is atomic number 92.

d. Suppose one kilogram of uranium is converted to one kilogram of nuclei with an atomic number of 46. How much energy is released? This process is called *nuclear fission*. Nuclear fission is how conventional nuclear reactors produce power. The actual process produces two nuclei which are unequal in mass, but the energy release is similar. The graph shows that splitting up large nuclei, with atomic numbers greater than 26, releases energy.

e. Research the energy content in 1 kilogram of gasoline. How many kilograms of gasoline does it take to match the energy released by the conversion of one kilogram of uranium into lighter elements through fission?

f. Research the energy used per person in the United States in one year. How many kilograms of uranium need to fission to release this much energy? How many kilograms of gasoline need to be burned to release the same amount of energy?

g. Stars convert hydrogen into helium. The graph also shows that combining light nuclei into heavier nuclei also releases energy. Research the binding energy of hydrogen and helium. How much energy is released when one kilogram undergoes nuclear fusion to become one kilogram of helium?

h. Research the energy output of the sun in joules per second, or watts. How many kilograms of hydrogen are converted to helium in one second to make this much energy?

i. Research the mass of the Sun. Assume the new Sun contained about 78 percent hydrogen and 22 percent helium. Suppose all of the Sun's hydrogen was converted to helium. How many years would it take to use up all the hydrogen in the Sun?

j. Research the age of the Sun. How does this compare with your calculation in part i above? The Sun will actually become unstable long before it uses all of its hydrogen. Current astrophysical theories tell us that the life of the Sun will be over when it uses up around 20 percent of the hydrogen in its core.

Engineering Design Log

Why Keep an Engineering Design Log?

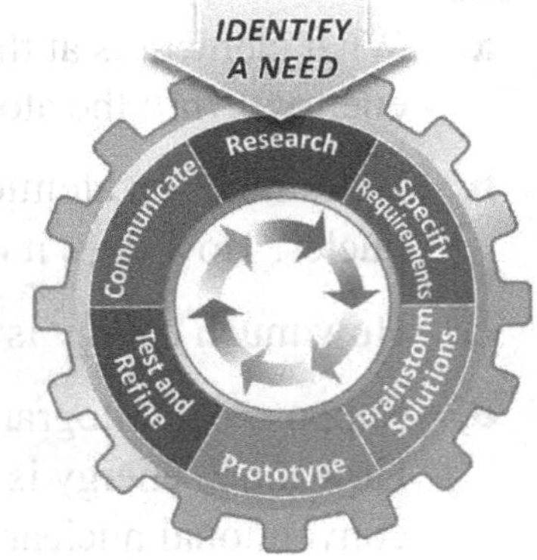

People who are trying to meet a need or solve a problem often follow an approach called the engineering cycle. The engineering cycle divides the design process into a series of stages that can be repeated as needed. Engineers often use a design log to keep track of their work. As you develop your own designs, keep a log to describe your ideas, gather information, ask questions, sketch plans, record results, and share findings. Your log should include the following stages:

1. Identify a Need

What problem are you trying to solve? What process are you trying to improve?

2. Research

Part of research is identifying what you know about a topic or closely related topics that may help you to design your solution. Then you may think about any additional information you need to gather and where you can find that information. What do you know about this topic? What additional information do you need?

3. Specify Requirements

The requirements of a design process include criteria, constraints, and trade-offs. Criteria are the standards a product or process must meet in order to be successful. Constraints are rules that limit the design of your product and include cost, availability of materials, and time. A trade-off is a compromise that exchanges one idea with another that may not be as good but still achieves the design requirements. For this design challenge, what are the design criteria and constraints? Can you identify any trade-offs at this stage? (Trade-offs often become more apparent during subsequent stages of the design process.)

4. Brainstorm Solutions

Describe the different ways you could solve the problem. Which solution should you explore first? Can you make a plan or blueprint of your solution?

5. Prototype

Now that you have chosen a solution, build your prototype. At this stage, you may need to gather some materials and/or tools. Did you learn something new as you constructed your prototype?

6. Test and Refine

How will you test your design and what data will you record? Once you have your prototype and other materials set up and you are ready to record your observations, conduct your tests. Construct a data table to record and analyze your data. Considering the results of your tests, what improvements should you make to your design?

7. Communicate

Scientists and engineers share their ideas within communities so that others may benefit from their work and review their conclusions. How will you share your design with others? What suggestions do others have for your design?

8. Reflect and Revise

The design process does not have to end when the results are communicated. What new information could you find that would help your design? Is there another type of test you could perform on your design, or an idea you weren't able to investigate in the first round of the process? Review your notes and reflect on these questions.

Engineering Design Log

The engineering cycle can also be thought of as a series of decisions. This diagram shows where these decisions are often made.

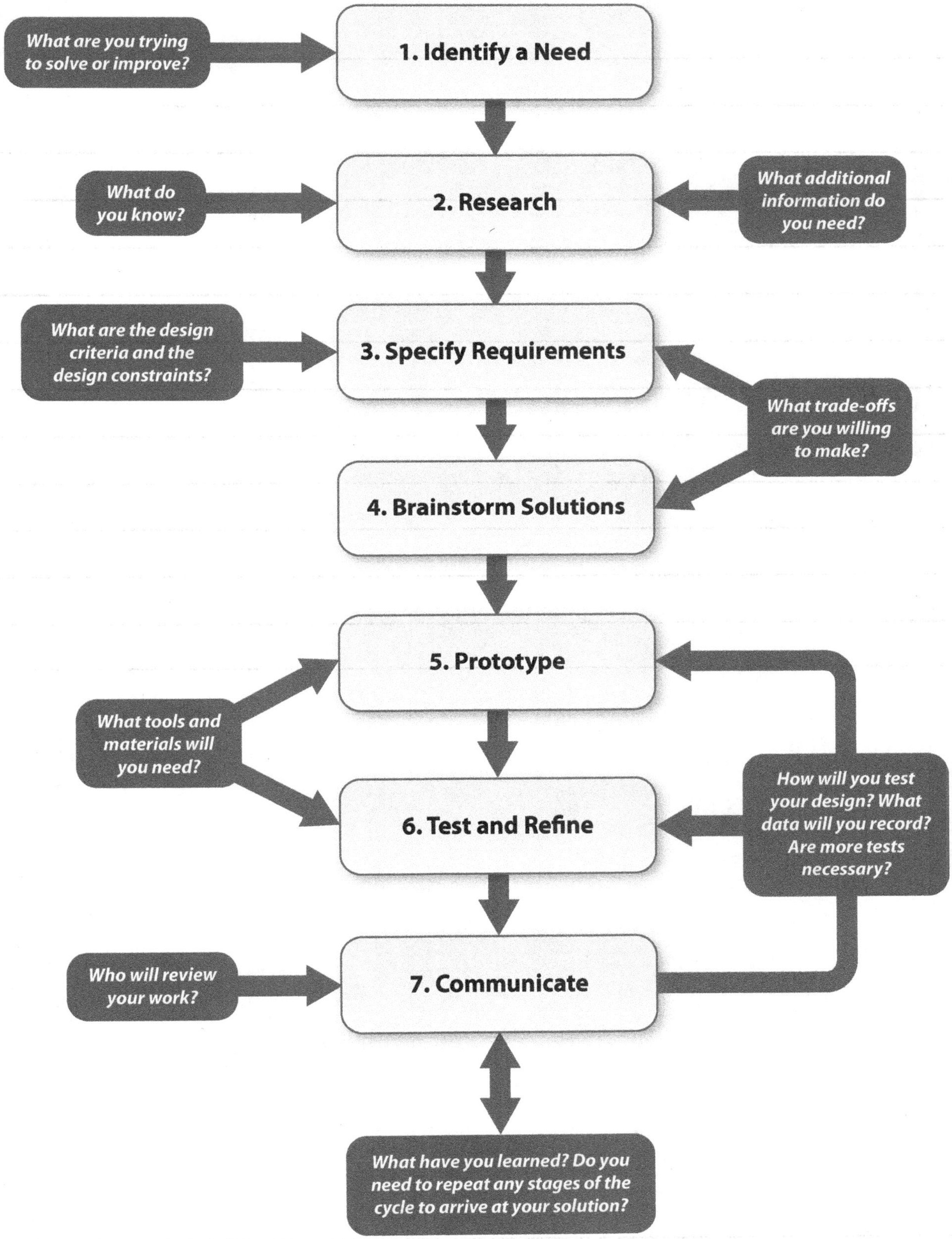

Name ______________ Date ________ Investigation [ ]

Engineering Design Log

1. Identify a Need

What problem are you trying to solve?
What process are you trying to improve?

Engineering Design Log

Name ____________________ Date __________ Investigation [ ]

2. Research

Part of research is identifying what you know about a topic or closely related topics that may help you to design your solution. Then you may think about any additional information you need to gather and where you can find that information. What do you know about this topic? What additional information do you need?

Name Date Investigation

Engineering Design Log

3. Specify Requirements

The requirements of a design process include criteria, constraints, and trade-offs. Criteria are the standards a product or process must meet in order to be successful. Constraints are rules that limit the design of your product and include cost, availability of materials, and time. A trade-off is a compromise that exchanges one idea with another that may not be as good but still achieves the design requirements. For this design challenge, what are the design criteria and constraints? Can you identify any trade-offs at this stage? (Trade-offs often become more apparent during subsequent stages of the design process.)

Engineering Design Log

Name ____________ Date ________ Investigation [ ]

4. Brainstorm Solutions

Describe the different ways you could solve the problem. Which solution should you explore first? Can you make a plan or blueprint of your solution?

Engineering Design Log

Name ____________________ Date __________ Investigation [ ]

5. Prototype

Now that you have chosen a solution, build your prototype. At this stage, you may need to gather some materials and/or tools. Did you learn something new as you constructed your prototype?

Engineering Design Log

Name ____________________ Date __________ Investigation [ ]

6. Test and Refine

How will you test your design and what data will you record? Once you have your prototype and other materials set up and you are ready to record your observations, conduct your tests. Construct a data table to record and analyze your data. Considering the results of your tests, what improvements should you make to your design?

Engineering Design Log

Name ____________________ Date __________ Investigation [ ]

7. Communicate

Scientists and engineers share their ideas within communities so that others may benefit from their work and review their conclusions. How will you share your design with others? What suggestions do others have for your design?

8. Reflect and Revise

The design process does not have to end when the results are communicated. What new information could you find that would help your design? Is there another type of test you could perform on your design, or an idea you weren't able to investigate in the first round of the process? Review your notes and reflect on these questions.

Engineering Design Log Sample Grading Rubric

Stages of the Engineering Cycle	Exceeds expectations	Meets expectations	Does not meet expectations
Research and Specify Requirements	• Problem statement identifies needs • Statement of design requirements includes **at least 3** constraints and **at least 3** criteria	• Problem statement identifies needs • Statement of design requirements includes some constraints and criteria	• The problem statement is incomplete or the statement of design requirements does not include criteria and/or constraints
Brainstorm and Prototype	• Includes a diagram or plan of the prototype that indicates each of the design elements and their purpose or function	• Includes design elements indicated by a diagram or plan	• Does not include design elements indicated by a diagram or plan
Test and Refine	• The data table indicates which design was tested (ex. "1, 2, etc.") and the performance of each prototype • The data table lists trade-offs in each design • Accompanying the data table are diagrams of each design. Each diagram labels refinements made on the pervious design.	• The data table indicates which design was tested (ex. "1, 2, etc.") and the performance of each prototype • The data table lists trade-offs in each prototype	• The data table does not indicate which design was tested or the performance of that design • The data table does not list the trade-offs in each prototype • OR, a data table is not included

Name ______________________________ Date ______________

STEM

Collisions and Restraints

What is the best way to minimize forces during a collision?

Imagine yourself riding in a car. The engine starts, the car begins to accelerate, and you feel yourself pressed into the back of your seat. The car turns and you feel your body lean away from the turn, pressed into the door and your seat belt. Finally, the car stops and you feel the seat belt dig into your waist. Throughout the ride different parts of the car have pushed and pulled on your body.

Materials:

- ✔ Energy Car and Track
- ✔ Physics Stand
- ✔ DataCollector and 2 photogates
- ✔ Various materials such as cardboard, rubber bands, tape, foam, etc.
- ✔ Electronic scale (or triple-beam balance)
- ✔ Hard-boiled egg

❶ Designing with the Engineering Cycle

Controlling the strength and distribution of those pushes and pulls exerted on your body is an important part of every car's design. If your car has a collision, those forces may be increased hundreds of times. Seat belts and car seats, among other parts, are designed to reduce and control those forces.

When engineers design cars, they test and refine each design many times. Together, these practices make up the **engineering cycle**.

The Engineering Cycle

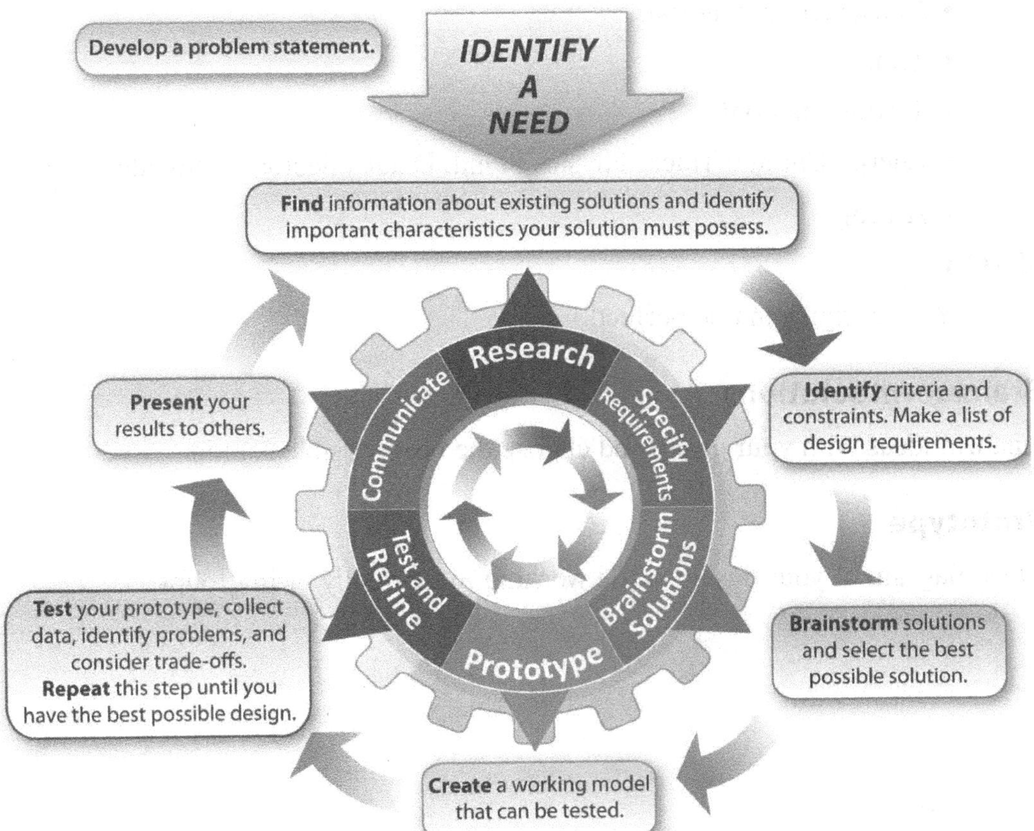

Name ______________________________ Date ______________

❷ Identify a need

Your restraint system must protect an egg in a collision.

❸ Research

Investigate different technologies used in automobiles to protect passengers in the event of a collision. Use this research to plan your design and think about your testing conditions. Record a summary of this in your design log.

❹ Specify requirements

Discuss the design requirements. You may find different design requirements than those listed below. Here are a few to consider:

Criteria

- The restraint system must secure an egg to the Energy Car
- The egg must survive impacts after achieving a velocity of at least 200 cm/s
- The design must include a way for the photogate to measure the speed of the car since the original fin will be blocked by the egg

Constraints:

- Total allowed mass of materials: __________ (your teacher will provide this figure)
- Construction materials
- Time
- Testing materials
- Energy Car and Track, Physics Stand, DataCollector, Photogates
- An egg

Trade-Offs

Your design can't be perfect!

❺ Brainstorm solutions

Brainstorm ideas with your group and choose the best possible solution.

❻ Prototype

Create a diagram of your design while working individually or in groups.

Name ______________________________ Date ______________

❼ Test and Refine

Test and refine your design using the Energy Car and Track. You may create different collision forces by changing the height of the track. Be sure to check your grading rubric and make sure you understand what you need to observe and record! Use the following procedure to test your design.

1. Set up a photogate at 80 cm on the track.
2. Roll the car down the ramp, and record the interval of the car passing through the catcher.
3. Stop the DataCollector and record the results of the trial. Find the final velocity of the car by dividing the width of prototype by the interval during which the photogate was blocked.
4. Observe and record the condition of the egg.
5. Repeat steps 1–3 of the procedure to achieve a total of five trials. Use your observations from these five trials to refine your design and change the prototype.
6. Record your results in data tables like the ones below.
7. Test the updated prototype. Use your observations from these five trials to refine your design and change the prototype.

Table 1: Data Tables

Trial	Final Velocity (cm/s)	Observations
1		
2		
3		
4		
5		

Trial	Final Velocity (cm/s)	Observations
1		
2		
3		
4		
5		

Name ______________________ Date ______________

8 Communicate

Discuss and communicate your results with your group and your class.

Investigation Design Log

Directions: Fill in the sections below as you complete the investigation. You may need additional space for some sections such as Test and Refine.

Identify a Need

____

____

Specify Requirements

1. What are the trade-offs or constraints for your project?
2. How will these affect your design?

____

____

Build Prototype

____

____

Test and Refine

Table 2: Test 1 Data Table

Trial	Final Velocity (cm/s)	Observations
1		
2		
3		
4		
5		

Name ______________________ Date ____________

✔ Based on your findings, how should your **prototype** change?

____

____

____

Table 3: Test 2 Data Table

Trial	Final Velocity (cm/s)	Observations
1		
2		
3		
4		
5		

Communicate

____

____

____

____

RUBRIC

	Unacceptable	Near expectations	Meets expectations	Exceeds Expectations
Specify Requirements	Trade-offs or constraints were not adequately identified or discussed	Trade-offs or constraints are identified but discussion is limited or illogical	At least 3 trade-offs or constraints are identified At least 3 trade-offs or constraints are discussed	More than 3 trade-offs or constraints are identified More than 3 trade-offs or constraints are discussed
Test and Refine	Observations were not recorded or are unacceptable No refinements were made to the design	Observations were recorded after each collision trial, but one of the following was incomplete: Final velocity Effects on restraint system Effects on egg At least one refinement of the design is described but the reasons for that refinement are unclear	Observations were recorded after each collision trial including: Final velocity Effects on restraint system Effects on egg At least one refinement of the design is described and justified by observations	Observations including high-quality diagrams, descriptions, or images recorded after each collision trial including: Final velocity Effects on restraint system Effects on egg Two or more refinements of the design are described and justified by observations
Final Design Performance	Two or more performance expectations were not met	The design did not meet one of the performance expectations (see next column) at velocities higher than 200 cm/s	At velocities higher than 200 cm/s: Egg remained inside restraint system Egg remained intact Restraint system uses ____ grams of materials	At velocities higher than 350 cm/s: Egg remained inside restraint system Egg remained intact Restraint system uses ____ grams of materials

Chapter 3 *Engineering Design Challenge* Assessment

Name ________________________________ Date ________________

Brainstorm	Constraint	Test and Refine
Criteria	Prototype	Communicate
Force		

1. A push or a pull is a ________________.

2. In the investigation, keeping the egg from cracking during a collision was an example of a

   ________________.

3. In the investigation, limiting the design process to a selection of classroom materials is an example of a

   ________________.

4. The first assembled restraint system you constructed in the investigation was a

   ________________.

5. The last step in the engineering cycle is ________________.

Name ______________________________ Date ______________

Projectile Motion

STEM

How can we control the lateral movement of a launched marble?

In previous investigations, you determined how spring setting and launch angle affect the horizontal range of a marble flung from the Marble Launcher. In this investigation, you will apply what you have learned in a design challenge.

Safety

- ***Wear safety glasses or other eye protection when launching marbles.***
- ***Launch only the plastic marbles that come with the Marble Launcher.***
- ***Never launch marbles at people.***

1 How engineers work

For this investigation, you and your group will use a process called the **engineering cycle** to design a technique for measuring and controlling the lateral, or sideways, motion of the marble during flight. You have learned that the horizontal, or forward, motion of the marble is affected by launch angle and launch speed. Lateral motion, on the other hand, is generally controlled by the long cylinder shape of the Marble Launcher. In your experimental data from Investigations B1 and B2, you probably noticed that the marble's landing site often varies along a plane perpendicular to the range of the marble (see the diagram below). This is due to the marble's lateral motion.

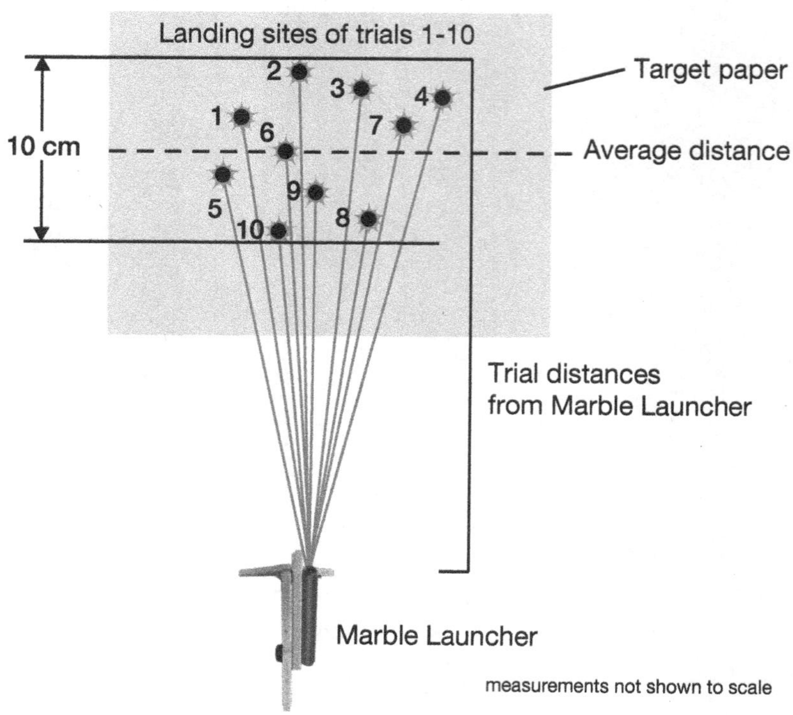

Name ______________________________ Date ______________

How could you control this variation in the marble's flight? This is the challenge of Investigation B4. **Engineers** design devices and procedures to perform tasks. This design process is accomplished using the engineering cycle. The steps in the engineering cycle are shown in this diagram:

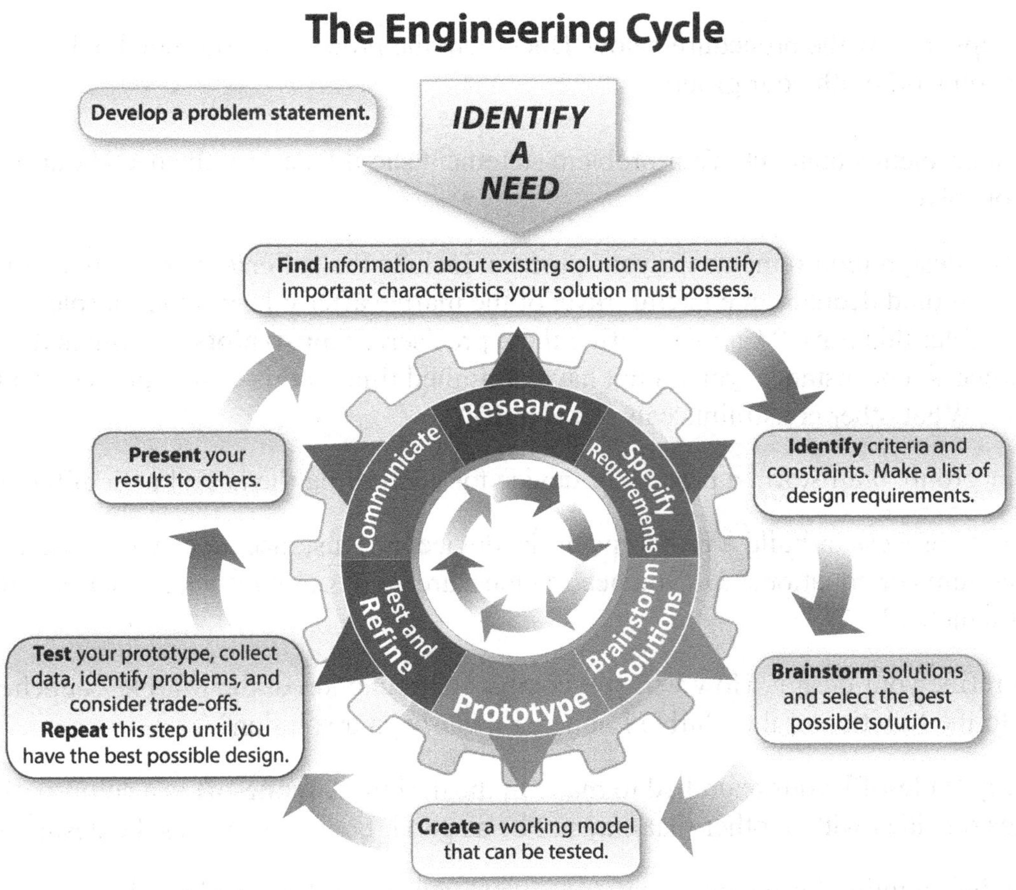

❷ Stop and think

✓ **a.** How might using the engineering cycle help your team design a successful process or device for controlling the lateral motion of the marble upon leaving the Marble Launcher?

____

____

____

____

Name ______________________ Date ______________

❸ Using the engineering cycle to improve your design

The goal of this investigation is to design a process or device for controlling the lateral motion of the marble upon leaving the Marble Launcher. This solution may take a wide range of forms.

Working in groups, follow the procedure below. Document the process in your notebook or Engineering Design Log as you work with your group.

1. Develop a problem statement. Your problem statement should identify the need your design will address or solve.
2. Specify the design requirements for your process. Include the **criteria**. For example, the solution you design should decrease the lateral travel of the marble after it leaves the Marble Launcher. What variables affect this travel? Can you define them precisely? **Constraints** are things that limit the design process. For instance, you do not have unlimited time in this design process, so time is a constraint. What other constraints can you identify?
3. With your group, brainstorm a process or device for controlling the lateral path of the marble's flight.
4. Describe the process or build a prototype of the device. For instance, does your process require the use of diagrams or equations? Do you need to measure angles or construct a platform to level the Marble Launcher?
5. Test and refine your design. How will you compare the function of the Marble Launcher *with* your solution to the function of the Marble Launcher *without* your solution?
6. Record any trade-offs your team had to make in the design. A **trade-off** is a compromise that exchanges one idea with another that may not be as good, but still achieves the design requirements.
7. Prepare a presentation for the class that documents your procedures and results.

Name ______________________________ Date ______________

❹ Extending your knowledge

✓ How did your final design compare with other designs in your class? What were some of the similarities and differences?

Chapter 7 *Engineering Design Challenge* Assessment

Name ______________________________ Date ______________

1. What were the constraints of your design process?

____

____

____

____

____

2. What is a trade-off?

____

____

____

____

____

3. What was a trade-off you made or considered in your design process?

____

____

____

____

____

For questions 4 and 5, complete each sentence with the correct stage of the engineering cycle.

Communicate	Prototype
Research	Test and Refine
Specify Requirements	Brainstorm Solutions

4. The stage in which the timer is used is ______________________.

5. The stage in which criteria and constraints are defined is

______________________.

Name ______________________________ Date ______________

Electric Circuit Game

How can the engineering cycle be used to build a game?

Do you have a steady hand? In this project, you will use the engineering design cycle to create a game to challenge a player's manual dexterity. The game will involve moving a small loop of wire over a complicated maze, and a mistake will make the light bulb light up.

❶ The engineering cycle

The **engineering cycle** is a process that is used to create or improve technology, including products and processes. Versions of this process are used by all types of engineers, from automobile designers to environmental engineers who are finding ways to reduce air pollution from factories.

The Engineering Cycle

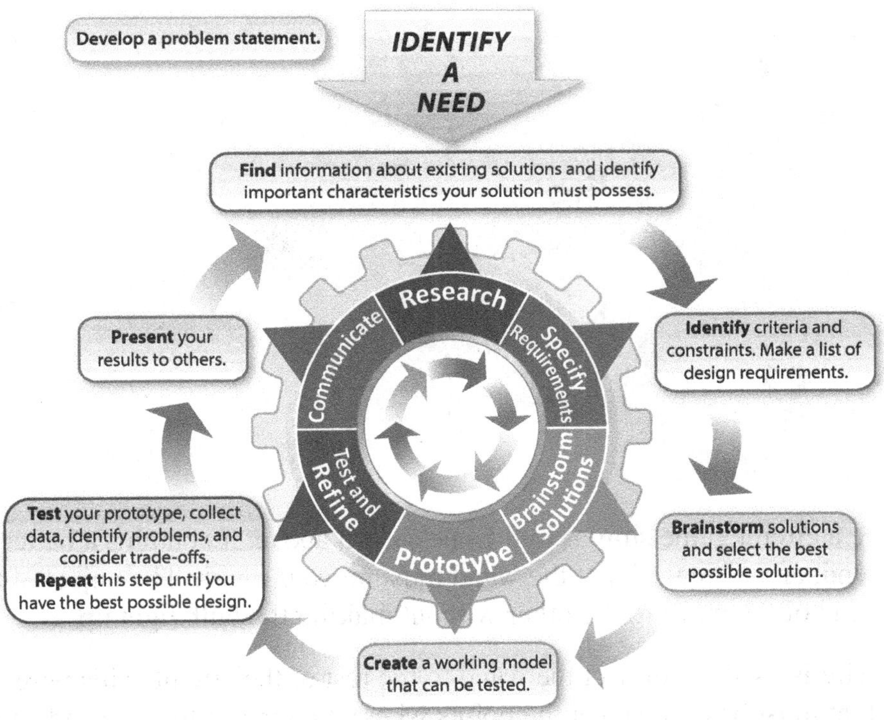

When developing a design, engineers often must consider criteria, constraints, and trade-offs. **Criteria** are requirements for the function and performance of a product. **Constraints** are rules that limit the design of a product and include cost, availability of materials, and time. Constraints sometimes lead to a need to make trade-offs in the criteria. A **trade-off** is a compromise that exchanges one idea with another that may not be as good, but still achieves the design requirements.

Name ______________________________ Date ______________

After possible ideas are brainstormed, a **prototype**, or model, is built. The prototype is tested and data are collected about its performance. Changes are then made to the design. The prototype is tested and changed as many times as needed until a final solution is reached. The final product is then shared with others.

❷ How the game works

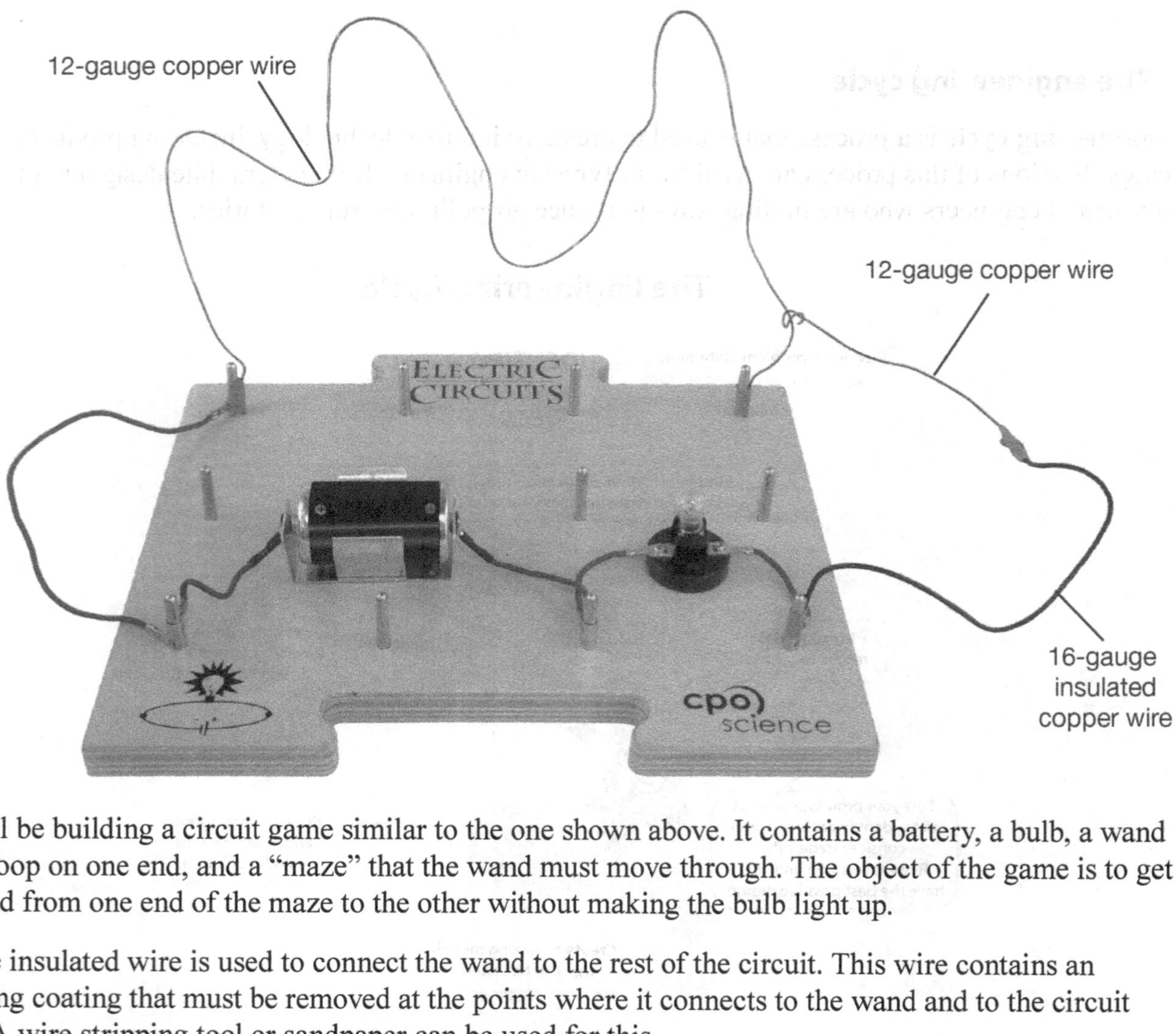

You will be building a circuit game similar to the one shown above. It contains a battery, a bulb, a wand with a loop on one end, and a "maze" that the wand must move through. The object of the game is to get the wand from one end of the maze to the other without making the bulb light up.

Flexible insulated wire is used to connect the wand to the rest of the circuit. This wire contains an insulating coating that must be removed at the points where it connects to the wand and to the circuit board. A wire stripping tool or sandpaper can be used for this.

The game pictured above is simply an example of what your game could look like. You can get more creative, and you might even want to add other components, such as a buzzer that sounds or a second bulb that lights if the player makes a mistake along a certain part of the maze.

Name ______________________________ Date ______________

❸ Designing and building your game

a. Discuss the criteria for the game as a class or with your group members. List the criteria you have decided on.

____

____

____

____

____

____

✓ **b.** What are the constraints you must consider when building your game?

____

____

____

____

____

____

c. Now you are ready to brainstorm ideas for your maze and begin building. Use the space below and on the next page to record your ideas. Follow the steps of the engineering cycle, making changes to your game as needed, until you have a final product. Then allow other members of the class to play your game.

____

____

____

____

____

____

Name ______________________________ Date ______________

4 Reflecting on your work

✓ **a.** Did you have to make any trade-offs when creating your game? Explain.

____

____

____

____

____

____

b. How would you change your game if you had fewer constraints?

____

____

____

____

____

____

Name ______________________________ Date ______________

1. List the steps of the engineering cycle.

____

____

____

____

____

____

____

2. Suppose you were an engineer working on designing a bridge for cars that spans from one side of a river to the other. List some of the criteria and constraints for your bridge.

____

____

____

____

____

____

____

____

Name ______________________ Date ____________

STEM

Designing Electric Motors

How does an electric motor work?

Materials:

- ✔ Electromagnet
- ✔ Electric Motor kit with all commutator disks (yellow, orange, pink, green, and blue), permanent magnets, and a battery pack
- ✔ 4 D-cell batteries
- ✔ Timer with one photogate

Electric motors are everywhere. You find them in locomotives, washing machines, cars, tools, spacecraft, and many other places where we use powered machines. All electric motors use one or both of the two kinds of magnets we just explored, **permanent magnets** and **electromagnets**. Permanent magnets are important because they create the magnetic field without needing any electricity. We will discover that electromagnets are necessary because the north and south poles can be reversed. This reversal of poles creates the motion that causes electric motors to run.

❶ Getting the rotor to spin

Electric motors spin because of the action of magnetism. Try to get the **rotor** to spin by manipulating magnets.

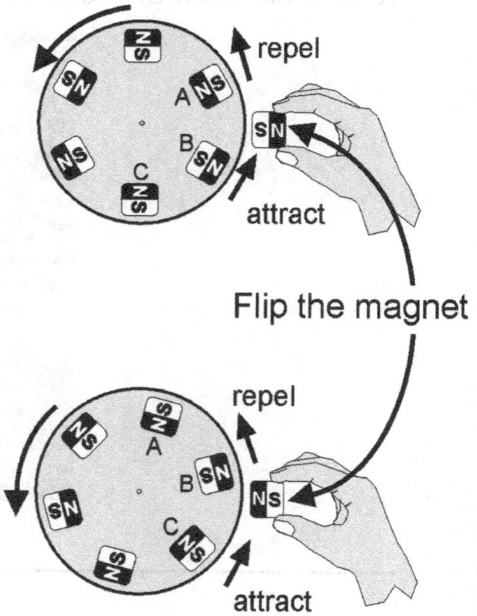

1. Take the motor apart and put 6 magnets in the rotor so they are evenly spaced and alternate north-south facing outwards.
2. Bring a stack of two or three magnets close and try to repel one of the magnets in the rotor. The rotor should spin a little.
3. As soon as you move one magnet, reverse the magnet in your fingers to attract and then repel the next magnet on the rotor.
4. By sequentially reversing the magnet in your fingers, try to push and pull on the magnets in the rotor to get the motor to spin.

You will see that reversing the poles of the magnet in your fingers is the key to making the rotor spin.

a. When is the right time to reverse the magnet in your fingers? Think about where the magnets are in the rotor.

____

____

____

____

Name ______________________ Date ____________

a. How could you make the rotor spin the other way?

____

____

____

____

❷ Building a 4-pole electric motor

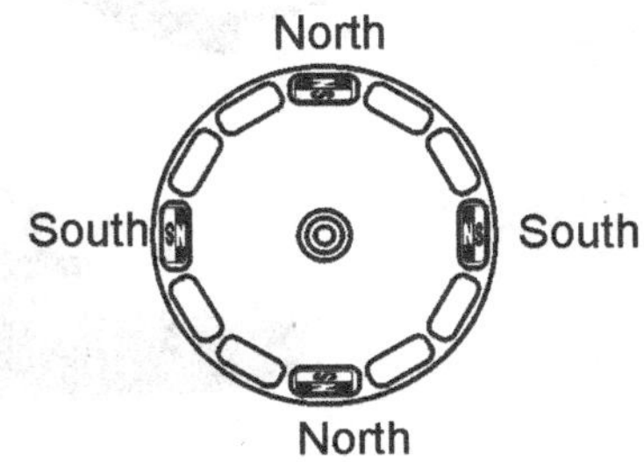

The key to understanding how electric motors work is learning how electromagnets are used to alternately attract and repel other magnets. If we arrange the electromagnets and permanent magnets just right, the rotor will turn when electricity is connected. The first motor to build is called a 4-pole motor because you are going to use 4 magnets to make 2 north poles and 2 south poles in the rotor.

All electric motors use some kind of switch to change the orientation of the poles of the electromagnet at the right time. A device called a **commutator** is used to switch the orientation of the north and south poles of the electromagnet. A plastic disk in the electric motor you are building acts as a commutator and switches the electromagnets using light.

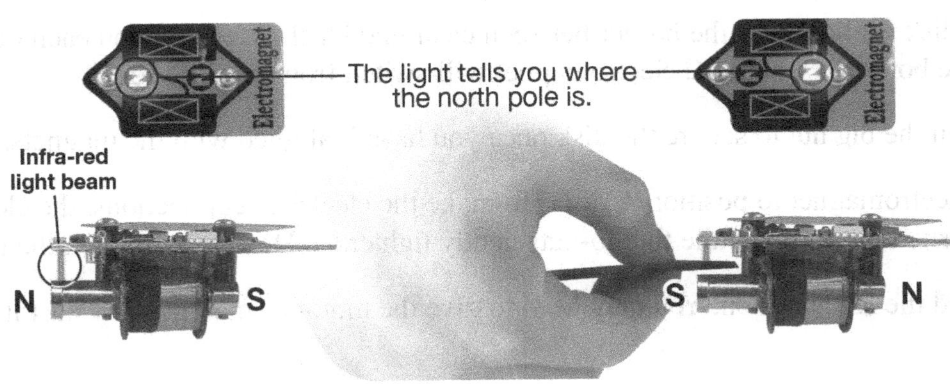

When the light is not blocked (see diagram above) the current flows in a direction to make the north pole at the front of the electromagnet. Blocking the beam causes the current to reverse, making a north pole at the back, and a south pole at the front. Two green lights indicate where the north pole is. The commutator (or switching) disks have alternating black and clear sections around the edge that switch the electromagnets by blocking the light beam. Aligning the clear/black edges with the centers of the magnets ensures that the switching happens in the right place.

Name ______________________________ Date ______________

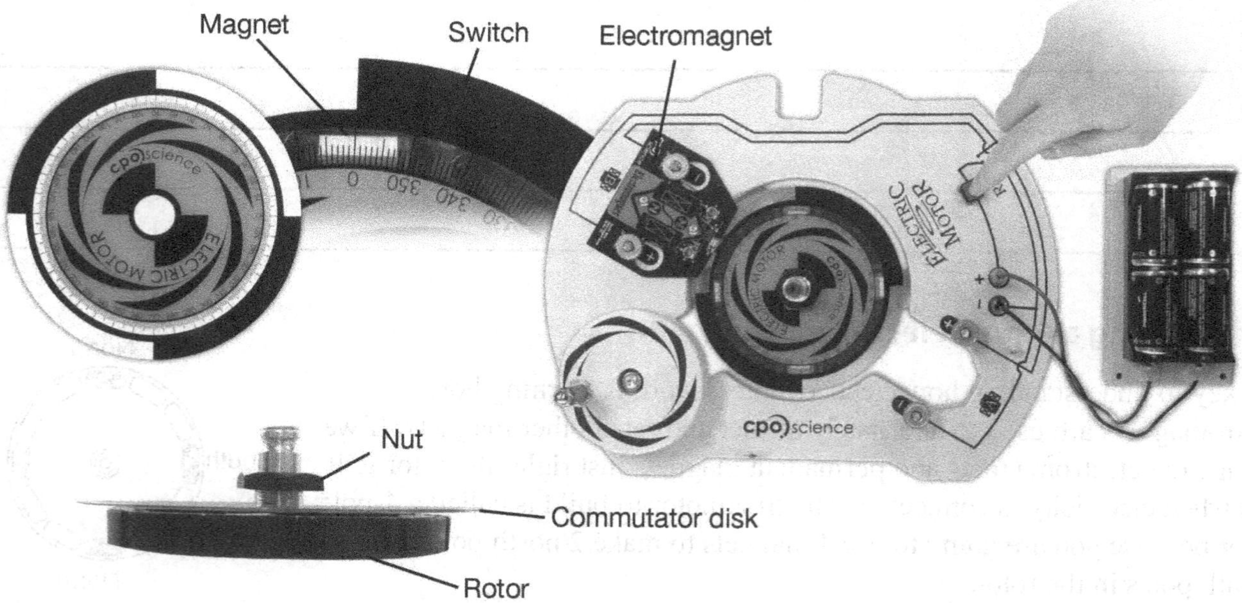

1. Find the pink four-pole commutator disk shown in the picture.
2. Arrange 4 magnets so the north or south poles alternate.
3. Be sure the disk is aligned so the border between clear and black is centered on each of the four magnets. The border is where the electromagnet will switch from north to south.
4. Finger-tighten the big nut to secure the disk once you have it aligned with the magnets.
5. Attach the electromagnet to position A or B. To make the electrical connections, the electromagnet should be pushed forward and the thumb-nuts gently tightened. Don't over-tighten the thumb nuts.
6. Push and hold the RUN button. You may need to give the motor a small push to start it.

❸ Designing and testing different electric motors

1. Using the procedure from Part 2, design and build a working electric motor for each of the five commutator disks.
2. You can change the number, placement, and direction of permanent magnets on the rotor. Keep changing things until you get the best design that works for each commutator disk.

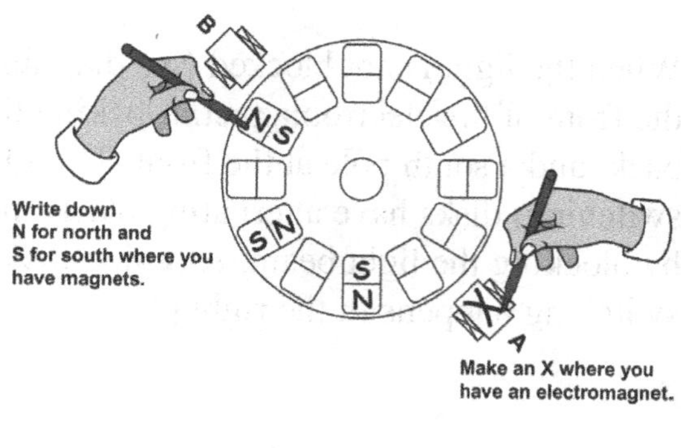

Chapter 23 *Engineering Design Challenge*

Name ______________________________ Date ____________________

1. For each motor you build, test and evaluate the design. Keep careful notes in your lab notebook on each design. Your notes should include:
   - Which color commutator was used
   - How many permanent magnets were used
   - How you spaced those magnets around the rotor
   - Which direction each magnet's poles are facing
   - How well the motor runs
2. You will discover that more than one design might work for each of the plastic disks. Choose the design that works best.
3. Use the diagrams to record your best designs. Here's how: Record the direction (north and south) and position of all magnets. Put an X where you placed the electromagnet. The electromagnet can be placed either at position A or B.

Pink Disk

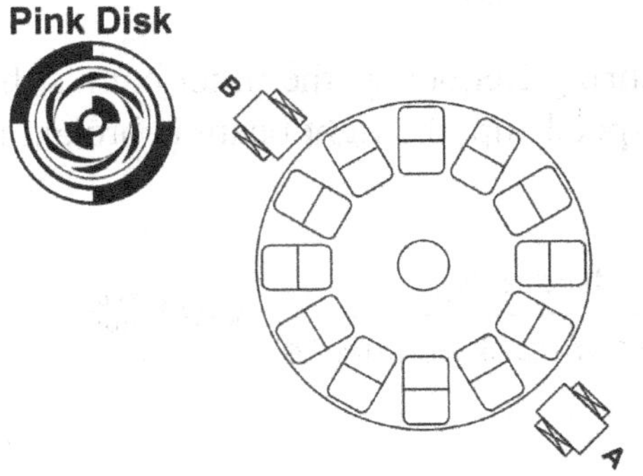

Orange Disk

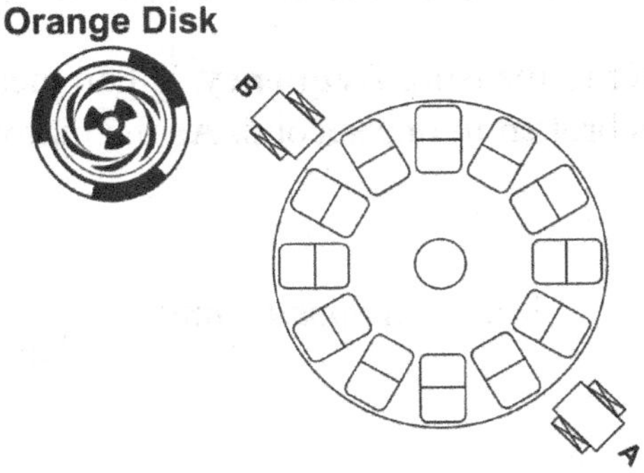

Blue Disk

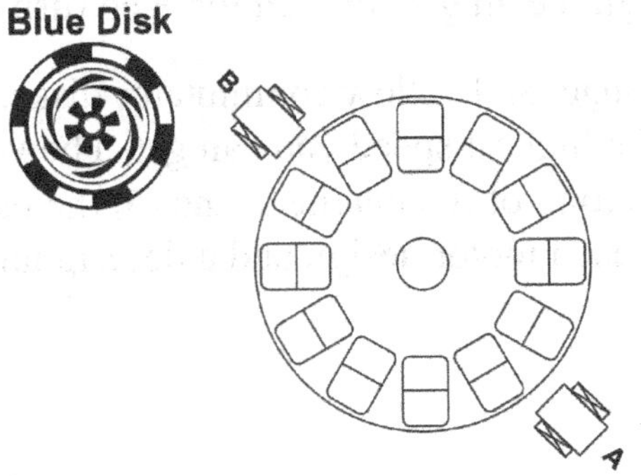

Yellow Disk

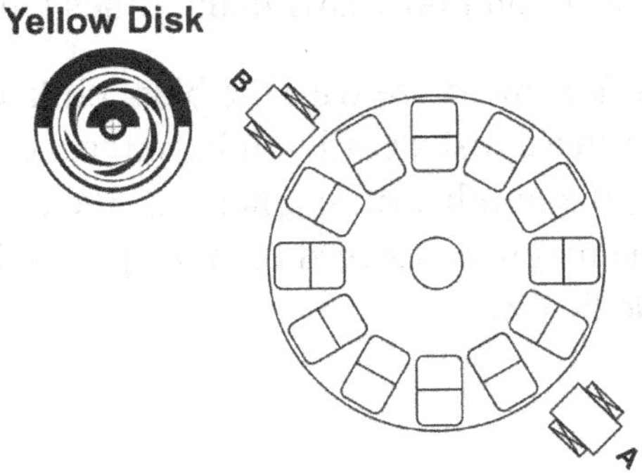

Green Disk: Version 1

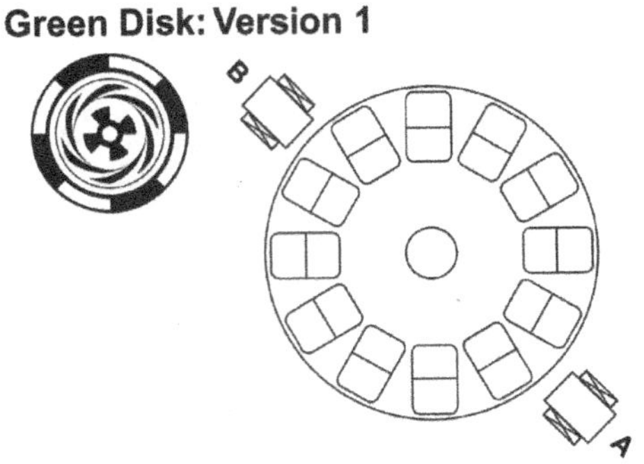

Green Disk: Version 2

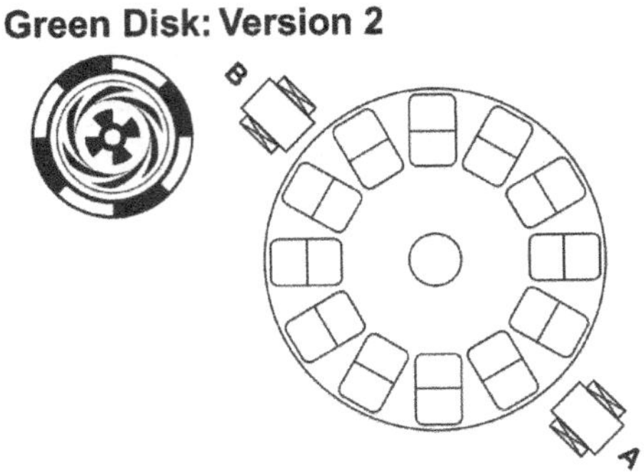

Name ______________________ Date ____________

❹ Testing for performance

Engineers want to build the best possible machine. In the design process, it is important to define what 'best' means. The word 'best' might mean fastest, least expensive, lightest, strongest, most shock resistant, or most attractive. It is usually impossible to be the best in all categories. The engineer chooses which categories are the most important depending on the machine's function. The speed of the rotor is one way to evaluate the electric motor. We will consider as the best design the one that goes fastest.

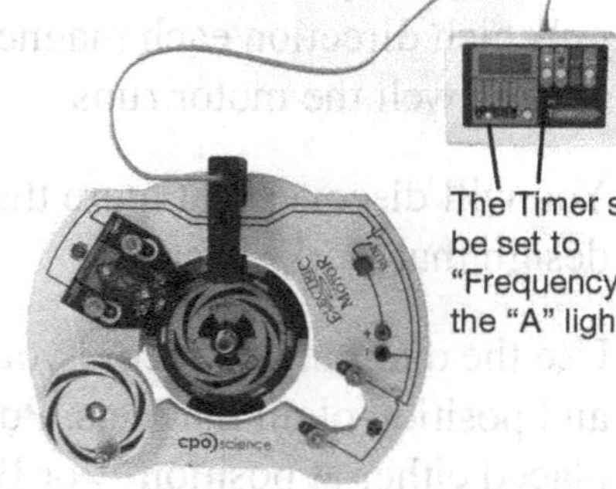

The Timer should be set to "Frequency" with the "A" light on.

1. Set up the timer with one photogate attached to the motor. When the photogate is all the way in the slot, the black and clear segments of the disk break the beam as the motor spins.

2. Set the timer to measure **frequency**. When measuring frequency, the timer counts the number of times the beam is broken in one second. As the motor speeds up, the light beam is broken more times per second.

$$\text{Rotation speed (rpm)} = \frac{\text{frequency}}{\text{Number of black segments}} \times 60\ \frac{\text{sec}}{\text{min}}$$

We want to calculate the speed in revolutions per minute (rpm). The equation above shows you how to calculate the speed if you know how many times the light beam gets broken for each turn of the disk.

Now, you will work some more with the blue, pink, orange, and yellow commutator disks. Adjust the motor design for each of these disks until you have reached the highest speed you can get. The variables that you can adjust are: (1) the number of magnets, (2) the orientation of the magnets, and (3) the placement of the disk relative to the magnets. Record the top speeds for each motor design and a description of the optimal design in the table below.

Name ______________________________ Date ____________________

Disk	Description of motor design	Highest frequency (# of times beam is broken/second)	Rotation speed (rpm)
blue			
pink			
orange			
yellow			
green, v. 1			
green, v. 2			

Name ________________________________ Date ____________________

✓ **a.** Which motor design gave you the highest speed and why do you think it was higher than the others? Your answer should identify the commutator disk and number of magnets you used.

✓ **b.** Give an explanation for why each variable has an effect on speed.

Name ______________________________ Date ________________

a. What would happen to your speed comparisons if the batteries lost their power during the experiment?

____

____

____

____

____

b. Write a procedure for checking the batteries to see that they stay the same throughout the experiment.

____

____

____

____

____

____

____

____

____

Name ______________________________ Date ______________

1. Would the motor work with a disk that was entirely clear? Why or why not?

____

____

____

____

2. Why must the electromagnet reverse polarity in order to spin the rotor?

____

____

____

____

3. Would the motor shown at right spin? Why or why not?

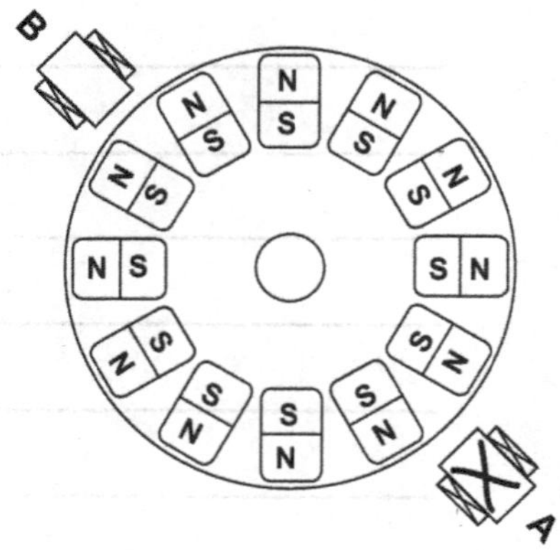

____

____

____

____

____

4. Draw the disk you would use if the magnets were arranged as in this diagram.

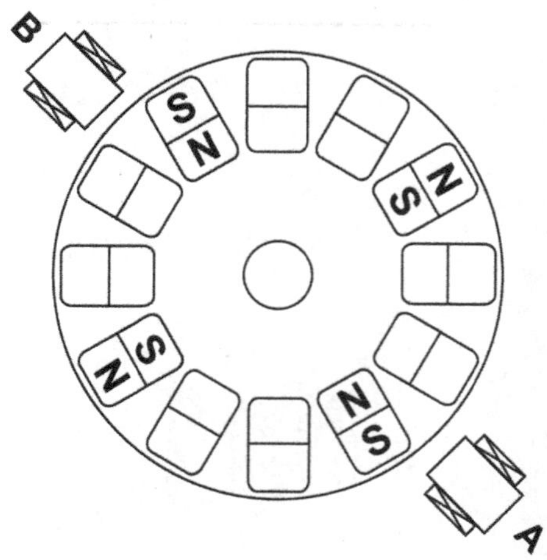